商用英文

張錦源 著

學歷：國立臺灣大學經濟系畢業
經歷：國立政治大學國際貿易研究所教授
　　　司法官訓練所講座
　　　中央信託局貿易處經理
　　　中央信託局副局長
現職：國立交通大學經營管理研究所教授
　　　貿協貿易人才培訓中心顧問

三民書局印行

國家圖書館出版品預行編目資料

商用英文／張錦源著．－－修訂三版十刷．－－臺北
市；三民，2007
　　　面；　公分
　　每章末附習題
　　ISBN 957-14-0021-1　（平裝）

　　1. 英國語言－應用文

805

Ⓒ　商　用　英　文

著作人　張錦源
發行人　劉振強
著作財
產權人　三民書局股份有限公司
　　　　臺北市復興北路386號
發行所　三民書局股份有限公司
　　　　地址／臺北市復興北路386號
　　　　電話／(02)25006600
　　　　郵撥／0009998-5
印刷所　三民書局股份有限公司
門市部　復北店／臺北市復興北路386號
　　　　重南店／臺北市重慶南路一段61號
初版一刷　1978年9月
修訂三版十刷　2007年10月
編　　號　S 800150
定　　價　新臺幣420元
行政院新聞局登記證局版臺業字第○二○○號

ISBN　957-14-0021-1　（平裝）

寫作技巧與能力。

　　由於本書是按進出口貿易過程先後編排，取材新穎，內容充實，講解親切，而且是配合我國對外貿易實況撰寫而成，讀者若能仔細研讀，勤加練習，相信不但可學習到商用英文的寫作技巧，同時還可領悟到做貿易的竅門，進而成為優秀的國際貿易從業員。

　　本書定稿後，即承三民書局劉經理振強欣然同意出版，在此敬申衷心的謝忱。

中華民國七十八年八月於臺北市

著者　張錦源　謹識

序

第二次世界大戰以後，國際貿易發展一日千里，各國爲繁榮經濟，提高人民生活水準，莫不致力於國外市場的開拓。我國在政府與民間的共同努力下，近年來對外貿易方面也已有飛躍的進展，頗受世人的注目。從事國際貿易也成了青年人嚮往的職業。

做國際貿易，與外國人打交道，須靠共同的語文來溝通。而在各種語文中，無疑英文已成爲國際性語文。因此，英文很自然地成爲國際商務往來的共同語文。商用英文(Business Engligh)就是應用於商業方面的英文，有志從事國際貿易工作的學子，爲求有效處理國際貿易事務，自需通曉商用英文。

想成爲一個能獨當一面的傑出國際貿易從業員，僅僅通曉英文是不夠的。因爲國際貿易作業程序錯綜複雜，從事國際貿易工作者，若不具備豐富的國際貿易實務知識，儘管英文程度很好，寫出來的英文，難免隔靴搔癢，不得要領。所以有志從事國際貿易的學子，不但需要學習商用英文的寫作技巧，同時還要勤加研習國際貿易實務。鑒於商用英文爲國際貿易從業員日益重視，著者乃藉日常工作之便，蒐集有關的資料，配合我國對外貿易，精心編了本書，以供有志從事國際貿易的青年學子參考。

在撰寫方法上，根據實務上的經驗，以「學英文做貿易，談貿易學英文」的方式，一邊談商用英文的寫作要領，一邊講解貿易的做法。在內容安排方面，則按進出口貿易過程的先後，循序漸進，採用現行我國進出口貿易實例，做親切的說明。並在每章之後，搜集了豐富的例句，以供隨時參考選用。

爲使讀者在研讀本書之餘，有機會自行測驗學習成果，在每章之後附有相關的習題，讀者如能一一練習，自我考核，深信必能增進商用英文的

商用英文　目次

第一章　緒　論

第二章　商用英文信的結構

第三章　撰寫商用英文信的七要

第四章　舊式文體與新式文體

第十一章　交易條件

第十二章　詢　　價

第十三章　答覆詢價

第十四章　報　　價

第十五章　接　　受

第十六章　推銷與追查

第十七章　訂　　貨

第十八章　答覆訂貨

第十九章　買賣契約書

第二十章　信用狀交易

第二十一章　貨物水險

第二十二章　海上貨物運輸

第二十三章　貨運單證

第二十四章　滙　票

第二十五章　收款與付款

第二十六章　索賠與調處

第二十七章　挽回客戶

第二十八章　代理關係的建立

第二十九章　招標與投標

第三十章　貿易電報與電報交換

第三十一章　通　　函

第三十二章　求才與求職

第三十三章　商業社交信

第三十四章　標點符號的用法

第一章 緒 論

(Introduction)

第一節 商用英文的定義

「商用英文」(Business English, Commercial English)就是應用於商業方面的英文。正如英文應用在醫學或化學方面一樣，在「商用英文」裡也使用各種商業術語和慣用語，例如 FOB, CIF, L／C, D／P, D／A 或 B／L 等，但是在文法和修辭方面，「商用英文」卻和普通英文並無兩樣。

想學好商用英文，必須：

1. 要有良好的英文基礎：如果英文根基不好，就無法寫出好的商用英文。所以，想學好商用英文，首先應具備一般英文文法的常識與熟諳英文的基本語法。

2. 要通曉商業術語：商用英文需使用各種商業術語才能達成簡潔的目的。如不通曉商業實務上的各種術語，必無法寫好商用英文。

3. 要熟諳商業慣例與實務：從事國際買賣的當事人往往居於不同風俗、習慣的環境裡，因此，想求交易的順利進行，就必須熟悉商業的慣例與實務。熟悉了商業慣例與實務，才能寫出理想的商用英文。

4. 要多閱讀別人所撰寫的商用英文：擷取可資模仿的各種表現(expressions)用例，以供適用於不同情形的需要。

第二節　商用英文的範圍

如上所述，商用英文是指運用於商業上的英文而言。然而，就非英語系國家而言，商用英文通常多應用於對外貿易，所以，在非英語系國家而言，將商用英文稱為貿易英文或許更恰當。對外貿易所涉及範圍相當廣泛，除貿易業務本身之外，尚與銀行、外匯、倉儲、運輸、保險、行銷等等有密切的關係。所以就我國而言，商用英文的範圍可包括：

1. 商用書信(Business Letters)
2. 貿易契約(Foreign Trade Contract)
3. 貿易單證(Foreign Trade Documents)
4. 保險與航運(Insurance and Shipping)
5. 銀行與外匯(Banking and Foreign Exchange Business)
6. 電報與電報交換(Cable and Telex)
7. 商業社交信(Social Letters in Business)
8. 求才和求職(Employment)
9. 其他(Others)：包括商情報告、廣告及貿易單證等。

商用書信裡，又包括下列各種：

1. 尋找交易對象(Looking for Customers)
2. 招攬交易(Trade Proposal)
3. 徵信(Credit Inquiry)
4. 詢價(Trade Inquiry)
5. 答覆詢價(Response to Inquiry)
6. 報價(Offer)
7. 接受(Acceptance)
8. 推銷與追查(Sales Promotion and Follow-up)

9. 訂貨(Order)

10. 答覆訂貨(Response to Order)

11. 收款與付款(Collection and Payment)

12. 索賠與調處(Claim and Adjustment)

13. 挽回客戶(Regaining Lost Customers)

14. 建立代理關係(Establishment of Agencyship)

15. 招標與投標(Invitations and Bids)

16. 通告(Announcement)等等。

習　　題

一、試說明商用英文的定義。

二、想學好商用英文，應具備那些條件?

三、試述在非英語系國家的「商用英文」課程應包括那些內容?

第二章　商用英文信的結構

(The Structure of an English Business Letter)

　　一封商用英文信是由信文(the letter)與信封(the envelope)組合而成。

第一節　商用英文信的構成部分

一、商用英文信的構成部分

　　商用英文信的構成部分，可分爲基本部分與附加部分。

　1.基本部分：這是一封信不可或缺的主要部分(main parts)，基本部分包括：

　　　(1) Letterhead　　　　　　　信銜

　　　(2) Date　　　　　　　　　發信日期

　　　(3) Inside Address　　　　　信內地址

　　　(4) Salutation　　　　　　　稱呼

　　　(5) Body of the Letter　　　本文，正文

　　　(6) Complimentary Close　　客套結束語

　　　(7) Signature　　　　　　　發信人簽署

　2.附加部分：一封信除上述基本部分之外，可能還有特別事項需附加的。例如下面幾項都是商用英文信中常出現的附加事項：

　　　(1) Reference Number　　　案號，參考文號

　　　(2) Attention　　　　　　　特定受理人，經辦人

⑶ Subject of the Letter　　主旨，事由

⑷ Identification Marks　　鑑別符號

⑸ Enclosure Marks　　附件符號

⑹ Carbon Copy Notation　　副本抄送單位記號

⑺ Postscript, P. S.　　附啓

⑻ Continuation Sheet　　續頁

現在就以上各項分別說明。

二、信銜

信銜是信箋上頂端事先印好了的有關發信人(Sender)的：

⑴　Firm's Name　　商號名稱

⑵　Address　　地址

⑶　Cable Address　　電報掛號

⑷　Telex Number　　電報交換掛號

⑸　Telephone Number　　電話號碼

⑹　Line of Business　　行業種類

⑺　Year of Establishment　　創業年份

⑻　Trade Mark　　商標

⑼　Bankers　　往來銀行

⑽　Capital　　資本額

⑾　P.O. Box　　信箱號碼

信銜的設計宜簡潔、美觀、大方，不宜多占地位，多用印妥信銜的信箋。

〈Illustrations of Letterhead〉

BURDA ENTERPRISES INC.

Exporters-Importers-Manufacturers
P. O. BOX 3557 TAIPEI
5TH FL., 26, SEC. 3, JEN-AI ROAD
TAIPEI, TAIWAN 106, R. O. C.

PHONE: (02) 705-9286 (10 LINES)
TELEX: 22615 BURDA
CABLE: BURDA TAIPEI

REFERENCE BANK:
FIRST COMMERCIAL BANK
BANK OF AMERICA

三、日期

1. 位置: 商用英文信應寫明發信日期，其位置在信銜最後一行的下面兩行至四行之間。至於其起寫地方，要看信的格式而定。

在 Full Block Form 時，日期自左緣開始。採用其他格式時，日期可以放在中央，也可從中央稍靠右的地方開始，但其最後一個字母須與正文〈body of the letter〉的右邊排整齊。

2. 寫法

⑴　October 12, 1989（美式）

⑵　12th October, 1989（英式）

⑶　12 October 1989（英美均用，尤其美軍）

注意事項:

①月份不宜用縮寫字，如 Jan., Feb., Mar., Aug.

Poor	Right
Apr. 2, 1989	April 2, 1989
9th Mar., 1989	9th March, 1989

②以下列方法表示日期，均易混淆，宜避免。

英式	美式
12/10/1989	10/12/1989
12/10/89	10/12/89
12/10/'89	10/12/'89
12-10-1989	10-12-1989
12-10-89	10-12-89

（以上均表示 1989 年 10 月 12 日）

四、信內地址

通常包括：

1. Name of Addressee　　　收信人姓名
2. Title of Addressee　　　收信人尊稱（頭銜）、職銜
3. Name of the Firm　　　商號名稱
4. Full Address　　　地址

但有時可能缺第 1，2 項。

1.位置：Inside Address 應寫在日期或案號下面二行的地方。

TAIWAN TRADING CO., LTD.

EXPORTERS & IMPORTERS

OUR REF. 123　　　　　　　　　　　October 12, 1989

MR. Charles H. Franklin

Vice President

Atlantic Trading Corp.

61 Broadway

New York, N.Y. 10021

U. S. A.

但也有將 Inside Address 放在信的左下角，低於簽署一、二行者。

2.注意事項：

⑴如寫給商號經理(Manager)、董事(Director)等，而不寫出其姓名時，應在商號名稱上面一行加上"The Manager"等字樣。例如：

The Manager (or The President, The Director…)

Taiwan Trading Co., Ltd.

100, Hung Yang Road

Taipei, Taiwan

(2)收信人地址寫在商號名稱下一行，通常分為三行，第一行為門牌
號碼及街名；第二行為城鎮及郵遞區號(ZIP　Code,為 Zone
Improvement Program Code 之縮寫)；第三行為國名。門牌
前面不加"No."或"#"。

Right	Poor
61 Broadway	No. 61 Broadway
New York, N. Y. 10021	New York, N. Y. 10021
U. S. A.	U. S. A

五、稱呼

稱呼為寫信人在進入正文前，對收信人的敬稱，等於中文書信中的「謹
啓者」、「敬啓者」。性質等於"How are you?"的問候語。

1.位置：在 Inside Address(或 Attention)下面兩行的地方，並與左
緣靠齊。

2.注意事項：

(1)稱呼的第一字母及頭銜必須大寫。例如：

Wrong	Right
Dear sir	Dear Sir
My Dear Sir	My dear Sir
Dear doctor Chang	Dear Doctor Chang

(2)如收信人為個人時，不可將其全名寫出。

Wrong	Right

Dear Mr. John C. Chang   Dear Mr. Chang

Dear John（但正式的商用信不宜
用暱稱）

⑶如果收信人為經理"The Manager"等, 而沒有人名, 在稱呼上只能用"Dear Sir", 不能用"Dear Sirs,"因為這是給經理一個人的。

⑷收信人為商號時, 用下列幾種稱呼：

Dear Sirs（英式, 對男人或男女組成的商號）

Gentlemen（美式, 對男人或男女組成的商號）

Mesdames ⎫
Ladies   ⎬ （對女人組成的商號, 英美均通用）

⑸ Mr., Messrs., Mrs.和 Miss 不能單獨用作 Salutation,後面必須有姓。

Wrong	Right
Dear Messrs:	Gentlemen:
Dear Mr:	Dear Mr. Chang:
My dear Mrs:	Dear Mrs. Chang:

⑹在 Salutation 一項內頭銜不可縮寫, 但 Mr., Mrs., Prof.,及 Dr. 除外。

Wrong	Right
D'r Sir:	Dear Sir:
Dear S'r:	Dear Sir:
Gents:	Gentlemen:
Mmes:	Mesdames:

茲將尊稱與 Salutation 的配合用法列表於下

收　信　人		尊　　稱	列　　示	稱　呼
男 性	單數	Mr. (Mister) Dr. (Doctor)	Mr. E. Hemingway Dr. R. P. Watson	Dear Sir Dear Mr....
	多數	Messrs. (Messieurs)	Messrs. Wilson & Co.	Dear Sirs Gentlemen
女 性	未婚單數	Miss	Miss Janet Parker	Dear Miss… Dear Madam
	未婚多數	Misses	Misses Lucy and Bessy Smith	Dear Ladies Dear Mesdames
	已婚單數	Mrs. (Mistress)	Mrs. Judy Ford（將 Mrs.冠以丈夫姓） Mrs. Maggie Krook（寡婦）	Dear Mrs.... Dear Madam
	已婚多數	Mmes (Mesdames)	Mmes. Lucy Cole and Jane Bennet	Dear Ladies Dear Mesdames
	不分已婚否	Ms.〔註〕（發音：miz）	Ms. Betty Ford	Dear Ms.... Dear Madam

【註】Ms.的多數形不詳，該是 Mses.吧！

六、正文

　　一封信的正文就是信的主體，也是一封信最重要的部分。其位置在稱呼或事由的下面兩行。正文視內容的繁簡，分為若干段。至於正文的排列形式，請參閱第四節。

七、客套結束語

　　客套結束語是寫信人在信尾客氣的自稱。有如中文信的「謹啓」、「順頌籌祺」、「謹請財安」。其性質等於"Good-bye"。位置則在正文下面兩行。如用 Full Block Form 應由左緣開始，其他格式則由信箋的正中央起，向右邊寫。現代商用英文信普通所用的客套結束語有下列三種：

　　Yours truly,

　　Yours very truly,

　　Very truly yours,

但

　　1.英國常用：　　　　　　　　2.有親交的商業社交信可用：

　　Yours faithfully,　　　　　　Yours sincerely,

　　Faithfully yours,　　　　　　Yours very sincerely,

　　　　　　　　　　　　　　　　Sincerely yours,

客套結束語必須與稱呼、正文語氣相稱。茲將配合稱呼的客套結束語列表於下：

Salutation		Complimentary Close
Male	Female	
Sir Sirs	Madam Mesdames	Yours respectfully, Yours very respectfully, Respectfully yours, Yours faithfully,

		Faithfully yours,
Dear Sir	Dear Madam	Yours truly,
Dear Sirs	Mesdames	Yours very truly,
Gentlemen	Ladies	Very truly yours,
My dear Sir	My dear Madam	Yours faithfully,
		Faithfully yours,
		Yours sincerely,
		Sincerely yours,
Dear Mr. Brown	Dear Mrs. Brown	Sincerely,
My dear Mr. Brown	Dear Miss Brown	Yours sincerely,
	Dear Ms. Brown	Sincerely yours,
	My dear Mrs. Brown	Very sincerely yours,
		Yours very sincerely,
		Yours cordially,
		Yours very cordially,
		Cordially yours,

八、簽署

　　簽署包括寫信人的姓名、職銜及簽字，信件寫好之後，必須由發信人簽字，以示負責。手寫簽字往往難於辨認，所以簽名的下面須將發信人姓名和職銜打出。Signature 的位置在 Complimentary Close 下面，左邊對齊(Indented form 除外)。例如：

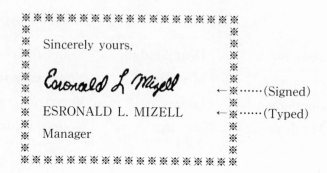

Sincerely yours,

ESRONALD L. MIZELL ←……(Signed)

ESRONALD L. MIZELL ←……(Typed)

Manager

注意事項:

1. 如發信人是女性，應將"Miss"，或"Mrs."加註於姓名之前，以便收信人回信時易於稱呼。

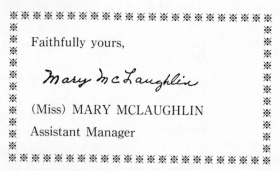

Faithfully yours,

(Miss) MARY MCLAUGHLIN

Assistant Manager

2. 商號名稱因已印在 Letterhead 裡，所以通常不必在簽名的上面再打上商號名稱。但比較正式和重要的信函，也打上發信人所屬商號名稱。這時，須將商號名稱用大寫打出。

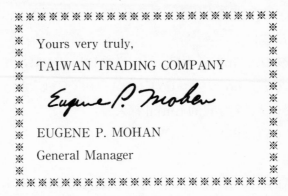

Yours very truly,

TAIWAN TRADING COMPANY

EUGENE P. MOHAN

General Manager

3. 爲表明有權代表商號正式簽名起見，往往在商號名稱前加上"Per Pro"或"P. Pro."或"P. P."字樣，這是拉丁文"Per Procuration"（授權代理）的縮寫。

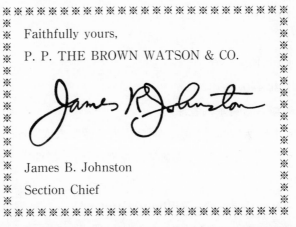

Faithfully yours,

P. P. THE BROWN WATSON & CO.

James B. Johnston

Section Chief

如無正式代理簽署權，但經商號負責人同意可在其職責範圍內代表商號行文者，應在商號名稱之前，加上"For"字樣，或在簽名之前面加打"By"字樣，以代替"P. P."。

九、鑑別符號

爲了便於查考起見，商用英文信中常將發信人及打字員姓名的第一字母(Initial)打在信箋左下面與簽署最後一行平行或下一、二行，緊靠左緣，其表現方式如下：

　　1.發信人的 Initial 大寫，打字員的 Initial 小寫：

　　　CYC: ob　　　　　CYC/ob　　　　　CYC ob

　　2.發信人、打字員的 Initial 都大寫：

　　　CYC:OB　　　　　CYC/OB　　　　　CYC OB

十、案號

　　現代的大商號，往來函件很多，因此為便於雙方將來查考之用，每一封信都編有案號，註明請收信人在覆信時提及該案號。案號的位置印在 Letterhead 或 Date 下面，靠左緣或右緣。其文字有下列幾種：

　　1. Please refer to:

　　2. In reply, please refer to:

　　3. In reply, please quote:

　　4. Our Ref. No.

　　5. File No.

十一、經辦人（特定人查照）

　　如果收信人對象為商號，而發信人希望某人或某部門特別注意到該信時，可加「經辦人某某」（或某某人查照）一行，這稱為 Attention line.其寫法有：

　　1. Attention: Mr. William Hunt

　　2. Attention of Mr. Charles C. Chang

　　3. Att: Mr. William Hunt

　　Attention line 通常位於 Inside Address 與 Salutation 之間，緊靠左緣，但也有放在中央低於 Salutation 一、二行或與 Salutation 同一行的。例如：

```
Irving Trust Company
1 Wall Street
New York, N.Y. 10021
U.S.A.
Attention: Letter of Credit Department
```

```
Irving Trust Company
1 Wall Street
New York, N.Y.10021
Gentlemen:            Att: Mr. Ernest Shaw
```

十二、事由（主旨）

為使收信人商號的收發人員能將信件迅速傳遞到有關部門或有關人員處理，可在信上標明事由或主旨。事由的內容不外兩種：

1.收信人來函的文號或日期或兩者

2.本信所討論的主題

事由應放在稱呼及正文之間上下各空兩行。其表現方式有三。

1.以"Subject"字樣開頭，緊靠左緣：

Gentlemen:

Subject: Your order No.123

We are pleased to inform you...

2.不註明"Subject"字樣，但放在中央，加底線：

Gentlemen：

Your order No.123

We are pleased to inform you...

3.加"Re"（Reference 的縮寫）字樣，不用底線：

Gentlement:

Re: Your order No.123

We are pleased to inform you...

第3種"Re"字樣，有些人認為關係陳腐的表現法。

十三、附件符號

如果信中有附件時，應在正文中提及，並在鑑別符號下一、二行的位置註明，如此一方面可提醒發信部門，以免遺漏；他方面可引起收信人的注意。這種說明有附件的記號稱為 Enclosure Mark，其表現方式有多種，茲列若干於下：

Enclosure
Enclosures
Encl.   } 用於不重要附件
Encls.

Enclosure: As stated
Enclosures:a/s
Encl.: As stated   } 同上 (As stated 或 a/s 為「附件如文」之意)
Encls.: a/s

Enclosure:1
Enclosures:2
Encl.:1   } 用於較重要附件
Encls:2

Enclosure:　A check ♯ 123 for US$100
Enclosures:　one cable copy and one catalog } 用於重要附件
Encl.: one price list

十四、副本抄送單位記號

如發信的同時，需將副本抄送有關單位時，以"C.C."(Carbon Copy

的縮寫）打在 Enclosure 下面的地方，這種副本通常須由發信人簽字。至於其表現方式如下：

CC to; CC:; cc:;c.c.-;cc-;ccs:（多數時）;Copy to

例：

CC: XYZ Company

CC to XYZ Company

Copy to XYZ Company

如副本寄送單位多時，爲使副本收信單位注意起見，在其名稱前面加" N "

CC:　ABC Trading Company

XYZ Trading Company

N Taiwan Marine Insurance Co., Ltd.

十五、附啓

當信已打好，忽然想加幾句話或補充一件事時，可以在信箋最下方也卽在副本抄送單位記號的下面寫上"P.S."符號，然後再將要追加的文字加上去。"P.S."爲 Postscript 的縮寫。"P.S." 有如中文信中的「附啓」，「再啓」，「又啓」，「附言」。

例：

CC: New York Corp.

P.S.: Your offer of August 10 has just been received. We will reply before August 15.cy.（cy.爲寫信人的草簽 [initial]）

十六、續頁

最理想的商用英文信爲打字部分的面積約占全頁的四分之三，這樣上下左右都留有適當的空白，形式美觀。如信文過長，不是一張信紙所能容納，則需用續頁。在這種情形，要注意：

1.續頁紙必須用白紙，不宜用印有 Letterhead 的信紙。

2.在各續頁紙的頂端，應將收信人的姓名或商號名稱、頁數、日期分
別打出：

例：

$-2-$

ABC Trading Co.   May 10, 1989

ABC Trading Co.   Page 2
May 10,1989

Page 2   May 10, 1989
ABC Trading Co.

第二節　　信封的寫法

一封信寫好後，便要寫信封。信封上記載的文字有：

1. Return Address　　　　發信人名稱、地址
2. Mailing Address　　　　收信人名稱、地址
3. Mailing Direction　　　郵遞指示
4. Others　　　　　　　　其他

信封上的收信人名稱、地址必須正確，以免誤遞，且須與 Inside
Address 完全相同。所用格式與標點也應一致。

收信人名稱、地址排列行數如不超過四行，則每行之間宜採用雙行間
隔(Double Spacing)，四行以上則宜採單行間隔(Single Spacing)。

信封的寫法並無標準格式可言，但下面的例子是比較普通的格式：

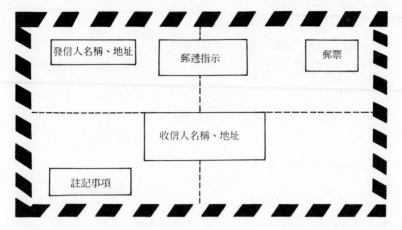

(1)的位置印有發信人商號名稱及地址。如以個人名義發信，則須加打個人姓名，以備無法投遞退回時，可退還個人。除此之外，也有在 Return Address 上面加印如下的文字。

If undelivered, please return to... （如無法投遞，請退還）

If not delivered within...days, please return to...

　（如於××日內無法投遞，請退還…）

(2)的位置為 Mailing Direction，可用於指示下列各事項：

Via Airmail （航空信）

Express （快信）

Special Delivery （限時專送）

(3)的位置為貼郵票處。

(4)的位置為收信人姓名、地址。其位置在信封下半部稍偏右方。其寫法、標點、格式須與 Inside Address 一致。須轉交(in care of, c/o)時，c/o 放在收信人姓名下一行轉交人姓名前面。

(5)的位置為註記事項，在信封左下方。有的航空信封在這裡事先印上 "By Air Mail"，"Par Avion"或"Correo Aereo"字樣。在此情形，(5)的位置稍向上升。

註記事項可能爲下列的一項或多項:

Attention of Mr. William Taylor	專陳威廉泰樂先生
Confidential	機密
Introducing Mr. Charles Chang	介紹張先生
Kindness(or By courtesy) of Mr. A	煩 A 先生轉交
Photo Inside	內有照片（請勿折疊）
Printed Matter	印刷品
Private	私函（限由收信人親拆）
Personal	親啓
Registered Mail（or Registered)	掛號郵件
Sample of No Commercial Value	無商業價值樣品
Sample of No Value	貨樣贈品
Second Class Airmail	第二類航空郵件
With Compliments of...	…敬贈
Strictly Confidential	極機密
Urgent	急件

也有將 Return Address 寫在信封蓋(Back Flap)上的。其形式請看下面的例示。

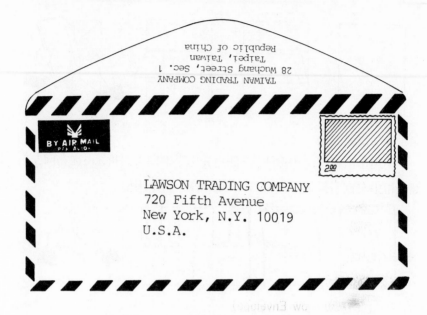

第三節　信紙的折疊法

　　英文信的折疊方式，雖是微末細節，但因爲它是給收信人一個最先的印象，因此不能忽略。

　　摺信紙的方法，依信封的大小而定。不過總是要摺得美觀，便於郵寄並使收信人易於拆閱。

一、大型信封

　　使用大型信封時，信紙的摺疊法爲先由下端向上端折⅓，然後再向上折與上端平齊，成三等分，然後裝入信封。

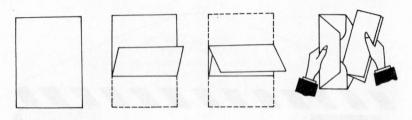

二、小型信封

　　使用小信封時，信紙的摺疊法爲先由下端向上對折，再從右向左折⅓，然後再由左向右折⅓，再將六折的信紙裝入信封中。

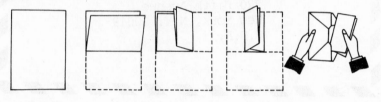

三、開窗信封(Window Envelope)

　　開窗信封即在信封的中央留一個窗口覆以透明紙。使用這種信封時，折疊信紙必須將收信人姓名、地址折於外面，使其裝入信封後，收信人姓名及地址可在窗口露出。使用開窗信封可免在信封上重打收信人姓名、地址，增加效率。

第四節　商用英文信的格式及標點

一、商用英文信的格式

　　商用英文信的各項構成內容已如前述。至於各部分的編排方式──即格式(forms or styles)──以及標點方法，視寫信人的喜愛而定。茲將常見的格式介紹於下：

　　1.全齊頭式(Full Block Form)：Date, Inside Address, Salutation, Body, Complimentary Close 及 Signature 等的每一行都

一律從左緣開始，排列整齊，不向右縮進。優點：打字不必考慮各段第一行縮進的問題，打字節省時間。缺點：失去平衡，不美觀，不便於閱讀。

2. 半齊頭式(Semi-Block Form)：Inside Address 及 Salutation 採齊頭式，Date 放在右上方與 Inside Address 並列。Body 採縮進式 Complimentary　Close 與 Signature 移到右下方。優點：Body 的每段第一行縮進若干字母，便於閱讀，而且 Inside Address 與 Salutation 在左上方，Complimentary Close 與 Signature 在右下方，看起來顯得平衡。因此這是目前最常用的一種格式。

缺點：因每段第一行須縮進若干字母，打字較費時。

3. 改良齊頭式(Modified Block Form)：Inside Address, Salutation 及 Body 採齊頭式，每一行都從左緣開始，Date 移到右上方與 Inside　Address 第一行並列，Complimentary　Close 與 Signature 移到右下方。

全齊頭式的優點爲打字省時，缺點爲不平衡，改良齊頭式則將此兩者加以調和，兼顧省時與平衡。

4. 縮進式(Indented Form)：Inside Address, Body 的每段開頭，向右縮進(Indented)形成梯形。此格式是傳統的形式，但現在用的人較少。優點爲各部分區分明顯，缺點爲打字效率較差。

5. 懸段式(Hanging　Paragraph　Form)：又稱爲 Hanging-indention Form)，卽 Body 的每段第一行從左緣開始，其餘各行則縮進若干字母。其他部分與半齊頭式及改良齊頭式同。

二、英文信上各部分的標點(Punctuations)

英文信上的 Date, Inside Address, Salutation, Complimentary Close 及 Signature 等各行末端所用的標點符號打法有閉式(Closed

Punctuation)、開式(Open Punctuation)及混合式(Mixed Punctuation)三種。

1. 閉式: Date, Inside Address, Salutation, Complimentary Close 及 Signature 各行後面加上適當標點符號, 如逗點 (Comma) (,)、句點(Period) (·) 或冒號(Colon) (:) 者, 稱為 "Closed Punctuation"。閉式標點多半配合 Indented Form, Semi-block Form 及 Modified Block Form 使用。

2. 開式: 前述各項每一行後面都不加任何標點符號者稱為 Open Punctuation。開式標點多配合 Full Block Form 使用。

3. 混合式: 前述各項除了 Salutation 後面用逗號 (,) 或冒號 (:) 及 Complimentary Close 後面用逗號 (,) 外, 其餘都不加標點符號 者稱為 Mixed Punctuation, 現在大都採用混合標點法。

Indented Form, Closed
Punctuation

Full Block Form, Open
Punctuation

Semi-Block Form, Mixed
Punctuation

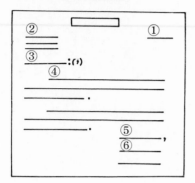

Modified Block Form,
Mixed Punctuation

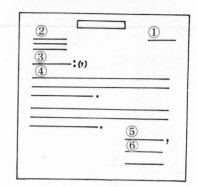

說明：

① Date　　　　　② Inside Address　③ Salutation

④ Body　　　　　⑤ Complimentary Close ⑥ Signature

習　題

一、試以圖解方式說明一封商用英文信的構成部分。

二、試將下列日期，以美國式及英國式寫出來：

　　1. 1990 年 3 月 5 日

　　2. 1990 年 6 月 12 日

三、試將下列地址，用 Full block with open punctuation 及 Indented form with closed punctuation 方式寫出來：

　　1.臺北市武昌街 2 段 37 號。

　　2.臺北市新生南路 1 段 1 號貿易大樓 3 樓 101 室

四、試將下列人名、地址及日期，分別用 Full block form with open punctuation 及 Indented form with closed punctuation 方式寫出：

　　1.印度，Calcutta,Gandhi Road,100 號,S. Jimoh & Son Co.

　　2.美國，郵遞區號 90848，加州，洛杉磯市，West Sixth St. 6204 號，Sears,

Roebuck & Co., Inc.

五、將下列用語譯成英文。

　1. 親啓

　2. 機密

　3. 內有照片

　4. 印刷品

　5. 附件如文

第三章　撰寫商用英文信的七要

(Seven C's of English Business Letter Writing)

一封完善的商用英文信應具備七 C(Seven C's)的要件，所謂七 C 是指：Completeness, Clearness, Correctness, Concreteness, Conciseness, Courtesy 及 Consideration 而言。茲分別說明於下：

第一節　完　備(Completeness)

所有的信都是爲某種目的而寫。那麼爲達成此目的，就應將其文意加以完備(Complete)的叙述，不可有所遺漏。假如一封信寫得不完備(Incomplete)，不但不能達成目的，反而可能引起反效果。例如 Dispute(糾紛)、Complaint(訴怨) 或 Claim（索賠）等等卽往往因爲不完備的通信而引起。

【例 1】

a.　（不完備）

Gentlemen,

We are returning the goods you shipped, as it arrived too late.

Yours truly,

這種惜字如金的信，除非賣方標榜"Return Goods Welcome"（歡迎退貨）另當別論外，就 Completeness 而言，可謂一文不值。謹愼的商人(careful merchant)起碼應改寫如下：

b.　（完備）

Gentlemen:

We are returning the cotton goods, your invoice No. 105, by s.s. "President" today.

This was received May 2, too late for our Spring Sale. You will find on reference to our order of January 3, that this was to be shipped to reach us not later than March 15.

Yours truly,

【例 2】

a.　（不完備）Shipment will be made *in due course*.

b.　（完備）Your order for 500 doz. umbrellas will be shipped at the end of this month. You should receive them early next month.

第二節　清楚(Clearness)

寫信時應推開窗子說亮話，措詞明白清楚，使閱讀的人能立即明瞭內容，不致引起誤會。商用英文最忌含糊(obscurity)、模稜兩可(ambiguity)、不明確(vagueness)。

一、避免使用意義不明確的語句

【例 1】

（含糊）*Fluctuations in the freight* after the date of sale shall be for the buyer's account.

"Fluctuations in the freight"一詞含有運費上漲或下跌之意，因此欠明晰，容易引起爭執。如改寫成如下，就清楚明白了。

（清楚）*Any increase* in the freight after the date of sale shall be for the buyer's account.

（出售日以後運費如有上漲，歸買方負擔。）

（清楚）Any change up or down（or Any increase or decrease）in the freight after the date of sale shall be for the buyer's account.

（出售日以後運費如有上漲或下跌，均歸買方負擔。）

【例 2 】

（含糊）As to the steamers sailing from Keelung to Sydney, we have *bimonthly* direct services.

"bimonthly"有「一個月兩次，即每半個月」或「兩個月一次」之意，因此欠明確。

（清楚）We have two direct sailings every month from Keelung to Sydney.

（……每個月二次）

（清楚）We have semi-monthly direct sailing from Keelung to Sydney.

（……每半個月一次）

（清楚）We have a direct sailing from Keelung to Sydney every two months.

（……兩個月一次）

二、修飾詞應放在適當的地方

【例１】

　　a) We shall be able to supply 100 cases of the item *only*.

　　b) We shall be able to supply 100 cases *only* of the item.

　　a) 為「以此商品為限」，b) 為「以一百箱為限」，由此可見修飾詞的位置的不同，其意思迥異。

【例2】

(含糊) The goods we received *contrary to our instructions* are packed in wooden cases without iron hoops.

原意是想說「木箱沒有鐵箍條,是違反我們的指示」,但由於"contrary to our instruction"放錯地方,變成「違反我們的指示而收到…」。

(清楚) The goods we received are packed in wooden cases without iron hoops *contrary to our instructions*.

三、使用修飾詞應注意其聯貫性(Coherence)

1. 由副詞、副詞子句、關係代名詞所構成的 Clause,應盡可能將 Clause 放在被形容之詞的旁邊,但非修飾文中特定語的 Sentence adverb,應盡可能放在文首或文尾。

(含糊) The proprietor's personality and sincerity have made, *in our opinion*, the firm what it is today.

由於"in our opinion"插在及物動詞與其受詞之間,使句子的意義不明,如將它放在文首就清楚了。

(清楚) In our opinion, the proprietor's personality and sincerity have made the firm what it is today.

(我們認為該店主的人格及誠懇使他的店有今日的成就。)

2. "accordingly"有"therefore"(因此)及"in agreement with what has preceded"(遵照……)兩種意義,如當作後者之意時,應放在句尾。

If you will give us 24 hours' notice, we will arrange accordingly.

(如於 24 小時前通知我們,則我們將遵照指示安排。)

3. 句子中避免使用數個關係代名詞。

(含糊) We have received with thanks your letter of May 15, *which* instructs us to purchase for your account 1,000 tons barbed wire, BWG ♯ 12, *which* we have today filled at US$190 per M／T.

第二個 which 以 barbed　wire 爲先行詞，結果變成 to　fill barbed wire 意思不通，因爲只能說 to　fill　order,而不能說 to fill goods.

(清楚) We received with thanks your letter of May 15 *instructing* us to purchase for your account 1,000 tons barbed wire, BWG ♯ 12,and have today filled *your order* at US$190 per metric ton.

四、注意代名詞、關係代名詞與其先行詞的關係位置

(含糊) Mr. Chang wrote to Mr. Ford that he had received his order.

(張先生寫信給福特先生說他的訂單他已收到了。) 這裡的"his" 是誰不明確。因爲可以是"Mr. Chang"也可以是"Mr. Ford"。

(清楚) Mr. Chang wrote to Mr. Ford that he had received Mr. Ford's order.

(張先生寫信給福特先生說福特先生的訂單他已收到了。)

第三節　具體 (Concreteness)

所謂"Concreteness"就是要言之有物，措詞、文意切中要領，據實直陳，切忌空泛、抽象。

一、避免抽象籠統

譬如說貨色好，不要單說 best, fine, highest grade, supreme (高

級）等抽象的形容詞。一定要將它的優點一一具體的臚列出來。抽象的字眼非但人家不會相信，而且意義上也不明確。

【例１】

(不具體) Our apples are excellent.

(具體) Our apples are juicy, crispy and tender.

【例２】

(不具體) Dear Sirs,

> We acknowledge with thanks receipt of your inquiry of May 2.

> We can supply you with *the types of Cigarette Lighters you described. If you send us your reply soon, we might be able* to meet the due date you mentioned.

> Awaiting your early reply.

> > Yours very truly,

這封信有不少不具體的敍述。除非 letter subject（案由）已將品種、數量等標示，否則收信人會懷疑寫信人對該品種及數量是否已正確了解。

"Send us your reply soon"一句中，"reply"是抽象的字，並未具體要求怎樣的答覆。"Soon"一字太籠統。"We might be able to..."有「也許可能，也許不可能」之意，使人覺得不可靠。

(具體) Gentlemen,

> We shall be pleased to supply you with *two dozen "Flamex"Cigarette Lighters according to the specifications* in your letter of May 2.

> *If you will telegraph us confirmation of the order by May 20, we will send them by air freight not later than May 30.*

We look forward to the opportunity of serving you.

Yours very truly,

二、日期、期間應明確表示

Early in July; late in August; begining part of May; middle part of April; later part of June; recent;in due course 等語句均含糊不具體。

【例 1】

(不具體) As requested in your *recent* letter we have *already* sent the samples to you.

這句中 recent, already 等字眼均不具體，應將日期具體寫出。

(具體) As requested in your letter dated October 12, we sent you the samples by air parcel on October 20.

【例 2】

(不具體) a) You will receive our reply about it *in due time*.

b) We enclose a copy of our latest catalog which you requested us to send *several days ago*

in due time（在適當時期），several days ago,等語句，均含糊不具體。

(具體)　a) You will receive our reply about it within one week.

b) We enclose a copy of our latest catalog , as requested in your letter of November 8.

三、明確寫出文件、單證的案號(reference)

(不具體) We have effected shipment under your L／C.

這句對於開狀銀行、信用狀號碼均未具體指出。

(具體) We have effected shipment under your L／C No. 123 issued by Bank of America.

第四節　簡　潔 (Conciseness)

寫商用書信應力求簡潔扼要，切忌冗長。須知商場中人，業務倥傯，最重視時間，無功夫閱讀連篇累牘的信。因此，商用書信，應在辭能達意的原則下，使用簡單淺明的辭句。

一、以簡單淺明的文字代替長而艱深的字

長而艱深	簡單淺明
apparent	clear
approximately	about
ascertain	find out
commence	begin
compensation	pay
conclusion	end
contribute	give
demonstrate	show
endeavour	try
equivalent	equal
expedite	hasten
facilitate	made easy
initiate	begin
modification	change
participate	take part
perform	do
procure	get
reimburse	pay

render	give
transmit	send

二、將數語變爲一語

at an early date	early, soon
at all times	always
at the present time	now
at this writing	now
be in a position to	can
by means of	by, with
by virtue of	by, with
costs the sum of	costs
due to the fact	as, because
for the reason that	as, because
in view of the fact that	since
during the course of	during
final completion	completion
for the purpose of	for
in accordance with	by, with
in spite of the fact that	although
in the amount of	for
in the case of	if
in the event that	if
in the meantime	meanwhile
in the near future	soon
in the neighborhood of	nearly, about, around
in this place	here

in view of	since, because
look forward to	await
on a few occasions	occasionally
on the part of	for, among
owing to the fact that	because
we would ask that	please
previous to	before
there can be no doubt that	doubtless
under separate cover	separately
with reference to	about

三、將子句變成片語

(冗長) Please read the third paragraph of our letter No. 125 of August 5 *so that you will get all the facts*.

將"so that you will get all the facts"的 clause 改爲 phrase 就較簡潔。

(簡潔) Please read the third paragraph of our letter No.125 of August 5 for all the facts.

四、將冗長句子化爲短句子

(冗長) *We take the liberty to approach you with request that you would be kind enough to* introduce to us some exporters of iron scrap in your city.

(簡潔) *Please* introduce to us some exporters of iron scrap in your city.

(簡潔) *We shall appreciate* your introducing to us some exporters of iron scrap in your city.

五、儘可能使用主動語態(Active Voice)藉以簡化句子

【例 1 】

(Passive):It is noted that the sales volume has been increasing.

(Active) :We noted that the sales volume has increased.

【例 2 】

(Passive):It is suggested that your consideration be given to the recommendations in this report.

(Active) :We suggest that you consider the recommendations in this report.

由上述例子，可知一般而言，主動語態較被動語態簡潔而且易於瞭解。

六、善用連接詞(Connectives)

簡潔並不是說亂將句子縮短，假如將短句子一連串並列，將顯得彆扭。爲避免這種情形，應善用連接詞。

(拙劣) Thank you for the samples. You sent them to us on June 20. We are pleased to place an order. It is specified on the enclosed order sheet.

這些句子是夠簡單了，但顯得很拙劣，不自然。

(較佳) Thank you for the samples which you sent to us on June 20. We are pleased to place an order as specified on the enclosed order sheet.

簡潔並不是簡陋，簡潔須兼顧 Clearness, Correctness, Courtesy 及 Completeness 等等。

第五節　正　確(Correctness)

商用書信是科學化的應用文學，正確乃爲不可或缺的要素。因爲商用

書信事關權利、義務、利害關係, 如內容不正確, 則易引起糾紛。所謂正確性包括下列各要件:

1. 敍述正確(Correctness of Statement)
2. 數字的精確(Accuracy of Numerical Expressions)
3. 商業術語的正確使用(Correct Use of Commercial Terms)
4. 正確的文法(Grammatical Correctness)

　　茲分述於下。

一、正確的敍述

　　1. 勿 Overstatement (誇張) 或 Understatement (抑制): 商業最重信用, 騙人只能騙一遭, 與其過甚其詞, 將來損失信用, 倒不如老老實實, 既不過份誇張, 也不過份含蓄或抑制。

【例 I】

(誇張) This stove is *absolutely the best* on the market.

　　　　這是老王賣瓜, 自說自誇的措詞, 既抽象又武斷, 不如將其性能、優點具體地向客戶介紹。

(正確) Our model TR 123 stove is designed on modern lives and gives, without any increase in fuel consumption, 25% more heat than the older models. So you will agree that it is the outstanding stove for economy.

【例 2】

(誇張) It is the *lowest* price *available* to you.

　　　　假如競爭者, 報出更便宜的價格, 豈不有欺騙對方的嫌疑?

(正確) It is the *lowest* price (that) *we can offer you now*.

(正確) It is a *modest* price (which) *we can quote you now*.

　　　　下列的用語都是加強語氣的字眼, 如亂用, 易構成 overstatement, 所以使用時應謹慎。

completely	strictly	entirely
exceedingly	extremely	
greatly	definitely	perfectly
quite	especially	utterly
very	very much	all
best	only	utmost

2. Understatement 與客氣的措詞: somewhat, nearly, as it were, in a measure, it seems to us 等字眼常使文章語氣顯得軟弱無力。如用得不恰當, 就易構成 understatement。如想表示客氣, 可使用如下的 Past Subjunctive (過去假設語氣)。

We *should* be glad to receive it.

You *might* report to us once a week.

二、數字的正確表示法

商用英文信中涉及數字的情形很多。數字的表現如發生錯誤易造成糾紛。尤其, 以「以上」,「以下」表示數量、金額; 以「以前」,「以後」表示日期時, 應特別注意, 以免收信人誤解。茲將其正確表現方法列記於下:

1.提及的數字包括在內者

50 打以上

50 dozen or up, 50 dozen and upwards

US$100 以下

US$100 or less, US$100 and below

六月卅日以前

on or before June 30

六月卅日以後

on or after June 30

從六月卅日起

$$\begin{cases} \text{on and from (after) June 30} \\ \text{as from June 30} \end{cases}$$

2.從……到……為止（包括最後數字在內）

40鎊為止

up to £ 40 inclusive

十月廿日為止

up to and including October 20

3.頭尾均包括在內

從三月一日到十五日

from the lst to the 15th of March both inclusive

從星期一到星期五為止

from Monday through Friday

三、使用正確的商業術語

　　使用商業術語可節省很多時間與說明。但使用時，不僅對該術語的意義須完全瞭解，而且必須是商界通行的術語。

　　1.慎用容易發生誤會的術語：例如FOB後面通常加以裝船港名稱，但在美國，FOB有六種不同的解釋，因此與美國人交易時，應在FOB與裝船港之間加"Vessel"等字樣，例如：FOB Vessel New York（紐約港船上交貨條件）。

　　2.不用意義寬泛的字語：例如，與其用寬泛的"economic"（經濟的）不如用"money-saving"（省錢的）或"time-saving"（省時的），更不如用"dollar-saving"或"minute-saving"；與其用寬泛的"say"不如用"tell"，"inform"，"assert"（主張說），"state"（陳述）。

　　3.應該用意義特定的字眼：例如「利息」應用"interest"，「利潤」應用"profit"，不要用普通的"advantage"（利益）。

四、正確的文法

　　文法是從語言文字的習慣逐漸形成的公認的法則。如果我們作文說話不講文法，別人將不瞭解我們的意思。因此撰寫商用英文信時，必須講求文法，而且還要顧到商用英文信中的習慣用法，關於此，學者宜研讀一般英文法書籍。

第六節　謙　恭(Courtesy)

　　謙恭有禮是商場上的重要法則。接待顧客固然要有禮貌，寫信時也不能缺少禮貌。有禮貌的信很可博得收信人的好感。所以書信中如能適當地應用"Kindly", "Please", "Thank you", "We have pleasure", "We regret that"等用語，必能引起閱信人的好感。但亂用恭敬的詞句，也是不對的。例如在"please find enclosed cheque"（敬請查收附上的支票）中，用"please"就不對。如沒有請求收信人賜惠，或沒有過份麻煩收信人時，在商用英文信中，"please"一詞大可不必濫用。「敬請」一詞，不能像在中文信中那樣隨便加上。

　　要如何寫出有禮貌的信呢？

一、善用被動式(passive)語氣

　　同樣一件事的敘述，用主動式表達則常有譴責對方的感覺，如改用被動式，則語氣就緩和的多。

（無禮）1. You made a very careless mistake.

　　　　2. You did not enclose the cheque with your order.

　　　　這些主動式的句子頗有嚴厲譴責對方之感，顯得欠缺禮貌。如將"you"省略，並改以被動式，則語氣委婉，不致刺激收信人。

（有禮）1. A very careless mistake was made.

　　　　2. The cheque was not enclosed with your order.

二、善用含有客氣語意的詞句

(無禮) 1. We demand immediate payment from you.

2. We must refuse your order.

"demand"一詞有不客氣地要求之意, 語氣太重, 如改用 "request"則顯得客氣, 使人看起來較舒服。"We must…"頗有威脅之感, "refuse"也語氣太強。

(有禮) 1. We request your immediate payment.

2. We regret that we are not in a position to accept your order.

三、以肯定(positive)詞句代替否定(negative)詞句

否定詞句常會引起對方不快, 如改以肯定語氣, 則可博得對方好感:

【例 1】

(否定) We do not believe you will have cause for dissatisfaction.

(肯定) We feel sure that you will be entirely satisfied.

【例 2】

(否定) It is against our policy to sell leftover goods below cost.

(肯定) It is our policy always to have full stocks of goods at fair prices.

四、善用婉轉句法(mitigation)

使用婉轉語法以免刺激對方, 此類常用詞句有:

We are afraid	It seems(would seem) to us
We would say	We(wound) suggest
We may say	As you are (may be) aware (如你所知)
We might say	As we need hardly point out (不用指出)

【例 1】

(無禮) It was unwise of you to have done that.

(婉轉) *We would say that* it was unwise of you to have done

that.

【例2】

(無禮) You ought to have done that.

(婉轉) *It seems to us that* you ought to have done that.

【例3】

(無禮) We cannot comply with your request.

(婉轉) *We are afraid* we cannot comply with your request.

使用婉轉句法時，應注意不可違反 Concreteness。

五、使用 Subjunctive mood

表達希望或意見時，可用含蓄語法，以示有禮貌。

例如:

1. *Would* you compare our sample with the goods of other firms?

2. We *should* be grateful if you *would* help us with your suggestions.

第七節　體　諒 (Consideration)

寫信人在寫信時，不能一味地只顧從自己的立場著想，而應設身處地，為對方著想，這就是"You　Attitude"或"You　View-point"或"The reader's point"。具體地說，提起任何事物，應在可能範圍內少用第一人稱的"I"，"We"，"My"，"Our"如何如何，而應多說些"You"，"Your business"，"Your profit"，"Your needs"如何如何，要把對方的利益放在讀者眼前。以「收信人的利害關係」為前提，是撰寫商用書信的重要原則。因此，書信中的每一段 (paragraph) 的開頭宜多用"You"而少用"We"。

"WE" expression	"YOU" expression
1. *We* are pleased to announce that...	1. *You* will be pleased to know that...
2. *We* follow the policy because...	2. *You* will benefit from this policy because...
3. As to *our* standing, *we* refer you to Bank of America...	3. As to *our* standing, *you* may address any inquiry to Bank of America...

然而少用"We"起頭並不是說要勉強地以"You"起頭，而故意不用"We"。假如譴責、責問、訴怨的信也滿篇"You"，"You"，不但不禮貌，反而易招致惡感。例如下面的句子是以"You"起頭，句子中也一再提起"You"，然而卻是很拙劣的句子。

You attribute your negligence to the fact that *you* are very busy, but it cannot be believed that *you* must take so long in reply.

由上面的例句可知，Consideration 與 Courtesy 是不可分的。

習　　題

一、將下列句子改寫，使其符合完備的原則。

　　1. Please send me two pairs of slippers for my customers at an early date.

　　2. Shipment will be made in due course.

二、將下列句子改寫，使其符合清楚的原則。

　　1. Mr. Chang wrote to Mr. Wang that he had accepted his proposal.

　　2. An instance may be recalled to your memory which will confirm our

opinion.

三、將下列句子改寫，使其更具體。

1. The ABC Company is one of our big customers.

2. We make the finest blankets.

四、將下列句子改寫，使其更簡潔。

1. In compliance with your request, we immediately contacted ABC Co., Taiwan, and now wish to inform you of the results as follows:

2. We expect you to send us your check by April 30 because we have bills of our own that must be met.

五、試以 Active Voice 將下列句子予以簡化。

1. It is desired by Mr. Swain that this be called to your attention.

2. It is occasionally found that one of our customers has been unintentionally missed by our representatives.

六、改錯。

1. Having received your cable of yesterday, our thanks in due for your kindness.

2. My Dear Sir:

七、試將下列文字予以簡化。

1. an actual fact

2. due to the fact that

3. first of all

4. in a most careful manner

5. many in number

第四章 舊式文體與新式文體

(Old-fashioned Styles v. Modern Styles)

第一節 舊派與新派之爭

關於商用英文信的用字，一直存在着舊派與新派無休止的論爭。新派主張商用英文信不應：

1.使用陳腔濫調的客套。2.使用深奧的詞語。3.使用詞意模糊的詞語。4.攙雜閒語。5.以分詞式作爲結尾。

然而舊派仍繼續使用一些新派所指責的「舊式用語」(Old-fashioned expressions)，並未向新派低頭。就實際情形來看，銀行、保險公司、船公司以及若干大型的公司或公營事業機構比較保守，仍繼續使用所謂的「舊式用語」，而較小型的公司或民營公司則有不使用「舊式用語」的趨勢。

第二節 舊式用語與新式用語的比較

茲將商用英文信中常見的舊式用語與新式用語作比較，讀者如能仔細比較對照，並瞭解其意義，則在運用時，當能作適當的選擇。

1. Above：舊派將"Above"當作形容詞，習見常用的有：

the above statement （上列的聲明）

the above address 　（上開的地址）

the above date 　　（上載的日期）

新派認爲不應將"above"當作形容詞用，而宜以"foregoing"，"preceding"或"above-mentioned"等字來代替，例如：

the foregoing statement

或將"above"正式地當做副詞用，例如：

the address given above

2. Advise: 舊派將"Advise"當作「通知」解釋，例如：

Please advise us of the shipping date. (請通知我們裝運日期)

新派認爲"Advise"的正確意義是「勸告」或「提供意見」，只能用於請教人家進告意見的時候。例如：

Please advise us how to proceed. (請告訴我們如何進行)

而不能用作"inform"，"notify"或"tell"解釋，因此：

Please advise us of the shipping date.應改爲：

Please inform(or notify) us of the shipping date.

3. Acknowledge receipt of; Acknowledge with pleasure; Acknowledging yours of:舊派認爲這些片語用起來堂皇鄭重，因此在很多書信中，尤其大的公司行號都樂於使用。新派則認爲這些片語，裝腔作勢，不平易，因此，應予摒棄不用。

例如：

We acknowledge receipt of yours of (June 3)…應改爲：

We thank you for your letter of (June 3)…或

We have received your letter of (June 3)…

4. And oblige: 舊派將"And oblige"這個詞放在信中的最末尾，接以"yours truly"以示謙恭，猶如中文書信中的「爲荷」、「爲禱」、「爲感」。新派認爲這個詞，顯得卑屈，毫無意義，而且減弱了結尾的語氣，應予刪除。例如：

Kindly ship the enclosed order and oblige. (請將訂貨卽行運下

爲感）應改爲：

Please ship the enclosed order immediately.

5. As per: 舊派將"As per"當做「依照」，「如...」解釋，"per"爲拉丁文，等於"by", "through"。例如：

As per your letter

As per your instructions

As per offer sheet enclosed

新派認爲這種用法太陳腐，而且在同一信內混用兩國文字不妥當。除了法律文字外，不要用，而應改爲：

According to your letter

According to your instructions 或 As instructed by you

As described in the attached offer sheet

6. At an earliest possible moment; At an early date; At your earliest convenience; At your convenience:舊派將這些片語用作「儘速」，「得便務請從速」，「有便卽請」，新派認爲不簡潔、不確定、陳腐、曖昧 (vague)。應改用更 specific 的詞，諸如"at once", "soon", "immediately", "promptly"或"as soon as you can"或寫出具體的日期。例如：

Please notify us at an early date.應改爲：

Please let us know immediately (or within 10 days).

We should appreciate hearing from you at your earliest convenience.應改爲：

We should appreciate hearing from you immediately.

7. Attached hereto; Attached herewith; Attached please find; Attached you will find: 舊派將以上片語用於「附上」，「檢附」之意。例如：

Attached herewith is an invoice.

新派認爲"hereto", "herewith", "please find"這些詞是多餘的, 應予刪除, 以求簡潔。例如:

Attached is an invoice.

8. At the present writing: 這是「在寫這信的時候」之意, 舊派很喜歡用這片語。例如:

At the present writing we have none in stock.

　　（目前我們存貨一點也沒有）

新派認爲商用書信在能表達意思的前提下, 應盡量運用簡短而有力的字句。"at the present writing"這個片語本身雖無弊病, 而且過去曾有一段時期被認爲是一種很時髦的用語, 不過在講求效率的現代商業社會, 還是直截用"now"一詞來得自然、簡潔、沉着有力。例如上面的句子改爲:

We have none in stock now.

9. Awaiting your letter (order, reply); Awaiting your further wishes: 這是「等待你的信（訂單、答覆）」及「等候你再惠顧」的舊式用法。新派認爲以分詞式開頭爲過時的用法, 而且"Awaiting your further wishes"的 expression 也未免迂迴, 不如開門見山, 改爲:

Please give us your further orders.

至於"Awaiting your letter"則改成:

Your early reply will be appreciated.

We hope to hear from you soon.

Please let us hear from you.

效果較佳。

10. Beg to acknowledge; Beg to inform; Beg to advise; Beg to assure; Beg to call your attention; Beg to confirm; Beg to state; Beg to suggest; Beg to enclose; Beg to say: 以上均在舊式英文信中習見。新派

認爲這些都是老古董，過份謙卑，除用於下級人員對於上級人員呈文外，不應用於普通商用書信。卽使用"Beg"一詞也該加上"leave"，"permission"等詞。例如：

Beg leave to say（敬告）

I beg permission to go（請准放行）

現在是民主時代，大家平等，像 Beggar 那樣拼命 Beg，未免卑屈，反而使人瞧不起。所以應改成：

We acknowledge（we have received…）；We inform you; We assure you; We call your attention; We confirm; we write（用以代 beg to state to you）; We suggest you; We enclose you

11. By return mail; By return of mail; By return post; By return of post; By return：「請卽回示」「收到卽覆」之意。

例如：

Please answer by return of post.

這些語法，在郵政不發達時，很習見。新派認爲這種語法已不合時代，而且與"at an early date"同樣是濫調，不夠 definite，應以"immediately"，"promptly"，"at once"等詞代替或明確指出日期。例如：

Please answer promptly.

12. Contents noted; Contents have been noted; Contents daly noted; Contents carefully noted：舊派的客氣套語，其意爲「內容均悉」，「內容詳悉」或「敬悉一是」。例如：

Yours of June 25 to hand, contents of which have been carefully noted.（6 月 25 日函已收到，詳悉一切內容。）

新派認爲這些用語在信中是冗句，徒佔篇幅，毫無意義，犯了繁滯重複的毛病。不知道來信內容，如何覆信呢？所以應改爲：

We have received your letter of June 25.

　　　　a) which we have read carefully.

　　　　b) which we have read with interest.

　　　　c) regarding the shipment by s.s. "Peseus".

13. The captioned claim; The captioned matter：信函列有標題（主旨）時，在本文中，常有"the captioned..."的用語。其意爲「如主旨所列事宜」。美國的銀行或保險公司的文件中常出現這種用語。新派認爲此字太古老，在一般商用書信中不宜採用，而宜改用"the above claim"，"the claim above referred to"，"this claim"，"the above matter"，"this matter"。

14. Enclosed find; Enclosed please find; Enclosed you will find; Enclosed herewith：這些用語意指「茲檢附…請察收爲荷」，「茲檢附…」。例如：
Enclosed please find a copy of our price list.
新派認爲這種措詞未免太滑稽，旣然已隨函附上，難道還要發信人請求發現（please find），收信人才會去覓找？不要把收信人當傻瓜吧！
因此應改爲：
We enclose a copy of our price list.或
We send you enclosed a copy of our price list.
再者"enclose"後面不應再加上"herewith"字樣。

15. Esteemed favour："Esteemed"爲「可敬的」、「受尊重的」之意，"favor"爲「信札」（letter）之意。所以舊派將「大札」、「尊函」稱爲"esteemed favor"新派認爲"favor"爲「恩惠」之意。收到人家指責或索債的信，稱其爲"favor"已經不恰當，再加上"esteemed"，更顯得矯飾過份。業務上的通信也不必用此 overpolite expression,因此應改爲"kind letter"或只說"letter"就夠了。此外如"esteemed letter"；"esteemed inquiry"中的"esteemed"也應予刪除。

16. Hoping for an immediate answer; Hoping to hear from you; Hoping

to be favored with a reply:「希卽賜覆」,「希賜覆」之意。舊派商用英文,常將這些措詞作爲信尾結束之用。新派認爲這種 expression 充分顯示寫信人的矜持與無能。因爲只敢「希望」(hope)而已,足見沒有自信心。宜改爲:

We are looking forward to your prompt reply.或

We are looking forward to hearing from you soon.

17. I remain; I am:舊派常將這些用語作爲信尾結束之用,例如:I remain yours truly (or sincerely, etc.) (你永遠忠實的) 也卽中文信的「謹上」。因此"I remain"後面不應有逗點(comma)。舊派的信,雖然喜歡用此詞,卻在"I remain"後面加上逗點,這是畫蛇添足也。而且"I remain"係用於第二次以後的通信才能用,初次通信,頂多只能用"I am yours truly."新派認爲這種客套太陳舊。因此:

Hoping to hear from you in the near future, I remain.應改爲:

I look forward to hearing from you soon.

而:

Thanking you for your kind consideration, I am.應改爲:

Your consideration is appreciated.

18. In answer to same; Same; The same:舊派將"same"當作代名詞用,等於中文的「該」,「這」或「他,他們,它,它們」(he, she, they, them, it, this)等等,其所指的主體可能單數,也可能多數。例如:

We regret the delay, and hope that same has not caused you inconvenience.

(對於遲延一事,我們很抱歉,希望這一延擱,並未引起你們的不便。)

新派認爲"Same"所代表的主體不明,易引起誤會,除了法律文書外,不應使用,而應用具體的代名詞"it", "they", "them", "him"等。例如:

We regret the delay and hope it has not caused you inconvenience.

19. In due course; In due course of time; In due time: "In due course" 爲「適時」,「屆時」之意; "In due course of time"爲「在相當時間內」之意; "In due time"爲「在適當時間內」之意。例如:

We have received your order and it will be processed in due course of time.

新派認爲這些措詞均有含糊不肯定之嫌, 且語氣軟弱, 應說出具體的日期。例如:

Thank you for your order No. 123. We will ship it by September 30.

20. In reply to your favor; In reply, would state that; In reply, wish to say; In reply, we wish to state that: 都是「謹覆」之意。新派認爲寫商用書信不是投稿, 需按字計酬。因此應力求簡潔, 開門見山, 不要拖泥帶水, 廢話連篇。所以:

"In reply, we wish to state that the samples you requested have been airmailed today." 一句中的 "In reply, we wish to state that" 這一段應予刪除, 而直書爲:

The samples you requested have been airmailed today.

21. Inst.; Ult.; Prox.: 這三個字分別爲 "instant" (本月), "ultimo" (上月) 及 "proximo" (下月) 的縮寫。例如:

Your inquiry of the 4th inst...

Yours of the 9th ult. received.

Your order will be shipped on the 10th prox.

新派認爲現代的商用書信貴在明確, 用了這些字, 讀者必須在腦海中多打一個轉才能領悟實際的月份。何況 "ultimo", "proximo" 又是拉丁文，犯了用字生僻的毛病。因此應改爲：

Your inquiry of June 4...

We have received your letter of May 9.

Your order will be shipped July 10.

22 Kind favor; Kind order：「親切（或仁慈）的信或訂單」。例如：

Thank you for your kind favor (order).

新派認爲用此措詞時應小心，因爲不是所有的來信都是「親切」的。至於"kind order"等於中文的「親切的訂單」，這種措詞也顯得怪彆扭。因此應改爲：

Thank you for your letter (order).

23. Kindly advise; Kindly be advised; Kindly inform：分別爲「請賜告」，「敬告」，「請賜知」之意，例如：

Kindly advise (inform) us when our order will be shipped.

Kindly be advised that your order has been shipped.

新派認爲以"kindly"代替"please"是多餘的，而且以"advise"代替"inform"；"tell"也不合理。同時，"kindly be advised"是說寫信人的通知是"kindly"並非說收信人的動作是"kindly"，因此這種措詞也很滑稽。所以應改爲：

Please tell (inform) us when our order will be shipped.

We are pleased to inform you that your order has been shipped.

24. Our records show; According to our records; Our records do not show：分別爲「我們的紀錄顯示」，「依據我們的紀錄」及「我們的紀錄不顯示」之意。

新派認爲這種表現法犯了兩個毛病①夜郎自大，好像自己的紀錄最正確、最權威。②逗圈子。因此，不如改用：

We find... (我們發現…)

We do not find...或 We fail to find... (我們未發現……)

25. Permit us to say; Permit us to explain: 分別為「希准我們說」,「希准我們來解釋」之意。這種客套為舊派所樂用。新派認為在信裡要說什麼, 由寫信人自行作主, 不必要收信人允許。何況寫信時已說出在先, 又何必請求准許說出於後? 豈非挪揄? 所以這是多餘的。

26. Please be advised that: 舊派人士迄今仍很喜歡用此措詞以表示「茲奉告」之意。例如:

Please be advised that we shall execute your order in the near future.

新派認為以這種措詞作為敍述某一事的起頭毫無意義, 要說什麼直說好了, 何必多費筆墨。因此, 宜改為:

We shall fulfil your order by October 2.

27. Recent date; Recent favor: 分別為「近日」及「近日的信」之意。例如:

We acknowledge the receipt of your recent favor.

In reply to your letter of recent date, we wish to state the samples you requested have been airmailed today.

新派認為這種不明確指出日期的措詞, 係草率而不禮貌。應改為:

We have received your letter of July 3.

The samples you requested in your letter of June 3 have been airmailed today.

28. Referring to; Regarding; Referring to yours of...wish to say that: 「關於…」之意, 例如:

Referring to yours of June 6 wish to say that we are pleased to enclose a copy of our latest catalog in accordance with your request.

新派認爲以分詞做爲開頭的老套應盡量避免，也不要用"yours"來代替
「來信」。至於"wish to say"這種庸俗的措辭也應予摒棄。因此宜改爲：

We enclose a copy of our latest catalog as requested in your letter of
June 6.

29. Said; The said：「該」，「上述的」之意。例如：

    We have notified them of (the) said arrangement.

    新派認爲商用書信中大可不必賣弄法律或契約用語，應予摒棄。因此，
    應改爲：

    We have notified them of the arrangement.

30. Sorry to say：「抱歉」，「可惜」之意。例如：

    We are sorry to say that we do not have these goods.

    （我們沒有這些貨，抱歉之至）

    新派認爲應該寫成：

    We are sorry that we do not have these goods.

    否則意思混淆，到底那「所說的話」使你抱歉呢？還是那「事實本身」
    使你抱歉呢？總之，現代商業書信已進步得像自然科學一般，要講求
    邏輯及清晰，絕不能如舊式書信那樣含糊不清。

31. Take the liberty of：「冒昧」之意。例如：

    I take the liberty of requesting you to send me a copy of
    your latest price list.

    新派認爲"take the liberty of"這個曾經出盡風頭的謙虛措詞應該退
    休了，還是直截了當地說出吧！例如改爲：

    Please send me a copy of your latest price list.

32. Thank you in advance; Thank you in anticipation：「謹先致謝」之意（請
    託用語）。例如：

    Kindly mail us any information you may have for removing

oil spot. Thanking you in advance for the favor, we remain.

　(請寄下有關去油漬的資料，謹先致謝)。

新派認爲該謝的時候再致謝不爲遲，預先致謝會使收信人覺得非爲你做不可。這種客套語，與其說謙恭，不如說不禮貌，如一定要先致謝，也應用"appreciate"(感激)此字，等事成後再說"Thank you"也不爲遲。所以應改爲：

We shall appreciate any information you may have for removing oil spot.

　(如能示知去油漬的資料，將不勝感激)

33. Taking this opportunity: 「趁機」、「藉此機會」之意。例如：

Taking this opportunity, I wish to inform you that I shall visit your country next month.

　(藉此機會奉告我將於下月訪問貴國)

新派認爲何不直接談正題呢? 用此措詞頗有藉機作書之嫌。因此，應改爲：

I shall visit your country next month.

34. Under separate cover: 「另封」之意。例如：

We are sending you under separate cover a copy of our catalog.

　(另寄上本公司貨品目錄一份)

新派認爲：如要表明另外寄上，必需註明分寄的方法，或不說「另寄上」或只說「寄上」也可。因此，宜改爲：

We are sending you by air parcel post a copy of our catalog.

We are sending you a copy of our catalog.

35. The undersigned: 「下面的簽字人」，卽寫信人的自稱。例如：

The undersigned wishes to state that we take extreme pleasure in accepting your esteemed order.

這種表現法顯然想給收信人謙恭的印象而避免使用"I", "we"。然而，弄巧成拙，新派認爲不要怕使用"I", "we"，而應改爲：

I am delighted to accept your order.

36. We shall be in a position to:「我們將能…」之意。

新派認爲這種表現法不夠簡潔，應改爲：

We shall be able to...

37. Will appreciate; Will be glad; Will be pleased:「將感激」，「將高興」，「將樂於」之意。例如：

We will appreciate your giving us an opportunity to display our merchandise.

We will be glad to discuss this matter more fully with you.

這些表達方式雖常有人用，可是新派認爲將"will"當做"shall"用是錯的，應將"will"改爲"shall"。

38. The writer:「筆者」之意。例如：

The writer wishes to acknowledge receipt of your letter.

The writer appreciates your efforts.

這是爲了給收信人有禮貌的印象而避免使用"I", "We"。新派認爲與前述"The undersigned"有同工異曲之妙，由於這種表現法很不自然，以致顯得不親切，該用"I", "We"時就用，不要怕用"I", "We"，而勉強用"The writer"。因此，應改爲：

I have received your letter...

I appreciate your efforts.

習　　題

試將下列各句改寫成比較新式的句子。

1. Kindly ship the enclosed order and oblige.

2. We shall be pleased to talk with you at all times.

3. Please notify us at an early date.

4. We welcome your esteemed favor of the 9th inst.

5. Hoping to hear from you soon.

6. We will appreciate your giving us an opportunity to display our merchandise.

7. The captioned company has advised that your order was shipped already.

8. We are attaching hereto a copy of our contract covering prices on toys.

9. We will be glad to discuss this matter more fully with you.

10.Enclosed please find sample of our #123 black elastic ribbon.

第五章　商用英文信的開頭句與結尾句

(The Openings and Endings of a Letter)

第一節　開頭句

一、開頭句的重要性

在商用書信，開頭句最重要，因為一封信的開頭寫得好不好，會給收信人以深刻的印象。開頭句一定要寫得生動有力，如果第一句話能喚起收信人的興趣或注意，他自然會高興地讀下去，否則他就覺得索然乏味，不願再費時間看下去。因此一切陳腐的開頭句，應盡量避免使用，以免減少收信人的興趣及注意。

二、開頭句的要領

下列幾種提示，可以幫助學者怎樣應用適當的開頭句子。

1.開門見山，廢去虛偽客套：一封信的開頭，應該快鞭直入，立刻告訴收信人所要知道的事情。尤其我們寫商用英文信時切勿將中文書信的刻板開頭用語加以套用，只有直接、簡潔、明確的開頭句才能使一封信的文字生動有力，而受收信人的歡迎。例如：

We are very sorry to be obliged to inform you that we forwarded an order for cotton-shirtings on the 3 rd inst., and have your acknowledgment of the sixteenth, but the goods have not yet at hand.

（鄙號曾於本月三日奉上購買棉布疋訂單乙紙，當蒙覆稱於十六日收到，然迄今該貨尚未寄到，迫切上陳，曷勝惶悚。）

像這種囉嗦的"We are very sorry to be obliged to inform you that"（等于中文書信中「迫切上陳，曷勝惶悚」）等句子，在現代，使人看了頭痛。我們應直截了當地這樣說：

On January 6, we placed an order for cotton-shirtings, which was acknowledged on the sixteenth, but the goods have not yet arrived.

（在一月六日寄上購買棉布疋訂單乙紙，曾蒙覆信說已於十六日收到，但那批貨迄今仍未寄到。）

同樣，下列的一些套語，應避免用作一封信的開頭句。

I have the honor of informing you that...

We beg to inform you that...

I take the liberty to address you that...

We beg to notify you that...

We take this opportunity to inform you that...

We regret to advise you that...

在舊式商用英文信中的覆信中，更多陳腐而模糊不清的開頭句子，我們應盡量避免使用，例如：

In reply to yours of 30 th ult., we desire to say that we have sent you under separate cover a copy of our latest catalog.

（敬覆上月三十日尊函，茲啓者，敝公司已將最新之目錄，另套寄奉矣。）

譬如上月是五月三十日，為什麼要說是 30 ult. 呢？為什麼要囉哩囉嗦說"We desire to say"等廢話呢？請和下列用語自然、明晰，而又站在讀者的觀點上說話的開頭句，比較一下：

The catalog you requested in your letter of May 30 has been sent today. On pages 14 and 15 you will find a complete description of the sewing machine in which you are interested.

（你五月三十日來信索閱的目錄，已於今日寄上。在第十四和十五頁上，你可以找到您所關切的縫衣機的完全圖樣。)

2.避免用分詞式開頭句(participial openings)：分詞式開頭句用法不僅陳腐，而且顯得了無生氣。例如：

Referring to your order of January 21...應改爲：

Your order dated January 21 for 1000 sets...又：

Confirming our cable of today...應改爲：

We confirm we sent you the following cable on May 5:

3.可能的話，盡量以"you"開頭，以收"you-attitude"的效果：

We have received your inquiry on...宜改爲：

Your inquiry on...has been received 又：

We wish to ask you to send us your latest catalog.宜改爲：

Will you please send us your latest catalog.

但不是說每一封信均以 "You"開頭，事實上很多信都以 "We"開頭，在一封信中滿篇"We"當然不好看，但故意不用 "We"有時反而顯得不自然。尤其在 Claim Letter，更不宜用"You"開頭，例如：

You did not reply to our letter about...應改爲：

We have not received your reply to our letter about...又：

Your letter of May 15 failed to tell us when you will open L/C宜改爲：

We failed to learn from your letter of May 15 when you will open L/C.

4.盡量避免以消極或否定句開頭：以消極或否定句開頭，易使人產生緊張或不快，以致影響收信人看信的情緒。例如：

⑴ (拙劣) We are discouraged that you have not replied to our letter of July 15.

(較佳) Did you receive our letter of July 15...?

⑵ (拙劣) We have your complaint about...

(較佳) Thank you for writing us about...

一般而言，一開始就說抱歉的話似乎不妥，但事實上不得不這樣說的話——尤其向人家道歉、謝罪或表示婉惜時，還是直說，以表示係鄭重聲明的意思。例如我們不宜毫不表示歉意地說：

Yours of the 2 nd at hand and in reply would say that we cannot accept any more orders for blankets, as our supply is exhausted.

(頃接二日台函，茲覆者，敝號絨氈存貨已罄，不能再行承辦矣。)

我們應該寧可刪掉那些陳套，而把歉意鄭重的放在開頭，明顯的表達出來。

We regret that we are unable to fill your order of the 2 nd for blankets, as our supply is exhausted.

(敝號非常抱歉，不能承辦台端於二日來信中所定購的絨氈，因爲敝號的貨物，已經賣完了。)

5.在覆信開頭句中，應將所要答覆的那一封來信指出，以便收信人能立即瞭解覆信的本意，例如下面的幾種開頭句是值得模仿的。

⑴ We appreciate the information you have so kindly furnished us in your letter of January 10.

(敝號非常感激寶號正月十日來信中，承蒙惠示的情報。)

⑵ We are unable to give you the information you asked for in your letter of May 10 concerning the financial status of Mr. Lee.

(五月十日來信中，承詢有關李君的財務狀況，敝號無法作答。)

⑶ We shall be glad to send our new price list of automobile tires, which you requested in your letter of May 8, as soon as it is printed.

(五月八日來信索閱的汽車胎新價目表，當印好的時候，敝行當欣然立刻寄奉。)

三、開頭句用語

以下按照由「我方主動先寫信」時，以及「覆信」時，兩種情形。分別列舉若干開頭句，以供寫信時參考之用。

1. 由我方主動先寫信時

　　(1)寄奉、寄送、文件、目錄、物品時：

We
{
send
are sending
have today sent
are glad to send
will send
are pleased to send
shall be pleased to send
}
you
{
under separate cover
by separate mail (post)
separately
by airmail
by airfreight
by s.s....
}

　　(2)檢附⋯時：(*舊式用語)

Enclosed pleased find*
Enclosed you will find*
You will find attached to this letter
We are attaching to this letter
We enclose
We send you herewith
There is (are) enclosed
Enclosed is (are)
We are enclosing
}
...

　　(3)通知、證明或介紹：

This is to
{
announce
inform you
certify
introduce to you Mr....
}
that...

We have the honor ⎫
We are pleased ⎬ to inform you...
We have the pleasure ⎭

(4)「請寄…」時:

Please send us ⎫
Will you please send us
Kindly favor us with

We shall be ⎰Pleased⎱
⎩glad⎭ to receive ⎬ your →
happy

We shall be ⎰obliged⎱ if you will send us
⎩grateful⎭

We shall appreciate it if you will send us ⎭

→ ⎰catalog
revised catalog
latest catalog
estimate
price list
proforma invoice
quotation sheet⎭

(5)「從…得悉貴商號名字」時:

We ⎰hear⎱ from...that you are ⎰the leading importers of...
⎩learn⎭ ⎩reliable firm of...

Your name has been given by CETRA.

We have had your name and address given to us by Mr....

We are indebted for your name and address to CETRA...

Through the courtesy of American Embassy in Taipei, we...

⑹表示追述、補充或確認前寄函電、談話時：

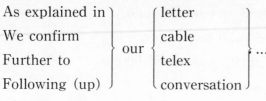

As explained in ⎫　　　　⎧ letter ⎫
We confirm 　　　⎬ our ⎨ cable 　⎬ ...
Further to 　　　⎩　　⎩ telex 　⎪
Following (up) ⎭　　　⎩ conversation ⎭

（Ⅰ）⎧ We confirm
　　⎪ We wish to confirm
　　⎨ We hereby confirm 　⎬ →
　　⎪ We have to confirm
　　⎩ This is to confirm

（Ⅱ）⎧ our telegram of (date) to you
　　⎪ our telegram sent you on (date)
　　⎪ having cabled you on (date)
　　⎨ having advised you by cable on (date) ⎬ →
　　⎪ cabling you on (date)
　　⎩ the exchange of telegrams

（Ⅲ）⎧ reading:
　　⎪ as follows:
　　⎪ which reads:
　　⎨ to the following effect: ⎬ → （Ⅳ） "(text of cable)", →
　　⎪ conveying to you that
　　⎪ advising you that
　　⎩ reading as follows:

which doubtless (no doubt) has had (has received)

→(Ⅴ)
- your attention.
- which we think is self-explanatory.
- to which we await your reply.
- as per confirmation attached hereto.
- as per copies attached.
- confirmation copy of which is attached.

as per attached confirmation copy.

【例】（Ⅰ）We hereby confirm

（Ⅱ）our telegram sent you on May

（Ⅲ）reading as follows:

（Ⅳ）"...(text of cable)...,",

（Ⅴ）which we think is self-explanatory.

(7)表示遺憾 (開頭句盡量避免使用，但並不排斥)：

We regret to $\begin{Bmatrix} \text{inform} \\ \text{advise} \\ \text{notify} \\ \text{remind} \end{Bmatrix}$ you that...

We regret to bring to your notice certain irregularities...

Much to our regret we have to advise you our inability to...

2.覆信的開頭句

(1)表示「來函敬悉，或來函已收到敬表謝意」時：

We {
have received (with thanks)
appreciate
are pleased to receive
are obliged for
acknowledge and thank you for
thank you for
}

This is to acknowledge (receipt of)

Please accept our thanks for

} your {
letter
inquiry
cable
telegram
order, etc.
} ...

(2)表示「敬覆者…」時: (*舊式用語)

In response to
Answering*
Replying to*
In answer to
We are glad to answer (reply to,
We are pleased to reply to
} your {
letter
inquiry
cable,etc.
}

(3)表示「關於…乙節」時:

In reference to
With reference to
Referring to
With regard to
As regards
Concerning
Regarding
In connection with
We refer to
Please refer to
This letter refers to
} { your / our } {
letter
inquiry
cable
telegram, etc.
} ...

(4)表示「遵照寶號×月×日來函（電）要求（指示、請求）」時：

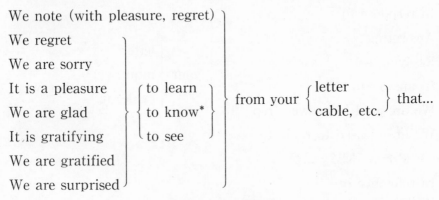

In conformity with ⎫ ⎧ the request
In compliance with ⎬ ⎨ the instructions
In accordance with ⎭ ⎩ the directions
According to   in your ⎧ letter
As requested   ⎨ cable, etc.
As desired
Following the instructions

(5)表示「拜讀×月×日大函，得悉……深感欣慰、遺憾」時：

We note (with pleasure, regret) ⎫
We regret
We are sorry
It is a pleasure   ⎧ to learn ⎫
We are glad   ⎨ to know* ⎬ from your ⎧ letter ⎫ that...
It is gratifying   ⎩ to see ⎭   ⎨ cable, etc. ⎬
We are gratified
We are surprised ⎭

＊"to know from…"的 expression 最好不用。

cf. I know right from wrong:我能辨別善惡。

第二節　結尾句

一、結尾句的重要性

任何事情的開頭與結尾給予人的印象最深，商用書信亦然。因此，一封信的結尾句與開頭句一樣，應以簡潔、強有力的文字表達，使收信人留下深刻良好的印象，從而達成寫信的目的。

要一封信前後聯貫，必須先從收信人的觀點，逐漸以合於邏輯的規則推進到寫信人的觀點。因此，開頭句應以收信人的立場開始，結尾句則應站在寫信人立場結束。換言之，開頭句應着重 "You"，結尾應着重 "Us"。

二、結尾句的要領

1.盡量避免以分詞式結束(participial close)：分詞式結尾，不僅顯得軟弱無力，而且俗氣、陳腐。

【例】

（乏力）Hoping that this information will be helpful to you.

（有力）We sincerely hope that this information will be helpful to you.

（乏力）Trusting that this will be satisfactory to you and hoping that we may be of further assistance to you.

（有力）If we may be of assistance to you in some other way, please feel free to write us.

以下各句都是常見的分詞式結尾句，應盡量少用：

Awaiting your reply, we beg to remain,（翹候回函）

Awaiting to hear from you at your earliest convenience...（敬俟復命，便祈即復）

Awaiting your further valued orders…（仍請源源賜顧為荷）

Awaiting the opportunity when our services may be of use to you...（俟機圖報）

以分詞式結尾時，有些人喜歡殿以"We beg to remain", "We remain", "We are", "We believe"等詞，以為這樣較客氣，其實都是陳腐的客套。

2.以具體明確的內容結束：結束句應有具體的內容才顯得有力。

Trusting that this is satisfactory to you.

We hope to hear from you soon.

上面兩句，既不明確又顯得乏力，下面的就顯得具體有力。

We shall do all in our power to continue to enjoy your confidence in us.

We are interested in hearing that this information fulfills your requirements.

3.以誠摯的語氣結尾：即用"Please..."或"Kindly..."的語法, 例如：

Please tell us whether we may expect to receive... （懇請惠寄……）

Please let us know about（whether, if）... （懇請賜知…）

4.「請求」時以疑問句結尾，可顯得客氣，例如：

Will you please reply us before Friday?

May we ask a favor of you to send us information available?

5.以被動句子結尾：既可避免使用"We"又可加強語氣，例如：

Your inquiry is appreciated, and we hope you will follow it with an order. （感激詢價，並盼接著就寄來訂單）

Your prompt reply will be appreciated. （迅速回覆，將不勝感激）

三、結尾句用語

1.表示「請賜覆、請見覆、請惠賜卓見」時：

（Ⅰ）

We hope to receive
We await (or wait for)
We are waiting for
We shall be obliged for
We shall appreciate
We shall be glad to have
Will you let us have
We look forward to

your
a, an

reply, answer.
early reply.
definite reply.
favorable reply.
satisfactory reply.
further news.
comment.
opinion.
cable reply.

（Ⅱ）

We trust that we shall hear
We hope to hear favorably
We shall appreciate hearing
We shall be glad to hear
Please let us hear
Will you let us hear

from you →

in this connection.
in this respect.
in this matter.
concerning this matter.
about this matter.

以上（Ⅰ），（Ⅱ）可分別接下列用語（有「儘速」、「及早」之意）。

$$
(\text{III})
\begin{cases}
\text{at your earliest convenience.} \\
\text{at an early opportunity.} \\
\text{as early as convenient.} \\
\text{without (the least) delay.} \\
\text{on receipt of this letter.} \\
\text{(very) soon.} \\
\text{at once (promptly, immediately).}
\end{cases}
$$

2.表示「歉意」時：

(1) We regret we have not informed you about this matter sooner.

(2) We are sorry for not having explained our position sooner.

(3) We regret that our reply has been delayed because of the necessity of making a thorough investigation in this matter.

(4) We hope that this delay will not cause you any inconvenience, and assure you that we shall give your further orders our prompt attention.

(5) We regret that we are unable to
$$
\begin{cases}
\text{meet your requirements.} \\
\text{be of assistance to you.} \\
\text{make use of your kind offer.} \\
\text{avail ourselves of your proposal.} \\
\text{be of service to you.}
\end{cases}
$$

3.表示「謝意、惠顧、合作」時：

(1) We are obliged to you for your kind attention to this matter.

(2) We thank you for your patronage in the past and solicit continuance of the same in the future.

(3) We thank you for the special care you have given to the matter.

⑷ We thank you for your generous cooperation.

　　4.表示「如有機會，必會報答」時：

⑴ It would give us a great pleasure to render you a similar service should an opportunity occur.

⑵ We hope to be able to reciprocate your good offices on a similar occasion.

⑶ We are always ready to render you such or similar services.

　　5.懇請惠予多加惠顧：

⑴ We solicit a continuance of your valued favor.

⑵ We hope we may receive your further favor.

⑶ We trust that on further consideration you may decide to give
one of these machines a trial.

　　6.以問候語結尾：

⑴ With the compliments of the season, (歲序更新，特此申賀)

⑵ With the season's greetings, (謹致時祺)

⑶ Wishing you a Merry Christmas and a Happy New Year, (謹
祝耶誕並賀新年)

⑷ We send you our cordial Christmas greetings and very best
wishes for a Happy New Year, (敬祝耶誕並賀新年快樂)

⑸ May our business relations in the coming year be more amicable and fruitful,

⑹ With kind remembrances and all good wishes for a Merry
Christmas and a bright New Year, (敬賀耶誕快樂與新年快活)

⑺ May all your plans and wishes come true in this New Year, (敬
祝您的計劃與願望能在今年實現)

⑻ With my most hearty greetings and sincere good wishes for

you for all the coming year, （謹祝新的歲序將是您及您全家最佳的
一年）

⑼ With best regards, （特此致意）

⑽ With kind personal regards, （謹致問候）

　　7.表示「希望以後有機會效勞（可奉告）、或將另函奉告等」時:

⑴ We hope that we may be of service to you in some other way.

⑵ We shall write to you further about the matter in a few days.

習　　題

一、將下列開頭句子改寫，使其符合「開門見山，廢去虛偽客套」的原則。

　1. We acknowledge receipt of your letter dated the 2 nd inst., stating that
　　your order has not arrived.　In reply we wish to inform you that we
　　shipped your order by air the 10 th inst.

　2. We are duly in receipt of your favor dated the 15 th May, and noted
　　its contents.　In compliance with your request, we take pleasure in
　　sending you herewith a price list of our Chinaware.

二、將下列分詞式開頭句子改寫，以免鬆懈無力。

　1. Complying with your request of even date, we have airmailed a copy
　　of our latest catalog to your Tokyo Office.

　2. Referring to your order of the 26 th, we beg to advise you that your
　　order for furniture will be shipped on 3 rd prox.

三、將下列句子改為"you attitude"措辭。

　1. We have received with many thanks your letter of November 6, and
　　we take the pleasure of sending you our latest catalog.　We wish to
　　draw your attention to a special offer which we have made in it.

　2. Through the courtesy of CETRA, we have obtained your name and

address.

四、將下列中文譯成英文。

1.敬請惠寄貴公司產品目錄為荷。

2.五月八日來函承索汽車輪胎(automobile tires)新價目表,一俟印就(as soon as it is printed)敝公司當即寄奉。

3.奉讀三月十日來函,承詢關於美國印書用紙(American printing paper)不勝感激。茲特由郵政包裹寄奉該項貨品之全部樣品(full line of samples)。

4.遲未作答,實感抱歉。

5.我們願無條件以供驅策(place ourselves unreservedly at your disposal),關於此事,並盼貴公司即賜回音。

6.謹致時祺。

第六章　進出口貿易步驟

(Steps of Import-Export Trade)

　　任何一筆進出口貿易，從市場調查開始，經業務關係的建立、詢價、報價、訂貨，而至賣方的交貨與買方的付款，完成交易，其間不僅須經過錯綜複雜的手續，而且很多都與商用英文有關。因此，貿易廠商對其進行的程序，必須先有充分的瞭解，而後才能寫出完好的商用英文。茲將以信用狀爲付款方式的進出口貿易進行步驟圖示如下：

1.尋找適當市場

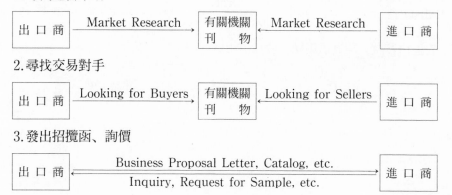

2.尋找交易對手

3.發出招攬函、詢價

4.徵信

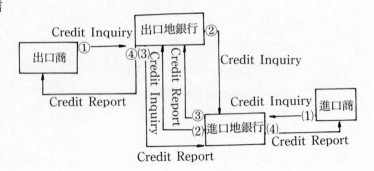

5.報價、還價、接受

6.訂貨、簽約

7.進口簽證、開發信用狀、收受信用狀

8.備貨

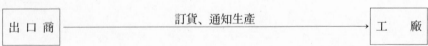

9.出口簽證

| 出 口 商 | EL Application
EL (Export Licence) | 國 貿 局
簽證銀行 |

10.申請檢驗、公證

| 出 口 商 | Application for Inspection or Survey
Inspection Cert., Survey Report | 檢 驗 局
公 證 行 |

11.洽訂艙位

| 出 口 商 | Booking Shipping Space
S/O (Shipping Order) | 船 公 司 |

12.購買保險

| 出 口 商 | Marine Insurance Application
Insurance Policy | 保險公司 |

13.出口報關及裝船

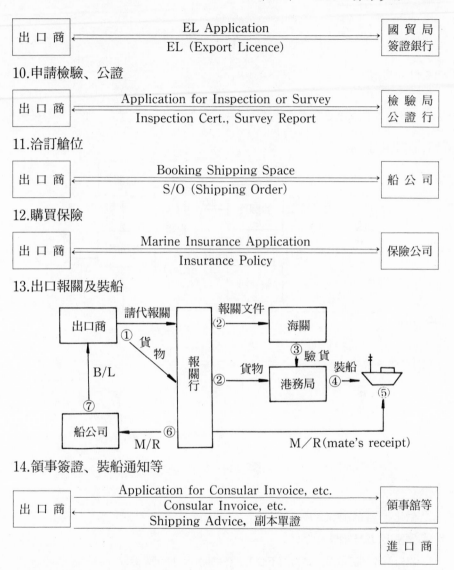

14.領事簽證、裝船通知等

| 出 口 商 | Application for Consular Invoice, etc.
Consular Invoice, etc.
Shipping Advice，副本單證 | 領事舘等 |
| | | 進 口 商 |

15.出口押滙

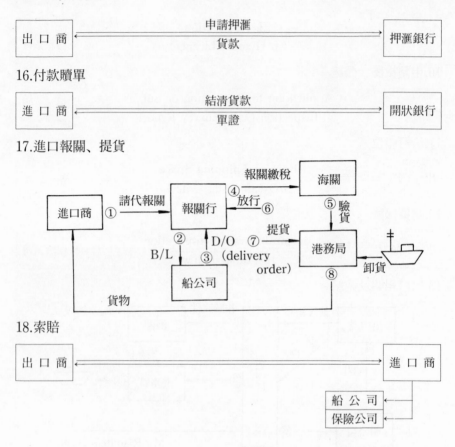

16.付款贖單

17.進口報關、提貨

18.索賠

習　題

一、試述進口貿易作業程序。

二、試述出口貿易作業程序。

三、出口商收到訂單後，是否可不等 L／C 的開達，即可先備貨?

四、在進出口貿易進行過程中，那些場合需要寫信?

第七章　尋找交易對象

(Looking for Customers)

第一節　尋找客戶信的寫法

進出口商經過市場調查之後，即可就獲得的資料比較分析，選定最有希望的市場作爲目標市場，再從這個市場尋找適當的交易對手。尋找交易對手的方法很多，而委託別人介紹交易對手時，難免要與這些人通信，這種通信的內容，因對象的不同而畧有出入，但通常大多包括下列幾項：

1.本公司希與該國客戶往來，請其介紹客戶

2.介紹本公司的營業項目

3.介紹本公司的優點——組織、經驗、資本

4.提供本公司往來銀行供徵信之用

請人介紹交易對手的信，可能由出口商採取主動(大多數)，也可能由進口商採取主動，茲舉例如下：

No.1　出口商函請國外商會介紹客戶

The Chamber of Commerce of New York
New York City, N.Y.
U.S.A.
Gentlemen:
　　We are desirous of extending our connections in

your country and shall be much obliged if you will give us a list of some reliable business houses in New York who are interested in the importation of Chinaware.

We are old and well-established exporters of all kinds of Chinese goods, especially of Chinese Typical Chinaware, and therefore, confident to give our customers the fullest satisfaction.

To justify our confidence in addressing you, we refer you to the following:

The Bank of Taiwan, Head Office, Taipei.

The Taipei Importers and Exporters Association, Taipei.

Your courtesy will be appreciated, and we earnestly await your reply.

> Your very truly,
>
> Taiwan Trading Co., Ltd.

【註】

1. desirous of extending our connections in your country:「擬拓展本公司在貴國的業務關係」, "connections"為"business connections"或"business relations"之意。

2. be much obliged:「很感激」, "to be much obliged to somebody"為「感激某人」; "to be obliged to＋verb"為"have to", "must"之意, cf. She was obliged to go back to work.:她不得不回去工作。

3. reliable business house:可靠的商號, 殷實商號。

4. who are interested in:對……有興趣 (的商號)。

5. in the importation of:輸入, 也可以"in importing"替代。

6. chinaware:陶瓷器。

7. old and well-established exporters:歷史悠久且殷實的出口商，因爲用"we"開頭，所以"exporters"用多數，假如用"I"開頭，就應用"exporter"。

8. give our customers...satisfaction:使顧客十分滿意。

9. to justify our confidence in addressing you:爲證明本公司所言不虛，也可用"to justify your confidence in us"。

10. typical:獨特的。

11. refer you to:請你向……查詢。

12. your courtesy will be appreciated:如承惠辦，則不勝感激。

"appreciate"爲"to place a sufficiently high value on"之意。此字本身已含有「深爲感激」之意，所以不必再加上"greatly", "deeply", "highly", "much"等副詞。但很多人仍加上這些字，這只好當作「禮多人不怪」吧！

No.2　出口商函請駐本地外國領事館介紹客戶

The Consul General,

The Korean Consulate-General

Taipei

Sir:

Would you be kind enough to give us the names and addresses of merchants and agents being Korean citizens, and any business organizations established by your countrymen, now having their offices in Taipei or Kaohsiung?

If possible, we should like to know what lines of business they are now carrying on.

Yours respectfully,

(Signed)

【註】

1. consul general:總領事, consul:領事。

2. consulate-general:總領事館, "consulate"為「領事館」。

3. Sir:因為信是寫給 Consul general，所以用單數，而不用多數的"Sirs"。

4. would you be kind enough to give us...? 「可否示知…?」也可以下面句子代替:

   a. we shall be much obliged, if you will give us...

   b. we shall be glad, if you would give us...

5. Korean citizens:韓國公民。

6. business organizations:公司行號。

7. we should like to know：「我們想知道」，「請賜知」。

8. if possible:如屬可能＝If it is possible。

9. lines of business:營業種類，也可用"lines of goods"（經營商品）。

10. carry on＝engage in「從事於…」。

11. yours respectfully:「謹上」，對官員或官署時用此詞。

No.3　出口商函請報社、雜誌社刊登介紹

Dear Sirs,

We have the pleasure of introducing ourselves to you as one of the most reputable electronics exporters in Taiwan, who has been engaged in this line of business since 1965; particularly we have been enjoying a good sale of TV sets and Transistor Radios and are now desirous of expanding our market to your district.

We would, therefore, appreciate it if you could kindly introduce us to the relative importers by announcing in your publication as follows:

"An Export Company of Electronics is now making a business proposal for TV sets and Transistor Radios which are said to have built a high reputation at home and abroad Contact them by addressing your letter to..."

We solicit your close cooperation with us in this matter.

Faithfully yours,

【註】

1. "We have the pleasure of..."以新式的"We are pleased to..."代替較佳。cf. We have the liberty of...冒昧地…

2. reputable:信譽良好的。

3. introducing ourselves to you as...:謹向您介紹我們（本公司）為…

4. line of business:經營項目。

5. expand our market:拓展本公司市場。

6. relative importers:相關的進口商。

7. announcing in your publication:在貴刊物中發佈……。

8. making a business proposal for:提議做……的生意。

9. are said:據說。這種用語有「不負責」之含義。

10. at home and abroad:在國內外。

No.4　進口商函請報社、雜誌社刊登介紹

The Economic News
555 Chunghsiao E. Road, Sec.4
Taipei, Taiwan
Dear Sirs,

We are one of the leading importers and dealers with long experience in international trading in all kinds of building materials, hand tools, hardware, readymade garments, sporting goods, hats & caps, watches and bands, umbrellas, footwear, imitation jewelry, stationery goods, sunglasses, handbags, electrical goods in Nigeria and we are seeking reliable suppliers of these items in your country.

We would appreciate your publishing our name in your newspaper so that any interested exporters and manufacturers in your country can contact us.

Our usual purchasing terms are full letter of credit, with each order, and our bankers are The Bank of America, Lagos from whom you will be able to obtain all the information you may require in regard to our business integrity and financial standing.

Your help in this matter will be appreciated.

Faithfully yours,

【註】

1. leading importers and dealers:主要的進口商及買賣商, "dealers"即 "traders"。

2. trading in...:經營… (某種商品)。

3. seeking:尋找, 物色。

4. items:項目,商品,可以"goods", "commodities", "merchandise", "articles" 等詞代替。

5. publishing our name in your newspaper:將我們 (本公司) 的名字刊登在貴

報紙上。類似措詞有:

placing our name in your publication:（出版物）

putting our name in your magazine:（雜誌）

advertising our name in your bulletin（刊物, 公報）

6. interested exporters:有意的出口商; 感興趣的出口商。

7. business integrity:業務上的誠實。意指做生意是否誠實。

8. 請報社、雜誌社刊登介紹客戶時, 有些需付刊登費用, 有些不必付這種費用。一般而言, 推廣貿易機構所發行的刊物, 都設有「貿易機會」(trade opportunities) 的專欄, 如請其在此專欄刊登, 通常都不收費。

No.5　出口商對於介紹顧客的謝函

Dear Sirs,

We have received with many thanks your letter of May 20, and wish to express our sincere gratitude for your kindness in publicizing our wish in your "Trade Opportunity".

We believe the arrangement you kindly made for us will connect us with some prospective buyers and bring a satisfactory result before long.

We thank you again for your taking trouble and wish to reciprocate your courtesy sometime in the future.

Faithfully yours,

【註】

1. wish to express our sincere gratitude:謹致謝忱。

2. publicizing:「廣為宣傳」, 以"publishing"代替較妥。

3. will connect us with...:使我們與…取得聯繫。

4. prospective buyers:未來的買主; 可能的買主。

5. before long:不久，可以"soon"代替。

6. taking trouble:費神，本句不如改爲"we thank you again for your assistance and wish to..."

7. reciprocate:報答；回報。

第二節　有關尋找交易對象的有用例句

一、開頭句

1. $\begin{Bmatrix} \text{As we plan} \\ \text{In order} \end{Bmatrix}$ to extend our business $\begin{Bmatrix} \text{connections in} \\ \text{to} \end{Bmatrix}$ →

→ your $\begin{Bmatrix} \text{country} \\ \text{city} \\ \text{area} \\ \text{district} \\ \text{market} \end{Bmatrix}$ → $\begin{Bmatrix} \text{would you please send us a list of} \\ \text{we request you to introduce to us} \end{Bmatrix}$ some →

→ reliable $\begin{Bmatrix} \text{business houses} \\ \text{firms} \end{Bmatrix}$ → in your $\begin{Bmatrix} \text{country} \\ \text{city, etc.} \end{Bmatrix}$ who are →

→ interested in $\begin{Bmatrix} \text{importing} \\ \text{exporting} \\ \text{handling} \end{Bmatrix}$ → $\begin{Bmatrix} \text{Christmas tree light bulbs.} \\ \text{household lamps.} \\ \text{porcelain.} \end{Bmatrix}$

$\begin{Bmatrix} \text{茲因計劃拓展本公司在} \\ \text{茲爲拓展} \end{Bmatrix}$ 貴 $\begin{Bmatrix} \text{國} \\ \text{市} \\ \text{地方} \\ \text{地區} \\ \text{地市場} \end{Bmatrix}$ 的業務關係→ $\begin{Bmatrix} \text{盼能提供} \\ \text{擬請推介} \end{Bmatrix}$ 貴→

→ $\left\{\begin{array}{l}國\\市\\\vdots\end{array}\right\}$ 若干有意 $\left\{\begin{array}{l}進口\\出口\\從事\end{array}\right\}$ $\left\{\begin{array}{l}耶誕樹裝飾燈泡\\家用燈泡\\瓷器\end{array}\right\}$ → 的可靠 $\left\{\begin{array}{l}商家\\商號\end{array}\right\}$。

2. $\left\{\begin{array}{l}\text{We would be very grateful to you}\\\text{We shall appreciate it}\\\text{We would be greatly obliged}\end{array}\right\}$ if you $\left\{\begin{array}{l}\text{will}\\\text{would}\end{array}\right\}$ kindly →

→ introduce to us → $\left\{\begin{array}{l}\text{proper}\\\text{reliable}\\\text{relative}\end{array}\right\}$ firms with whom we could →

→ establish a business relationship for the $\left\{\begin{array}{l}\text{export}\\\text{import}\\\text{import and export}\end{array}\right\}$ →

→ trade business.

如蒙推介本公司可與他們建立 $\left\{\begin{array}{l}出口\\進口\\進出口\end{array}\right\}$ 貿易業務關係的 $\left\{\begin{array}{l}適當\\可靠\\相關\end{array}\right\}$ 商號 →

→本公司 $\left\{\begin{array}{l}當感謝你\\當感激不盡\\當非常感激\end{array}\right\}$。

3. $\left\{\begin{array}{l}\text{We are desirous of expending our market to}\\\text{We are }\left\{\begin{array}{l}\text{seeking}\\\text{looking for}\end{array}\right\}\text{ new business connections in}\end{array}\right\}$ your →

→ $\left\{\begin{array}{l}\text{district}\\\text{area}\\\text{city}\\\text{country}\end{array}\right\}$ and would appreciate your $\left\{\begin{array}{l}\text{supplying}\\\text{furnishing}\\\text{providing}\end{array}\right\}$ us with

→ the names and addresses of $\begin{cases} \text{manufacturers} \\ \text{firms} \\ \text{importers} \\ \text{exporters} \\ \text{concerns} \end{cases}$ who are →

→ interested in $\begin{cases} \text{garments.} \\ \text{toys.} \\ \text{electronics.} \end{cases}$

二、自我介紹

1. We are one of the $\begin{cases} \text{most reputable} \\ \text{well organized} \\ \text{experienced} \\ \text{leading} \end{cases}$ trading firms in →

→ Taiwan, who have been engaged in $\begin{cases} \text{Tv Sets} \\ \text{garments} \\ \text{radio sets} \end{cases}$

business → since 1960.

2. We have close connections with outstanding manufacturers of various kinds of General Merchandise, and are very well-placed to supply our customers with high grade goods at → $\begin{cases} \text{competitive} \\ \text{reasonable} \\ \text{moderate} \end{cases}$ price.

＊ outstanding:著名的。　＊ general merchandise:雜貨。

＊ well-placed:居於有利的地位; 方便。

3. Our firm was $\left\{\begin{array}{l}\text{established}\\ \text{founded}\end{array}\right\}$ in 1950 and has enjoyed an

excellent reputation for these $\left\{\begin{array}{l}\text{lines of product.}\\ \text{items.}\\ \text{lines of business.}\end{array}\right.$

4. Being one of the biggest manufacturers and suppliers of... in Taiwan, we are confident to give full satisfaction to your people.

   * are confident to give full satisfaction to your people:深信必能令貴國人稱心滿意。

5. As we are long-established exporters of all kinds of porcelain, we have close connections with the leading manufacturers here.

三、提供備詢人

1. To justify your confidence in us, we refer you to the following references:

2. For information $\left\{\begin{array}{l}\text{concerning}\\ \text{regarding}\end{array}\right\}$ our $\left\{\begin{array}{l}\text{financial status}\\ \text{credit standing}\\ \text{financial position}\end{array}\right\}$, →

   → $\left\{\begin{array}{l}\text{may we refer you}\\ \text{please refer}\\ \text{please direct your inquiry}\\ \text{we wish refer you}\end{array}\right\}$ to the Bank of Taiwan.

四、請刊登介紹

$\left\{\begin{array}{l}\text{We would be greatly obliged}\\ \text{We shall appreciate it}\end{array}\right\}$ if you will kindly introduce us → to the $\begin{Bmatrix} \text{relative} \\ \text{interested} \end{Bmatrix} \begin{Bmatrix} \text{importers} \\ \text{exporters} \\ \text{suppliers} \\ \text{manufacturers} \end{Bmatrix}$ by →

$\begin{Bmatrix} \text{announcing} \\ \text{inserting} \end{Bmatrix}$ in → your publication as follows:

五、結尾句

1. We solicit your close cooperation with us in this matter.

2. Your $\begin{Bmatrix} \text{assistance} \\ \text{help} \end{Bmatrix}$ in $\begin{Bmatrix} \text{this} \\ \text{above} \end{Bmatrix}$ matter will be appreciated.

3. Your courtesy and $\begin{Bmatrix} \text{early} \\ \text{prompt} \end{Bmatrix}$ reply will be much appreciated.

   ＊"much appreciated"中的"much"是多餘的。

習　題

一、將下列中文譯成英文。

1. 茲因計劃拓展本公司在貴國之業務關係(business connections)，盼能提供紐約地區對進口聖誕樹裝飾燈泡(Christmas tree light bulbs)、手電筒燈泡(flash-light bulbs)及家用燈泡(household lamps)有興趣的若干可靠商號名單。

2. 企盼回函惠予同意。

3. 關於本公司的信用情況，請向下列銀行查詢。

4. 承蒙惠賜消息，謹此致謝忱。該消息對于我們與他們結成彼此有利的交易關係

(in forming mutually profitable trade relation)將大有助益(great assis-tance)。

5. 本公司從事瓷器品(porcelain)出口業務迄今已逾十年，並與本國第一流瓷器工廠素有廣泛(extensive)及密切的聯繫(close connections)。

二、試將下列縮寫字還原，並譯成中文。

1. CETRA
2. BOFT
3. BCIQ
4. KEPZ
5. CBC
6. MOF
7. FETS
8. MOEA

第八章　招攬交易

(Trade Proposal)

第一節　招攬信的寫法

貿易商經由各種途徑找到了可能的交易對手之後，卽應作成記錄，以便發出招攬信(Letter of Proposing Business)，提議建立業務關係。這種招攬信的內容，一般而言，包括下列各項：

1.獲悉對方的途徑：例如由銀行、外貿協會、進出口商名錄。

2.表示願意與其建立業務關係：這是招攬信的重心。

3.自我介紹及所經營商品的詳情。

4.扼要說明交易條件：主要是說明付款的方法。

5.提供信用備詢人。

若招攬信是由賣方主動發出者，可檢附商品目錄(Catalog)、價目表(Price List)，甚至另寄樣品(Sample)。凡此，均應在信中提及。

一、由出口商發出的招攬信

No.6　由商會獲悉

> Gentlemen:
>
> 　　Your neme was given through the...Chamber of Commerce as a firm doing an extensive trade in (商品) and we take the liberty to write you with the earnest

desire of having the opportunity to enter into business relation with you.

For more than twenty years we have been exporting Chinese (商品) and shipped considerable quantities to your country; but have not had the pleasure of doing any of this business with you. We therefore look forward with much interest to having the privilege of your requirements.

We are well organized with experienced men who have a thorough knowledge of the requirements and the taste of your market. In addition to this, our close connections with many of the leading factories and our investigation into the foreign traffic situation place us in a position to render the fullest satisfaction to our customers.

As to the terms of payment, we usually request our customers to open a letter of credit in our favor, under which we will draw at 60 days after sight.

In order to show you the excellence of our goods, we are sending you under separate cover our latest descriptive and illustrated catalogue. Our price-list is enclosed into this letter.

For any information you may desire in regard to our standing, we are pleased to refer you to the following bank:

(Bank's Name)

> If you are inclined to entertain our proposal mentioned above, please give us your definite inquiries upon which we shall be pleased to send you our best prices and deliveries.
>
> > Yours faithfully,

【註】

1. We take the liberty to write...「冒昧地寫信」。也可寫成:
   - We take the liberty of writing...
   - We take liberty in writing... (注意: "liberty"前面沒有"the")
2. to enter into business relations: 開始業務聯繫; 建立交易關係。
3. considerable＝large; no small
4. look forward to: 期待, "to"是介系詞, 因此應接 gerund 或名詞。
5. having the privilege of your requirements: 惠請賜顧。
6. a thorough knowledge: 精通。
7. the taste: 嗜好。
8. to render＝to give
9. terms of payment: 付款條件。
10. under which: 憑該信用狀之意。
11. under separate cover: 另寄。
12. descriptive and illustrated catalogue: 附有說明及插圖的目錄。
13. standing: 身分; 信用情況。
14. inclined to entertain: 俯允; 有意接受。
15. mentioned above＝above-mentioned
16. best prices: 最克己的價格。
17. deliveries＝delivery terms: 交貨條件。

No.7　由外匯銀行獲悉

10th March, 19...

Messrs. Carlton & Co., Ltd.

London, England

Dear Sirs,

We have lately requested our Bankers, Bank of America, Taipei, to give us names of reliable firms in London. They have recommended to us your firm in such high terms that we are very anxious to enter into relations with you in our soft goods.

Our firm has been in existence upwards of forty years as Exporters of Silk Piece Goods, in which, we are told, you are specially interested. Our Silks cover many different items, but the undermentioned are some of our chief exports:

　　　　Pongee Silk, Habutae, Crepe de Chine,

　　　　Tussore Silk Crepe, etc.

In addition to our specialities we have in recent years been shipping Cotton and Nylon Piece Goods, in which we have done a considerable business with some of the reliable houses on your side up to the present. It is, however, in these Silk Piece Goods that we wish to increase our business with your support and cooperation. The special facilities we have as shippers and our thorough familiarity with the tastes and requirements of your market place us in a favourable position to render you our services to your best advantage. Should you fall in with our proposal and make inquiries for your specific needs, we shall lose no time in sending you samples along with prices and pertinent details.

In reply to inquiries and in making offers we shall quote prices C. I. F. London or F. O. B. Keelung in Pound Sterling, whichever your prefer. Our usual terms are to draw a draft at 60 d/s under a Confirmed Banker's Letter of Credit to cover each order placed with us.

References: For information as to our financial stability, we would refer you to the following:

> Bank of America, Taipei
>
> Bank of Taiwan, Taipei

We trust that we may soon have the pleasure of opening up business with your respectable house and of proving that such a connection would yield you a considerable profit, while assuring you of our exceptionally careful and prompt attention to your interests.

Yours faithfully,

【註】

1. in high terms: 鄭重地。這裡的"terms"為「措詞」(mode of expression)之意。cf. He referred to your work in terms of high praise. (他對你的工作大加讚揚。)

2. enter into relations: 開始交易關係。相似的用例有:

　　　　to enter into (business) negotiations with...

　　　　to enter into connections (＝connexions) with...

　　　　to enter into correspondence with...

　　　　to open a correspondence with...

　　　　to open an account with...

3. soft goods: 織物類貨品。

4. in existence: 存在, 卽成立之意。

5. Pongee silk: 府綢紗。

　　crepe de Chine: 縐紗。

　　Tussore silk crepe: 山東綢紗。

6. speciality: 又寫成"specialty",特產品; 特製品。

7. shipping: 運出, 卽銷出之意。

8. on your side: 在你那邊, 卽貴地, 又可寫成"at your end"。

9. fall in with our proposal: 同意我們的提議。

10. lose no time: 馬上。

11. pertinent details: 相關的細節。

12. at 60 d/s＝$\left\{\begin{array}{l}\text{at sixty days after sight}\\\text{at sixty days' sight}\end{array}\right\}$見票後60天付款, "d/s"爲"days

　　after sight"或"days' sight"的縮寫。"at"表示期限, 例如"at sight"見票卽付。

13. under＝according to

14. cover: 抵付; 支付; 清償。此字在商業上意義很特殊, 茲舉例說明其用法:

　　(a) In order to *cover* our order, we have arranged with The Bank of
　　　　Tokyo a credit for U. S. $1,000.00 (a credit＝a letter of credit)

　　(b) We enclose a cheque for DM 10,000 to *cover* your invoice of the 30th
　　　　May.

15. yield: 產生。

16. 本信最後一段係從收信人利益爲着眼, 收信人看了之後, 一定很高興。這就是
　　　"You attitude"的寫法, 須知, "You attitude"未必以"You"開頭。只要以收

信人的利益爲前提所寫的信，就符合"You attitude"的原則。

No.8　由刊物獲悉

Gentlemen:

　　In a recent issue of the "Foreign Trade," we saw your name listed as being interested in making certain purchases in this country.

　　We take this opportunity to place our name before you as being a buying, shipping and forwarding agent. If you do not have anyone here to look after your interests in that capacity, we should be glad if you give us your kind consideration.

　　We inform you that we have been engaged in this business for the past 20 years. We, therefore, feel that because of our past years' experience, we are well qualified to take care of your interests at this end.

　　Further, as for references, we can give you the names of some concerns in your country and our bankers are The New York State Bank, 20 Wall St., New York 4, N. Y.

　　We look forward to receiving your reply in acknowledgement of this letter and with thanks in advance.

　　　　　　　　　　　　　Yours faithfully,

【註】

1. recent issue of the "Foreign Trade": 最近的「對外貿易」刊物。

2. being interested in ... this country: 有意在本國採購某些貨物。

3. We take this opportunity to place our name...agent：本公司冒昧向貴公司自我介紹擬代理貴公司在此地之採購、裝船及轉運業務。

4. We should be glad if you give...consideration：如惠予考慮，不勝感禱。

5. engaged in this business：從事此項業務。（必須用被動語態動詞）

6. We are well qualified to take care of your interests at this end：本公司深信在此方面具有爲貴公司提供服務的資格。

7. as for references：至於備詢人。

8. look forward to receiving your reply：盼望能收到貴公司的回音。

9. in acknow ledgement...in advance：承認收到本信並在此預先致謝。

No.9　未說明由何處獲悉

Dear Sirs,

As we are given to understand that you are interested in the Taiwan textiles, we take this opportunity of introducing ourselves as a reliable trading firm, established 10 years ago and dealing in the articles ever since with fair record especially with the Southeast Asian countries.

In order to give you an idea, we quote some of them without engagement as follows:

Sleeveless shirts, cotton & synthetic fibres mixed, white or colored, US$15 per doz. CIF your port.

We are sending you under separate cover our catalogues and free samples. Please give us your specific inquiries upon examination of the above as we presume they will be received favorably in your market.

We also handle...(name of goods)...which are selling

well in Latin America.

As to our standing, please refer to the following banks:

Bank of Taiwan, Taipei

Central Trust of China, Taipei

We hope to be of service to you and look forward to your comments.

Yours faithfully,

【註】

1. We are given to understand: 茲獲悉。

2. take this opportunity of...: 藉此機會 (向貴公司自我介紹本公司爲可靠…)。

3. dealing in: 經營。

4. ever since: 自…以來。

5. with fair record: 獲致良好成績。

6. In order...idea: 茲爲給予貴行獲得概念。

7. sleeveless shirts: 無袖襯衫。

8. cotton & synthetic fibres mixed: 棉與人纖混紡。

9. free sample: 免費樣品。cf. The sample is sent gratis.樣品免費贈送。

10. presume...favorably: 相信必能獲得好評。

11. handle: 經銷。

12. sell well: 暢銷。

13. as to our standing: 關於敝公司的地位，即指關於本公司的"reputation"。

14. to be of service to you: 對你有所助益；能幫助你。

15. your comments: 你的敎言；你的評語。

二、由進口商發出的招攬信

No.10　由商會獲悉

Gentlemen:

We are indebted to the Manila Chamber of Commerce for your name and address as one of the respectable concerns, offering Rubber Products of all descriptions.

We are not only buying for our own account, but also booking indent orders for our numerous clients. We are interested in general merchandise, but look specially for the supply of Rubber Products, such as Beltings, Hoses, Tires and Tubes, Athletic Shoes, Shoe Soles, etc.

Being extremely desirous of establishing business relationship with your house, we invite you to offer us the abovementioned items and send us a range of samples, together with illustrated literature, so that we may scout the market. In indent business, however, we hope you will include in your offer our commission of 3% on C. I. F. cost. Any overprice that we may be able to secure from our buyers is to be understood for our account.

Terms: Irrevocable letter of credit.

You may inquire our business standing and integrity through the China Banking Corporation, and the Kian Lam Finance and Exchange Corporation, both of Manila.

With nothing further by this mail, we look forward to your reply at your earliest convenience.

Sincerely yours,

【註】

1. We are indebted to...本公司承蒙…

2. respectable: 有聲望的。

3. buying for our own account: 自行買進，非代理。

4. indent order: 受託訂購。

5. belting: 調帶（機械用），hose: 軟管，shoe sole: 靴底。

6. illustrated literature: 有插圖的說明書。"literature"又解作「廣告、宣傳用的印刷品」(printed matter)。

7. overprice＝overage，溢價。

8. for our account: 算我們的收入。

No.11　由報紙、刊物獲悉

Dear Sirs,

　　Your name has been listed in the "Traders' Express" published by the China External Trade Development Council as one of the leading exporters of electronic equipment and a firm seeking the possibility of promoting your business in Taiwan.

　　As the enclosed pamphlet shows, we have been specializing in electronic equipment for over 10 years. Our experience in this line will surely add to your line of business. Furthermore, our broad range of business connections with many firms in Taiwan places us in a very competitive position.

　　We, therefore, feel that we are well qualified to take care of your interests at this end, and would like very much to open an account with you.

　　As for references, we can give you the names of

some concerns in your country and also our bankers are
Bank of Taiwan, Taipei.

We are looking forward to your early reply.

Yours faithfully,

【註】

1. has been listed in: 「刊登在…」, 也可以下列句子代替:

    has been placed in...

    has been published in...

2. Traders' Express: 「貿易快訊」, 係由外貿協會出版的日刊。

3. promoting your business: 推廣你的業務。

4. this line＝this kind of commodity

5. add to: 增加, 擴大。

6. to open account with: 開立往來帳戶, 卽建立業務關係。

No.12　由工商名錄獲悉

Dear Sirs,

We owe your name and address to Kelly's Directory,
from which we understand that you are general
exporters of Taiwan products.

Therefore, we take the opportunity to introduce ourselves as a reliable firm and inform you that we are very
interested in establishing trade relations with your good-
selves for the sale of products from your end. we are
mainly dealing as commission agents with very wide clientele and therefore it should be understood that if we
succeed in securing business, we shall be considered as

your "Sole Agents" and you will not deal direct with any of our clients except through our medium.

　　We are, at present, very keen on your offer for the following:

　　　　Canned Pineapples

　　　　Textiles

　　　　Chemicals and Fertilizers

　　Please let us have your quotations in Sterling currency on basis of CIFC$_5$ Port Sudan.

　　We are looking forward to your early news.

Yours faithfully,

【註】

1. sole agents: 獨家代理。

2. clientele (ˌklaɪənˈtɛl): 集合名詞，指顧客而言。

3. through our medium: 透過我們。

4. very keen on: very interested in, 對…很有興趣; eager: 渴望。

5. CIFC$_5$: C$_5$是指回佣(return commission)5%之意。至於5%究係按 CIF 計算或按 CIFC$_5$計算並不清楚。因此，爲避免糾紛，較謹愼的出口商在報價時多另加說明。例如: The price includes your commission 5% on CIF basis.但一般而言，多以 FOB 爲計算回佣較妥。

第二節　有關招攬交易的有用例句

一、開頭句

　　1. Your $\begin{Bmatrix} \text{good firm} \\ \text{name (and address)} \end{Bmatrix}$ has (have) been →

$\rightarrow \begin{cases} \text{recommended to us} \\ \text{brought to our attention} \\ \text{given to us} \end{cases} \rightarrow$

$\rightarrow \text{by} \begin{cases} \text{the Hongkong Chamber of Commerce} \\ \text{the American consulate at...} \\ \text{the China External Trade Development Council} \\ \text{our bank, Irving Trust Company} \\ \text{a certain reliable source (某一可靠來源)} \end{cases} \rightarrow$

$\rightarrow \text{as} \begin{cases} \text{a large exporter} \\ \text{one of the leading exporters} \\ \text{one of the greatest importers} \\ \text{a manufacturer} \\ \text{a distributor (經銷商)} \end{cases} \text{of...(goods)...}$

2. $\begin{cases} \text{From} \begin{cases} \text{Jerro Trade Directory} \\ \text{the American Chamber of Commerce} \end{cases} \\ \text{Through} \begin{cases} \text{the courtesy of CETRA} \\ \text{the medium of} \begin{cases} \text{Kelly's Directory} \\ \text{China News} \end{cases} \end{cases} \end{cases} \rightarrow$

$\rightarrow \text{we} \begin{cases} \text{have learned} \\ \text{understand} \end{cases} \text{that you are} \rightarrow$

* have learned：聞悉　* understand：獲知

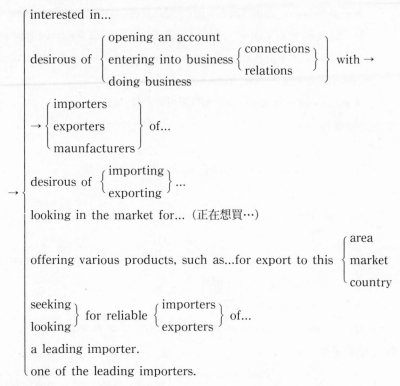

$$\rightarrow \begin{cases} \text{interested in...} \\[2pt] \text{desirous of} \begin{cases} \text{opening an account} \\ \text{entering into business} \begin{Bmatrix} \text{connections} \\ \text{relations} \end{Bmatrix} \\ \text{doing business} \end{cases} \text{with} \rightarrow \\[2pt] \rightarrow \begin{Bmatrix} \text{importers} \\ \text{exporters} \\ \text{maunfacturers} \end{Bmatrix} \text{of...} \\[2pt] \text{desirous of} \begin{Bmatrix} \text{importing} \\ \text{exporting} \end{Bmatrix} ... \\[2pt] \text{looking in the market for...（正在想買…）} \\[2pt] \text{offering various products, such as...for export to this} \begin{cases} \text{area} \\ \text{market} \\ \text{country} \end{cases} \\[2pt] \begin{Bmatrix} \text{seeking} \\ \text{looking} \end{Bmatrix} \text{for reliable} \begin{Bmatrix} \text{importers} \\ \text{exporters} \end{Bmatrix} \text{of...} \\[2pt] \text{a leading importer.} \\ \text{one of the leading importers.} \end{cases}$$

二、自我介紹

可參考前章，這裡再列舉若干供參考。

1. We are well established in manufacturing...（商品）...to which we have devoted years of experience and research. You may rest assured that whatever we offer you excels in workmanship and function.

   ＊ devoted to: 致力於… ＊ you may rest assured: 保證你

2. We $\begin{Bmatrix} \text{trust} \\ \text{believe} \end{Bmatrix}$ that our experience in foreign trade and intimate knowledge of international market conditions will entitle us to your confidence.

* entitle us to your confidence: 足以獲得你的信賴

3. We believe that many years of experience which we have had in this business world and the means at our disposal are sufficient for us to undertake any orders from you.

　　* means at our disposal: 可供運用的資力

三、付款條件

1. As to the terms of payment, we trade on $\begin{cases} \text{D/P.} \\ \text{D/A.} \\ \text{L/C.} \end{cases}$

2. Our usual terms are to draw a draft at sight under a banker's irrevocable L/C.

3. We make it our customs to trade on L/C to be opened concurrently with the placing of an order.

　　* concurrently with: 與……同時

4. Our terms are shipment within 60 days after receipt of L/C.

四、結尾用語

1. If you are $\begin{cases} \text{inclined to entertain} \\ \text{interested in} \end{cases}$ our proposal mentioned above, →

→ please $\begin{cases} \text{give us} \\ \text{let us know} \\ \text{let us have} \end{cases}$ your definite inquiries upon →

→ which we shall be $\begin{cases} \text{pleased} \\ \text{glad} \end{cases}$ to $\begin{cases} \text{send} \\ \text{cable} \end{cases}$ you our best →

→ $\begin{cases} \text{prices and deliveries.} \\ \text{terms and conditions.} \end{cases}$

2. We $\begin{cases} \text{trust} \\ \text{hope} \end{cases}$ that we may $\begin{cases} \text{soon} \\ \text{in the near future} \end{cases}$ have the →

$$\rightarrow \text{pleasure of} \begin{cases} \text{doing business} \\ \text{opening up} \\ \text{opening an account} \end{cases} \text{with} \begin{cases} \text{you} \\ \text{your reputable house} \\ \text{your firm} \end{cases} \rightarrow$$

→ and proving that such a connection would yield you a considerable benefit.

3. We are looking forward to $\begin{cases} \text{receiving your reply} \\ \text{your reply} \end{cases}$ and hope →

$$\rightarrow \text{we shall be able to} \begin{cases} \text{establish a connection} \\ \text{open an account} \\ \text{enter into business relations} \end{cases} \rightarrow$$

→ between our two firms which will be pleasant and mutually beneficial.

習　　題

一、將下列中文譯成英文。

1. 茲由外貿協會(CETRA)獲知寶號（及地址）係拉哥斯(Lagos)主要進口商之一。

2. 我們由 Kelly's Directory 得悉貴公司係電視機的主要製造廠商之一。

3. 因為本公司在進出口貿易方面有多年的經驗，而且又與本地主要製造廠商有很密切的聯繫，本公司相信，不拘訂購數量多寡，均可以廉價(at competitive prices)供應。

4. 我們特別強調「貨真價實」為敝公司永遠遵守的格言。

5. 本公司可以盡可能的最低價供售(offer)該貨。

二、試用下列片語造句。

1. to enter into business relations　2. mutually beneficial

3. as to　　　　　　　　　　　　　4. as for

5. be inclined to

三、試將本書 No. 8 的英文信譯成中文。

第九章　徵信與答覆

(Credit Inquiries and Replies)

第一節　委託徵信

接到招攬信後，在未弄清對方身分之前，不宜立卽進入交易階段。換言之，在正式進行交易之前，宜先做徵信(Credit Inquiry)，以免日後因對方不誠實而引起糾紛。但徵信費時，因此，在徵信期間，應先覆謝。

NO.13　收到招攬信後，正式建立關係前的覆函

Gentlemen,

We are grateful to you for your proposal, expressed in your letter of May 10, to open an account with us.

From your letter we are glad to learn that you are interested in Taiwan-made nylon piece goods, and in these we may say that we are specialists. It is our sincere wish to become connected with your establishment, but before we accept your proposal and go into negotiations in a definite way, it is usual with us to take up the reference named in your letter. We shall, therefore, do ourselves the pleasure of communicating with you further as soon as we hear from the bankers you named.

> We thank you for your courtesy in making the proposal and hope we may soon be able to do a good business with you.
>
> > Very truly yours,

【註】

1. establishment: (工廠、商號等) 機構, 公司, 行號。

2. it is usual with us＝it is the custom with us＝it is customary with us

3. take up the reference:向備詢人查詢。

4. do ourselves the pleasure:比"have the pleasure"鄭重。

如因某種原因, 不願意與其建立往來關係也應以委婉的措詞覆函拒絕。下面就是一例:

No.14　不願與其往來的覆函

> Dear Sirs,
>
> Thank you for your letter of May 4 proposing to establish business relations between our two firms.
>
> Much as we are interested in doing business with you, we regret to inform you that we are not in a position to enter into business relations with any firms in your country because we have already had an agency arrangement with Taiwan Trading Co., Ltd. in Taipei. According to our arrangement only through the above firm, can we export our products to Taiwan.
>
> Under the circumstances, we have to refrain from transacting with you until the agency arrangement expires. Your letter has been filed for future reference.

Thank you again for your proposal and your understanding of our position will be appreciated.

Faithfully yours,

【註】

1. agency arrangement＝agency agreement＝agency contract:代理契約；代理協議書。

2. much as:雖然很（有興趣）。

3. are not in a position:比"can not"委婉。

4. only...,can we...:這是加重語氣的措詞。

5. under the circumstances:「在此情形下」，注意"circumstances"必須用多數形。

6. refrain from:抑制，強調抑制一種衝動，自動地不做所欲做或所願做的事。與"abstain from"同義，但"abstain from"係強調以意志力克制自己，也指自動戒除某些所欲的或有害的東西，尤指享樂及飲食。例:He is abstaining from pie. (他在克制自己不吃〔他愛吃的〕餡餅。)

7. transacting:可以"doing business"代替。

8. position:立場，也以"situation"代替。

調查對方信用的方法，最常用的有下列三種:

1. 委託本地外滙銀行代爲調查。

2. 逕函對方所提供的備詢銀行(reference bank)或備詢商號(house reference, trade reference)查詢。

3. 委託徵信所(mercantile credit agency)代爲調查。

委託徵信時，其所需徵信事項宜具體寫出，通常一封委託徵信函應包括下列各項:

1. 被調查商號名稱、地址。

2. 調查理由。

3. 調查事項:

   ⑴ Character（品性）: 指負責人的 Integrity, Reputation, Willingness to meet obligation（履行債務的意願）, Attitude toward business（對業務的態度）。

   ⑵ Capacity（才能）: 經營者在商場上的經營能力、經學歷。

   ⑶ Capital（資本）。

以上稱爲3C，此外宜包括: 創業日期(date of establishment)、營業項目(line of business)、過去三年營運量(business volume for the past 3 years)、過去三年損益(profit/loss for the past 3 years)、員工人數(number of employee)等。

4. 表示對所提供消息，保證絕對保密不外洩。

5. 表示惠請協助將感激不盡。

No.15　請求本地外滙銀行代爲徵信

委託本地外滙銀行徵信時，可用本國文:

受文者: 臺灣銀行徵信室

主旨: 請代查報 ABC Co., Ltd., 123 Wall Street, New York, N.Y. 10012, U. S. A.信用狀況。

說明: 1.本公司爲貴行進出口及存款多年客戶，開有支票存款第1234 號戶，每月結滙多筆。

    2.茲正與上開紐約新客戶洽商交易中，據悉該商之備詢銀行爲 Irving Trust Company, New York.

           台灣貿易公司　謹上

中華民國 79 年 8 月 10 日

No.16 逕向對方所提供備詢銀行查詢

Dear Sir,

Anderson Co., Inc., 5000 Market Street, San Francisco, who have recently proposed to do business with us, have referred us to your Bank.

We should feel very much obliged if you would inform us whether you consider them reliable and their financial position strong, and whether their business is being carried on in a satisfactory manner.

In addition to the above, please, if possible, also furnish us with the following information:

1. date of establishment
2. name of responsible officers
3. line of business
4. business volume for the past three years
5. profit/loss for the past three years.

Any information you may give us will be held in absolute confidence and will not involve you in any responsibility.

We apologize for the trouble we are giving you. Any expenses you may incur in this connection will be gladly paid upon being notified.

Faithfully yours,

【註】

1. carry on:經營。

2. furnish...with:提供, 同類用語為"supply...with", "provide...with"。

3. responsible officers:負責人。

4. Any information...的相似用語有:

(a) Any information you may favour us with will be held in strict confidence.

(b) Your communication on the subject will, you may rest assured, be treated confidentially and discreetly.

(c) We will make a discreet use of any report you may give us concerning the subject.

(d) Any information with which you may favour us will be used in absolute confidence.

5. incur:發生。

6. upon being notified:一經通知。

　　銀行有代客保密之責, 因此逕向對方所提供備詢銀行調查客戶信用時往往得不到結果。所以, 還是委託自己的往來銀行徵信較妥。

No.17　受託銀行轉請國外銀行徵信

Request for a report on
XYZ Co., Ltd....

Dear Sirs,

　　At the request of our valued customer, we would appreciate your furnishing us by airmail with a detailed report on the business, means and credit standing of the above firm.

　　Any information you supply will be treated in strict confidence and without any responsibility on your part.

　　We thank you in advance and assure you of our

readiness to serve you in a similar manner.

Faithfully yours,

【註】

1. readiness:願意。

2. detailed report:詳細報告。

3. means:資力，即財力。

4. We thank you in advance 的措詞宜少用，可將"in advance"刪除。

5. to serve you in a similar manner:爲您做同樣（類似）的服務。

第二節　徵信報告

No.18　國外銀行逕向商號提出的徵信報告─favorable（有利）

Gentelmen:

The subjoined is a report on the firm referred to in your letter of March 17. Please note that this information is supplied in strict confidence and without any responsibility on the part of this Bank or any of its officers.

We thank you for your offer to reimburse the expense incurred, but the cost was too trifling to bring into account.

Yours very truly,

Anderson & Co., Inc.

500 Market Street, San Francisco

The above-mentioned firm is one of the oldest and best esta-blishments in this city, doing wholesale and import and export business in general merchandise. They have their branch houses in Portland and Seattle and have been doing a business of considerable volume.

The firm has maintained an account with us for the past twenty years and has been one of our best clients. We have often made loans for good amounts, and all these obligations have been met as agreed upon. We are quite satisfied to see that our confidence in the firm has never been misplaced.

Mr. W. Anderson, President of the firm and all other directors are everywhere held in the highest esteem, both for their business ability and for their integrity. From our records we do not hesitate to say that the firm may be rated as Al.

S.J.

【註】

1. the subjoined:另附的東西，卽「另紙」。

2. reimburse:歸償; 歸墊。

3. too trifling to bring into account:微不足道，所以不算帳。

4. made loans for good amounts:貸給相當數額的款項。

5. these obligations...upon:債務各部均依約付清。

6. our confidence...misplaced:我們信任該商號並未錯誤。

7. rated as AI:信用等級爲最佳，AI(éi wʌn)＝first rate(class)。

8. 銀行對於徵信的回答，總是加上免責條款，相似的條款尚有:

(a) It is understood that the information contained in this letter is given in absolute confidence, and entirely without prejudice or the assumption of responsibility by us.

(b) All persons are informed that any statement on the part of this bank, or any of its officers, as to the responsibility or standing of any person, firm, or corporation, is a mere matter of opinion and given as such, and solely as a matter of courtesy, and for which no responsibility, in any way, is to attach to this bank or any of its officers.

No.19　銀行對銀行的徵信報告—favorable

Dear Sirs,　　　　　Credit Report

In answer to your inquiry of October 14, 19··· regarding Anderson & Co., London, we are pleased to send you the following information.

This firm was established by Mr. Jones Anderson and his family in 1955. It operates as an importer of ladies' handbags which are distributed through the U. K. in a popular price line.

We have an account of the subject for a good many years and the relationship has been monthly openings of letters of credit in low six figures with deposit balances fluctuating between low five and moderate five figures.

This firm meets all obligations in a prompt manner. The principal is well-known to our officers and is highly regarded by us.

The financial statement as at the end of 19··· indicates that the net worth is Stg. £200,000.

　　We recommend this firm to your customers for their normal business engagements.

Yours very truly,

【註】

1. in a popular price line:以大衆化價格（出售）。

2. fluctuating between...figures:在低五位數與中下五位數之間變動。

"low five figures"又可寫成"low five figure proportions"或"low five fig-.ure category"，均可譯爲低五位數。銀行基於業務保密理由，不明示實際餘額，而以位數來代替。茲將其代字的意義列示如下：

Low...1-2

Moderate...3-4

Medium...5-6

High...7-9

〔例〕： low six figures in dollar:$100,000—$200,000; low to moderate five figures: 10,000—40,000; balance in medium seven figures:餘額爲 5,000,000—6,000,000

3. highly regarded:很受重視。

4. net worth:淨值，卽資產減負債的差額。

5. normal business engagements:普通程度的債務。

No.20　無法提供信用資料時

Your credit inquiry of June 6
on ABC Co., Ltd.

Dear Sirs,

　　The subject firm does not maintain any account

relationship with us, therefore, no information regarding it is available on our file.

Although we have contacted some of our banking friends here trying to obtain information on the subject firm, we failed.

Such being the case, we are not able to furnish you with the credit standing of the subject firm. We regret that we have been unable to assist you in this matter.

Yours very truly,

【註】

1. on our file。在我們檔卷中。也可用"in our file"。

   cf. We have placed the correspondence on our files:我們已將該信件歸檔了。

2. contact:接觸; 聯繫。當做動詞用時, 係他動詞, 因此不能說"contact with..."。如當做名詞用, 則須加"with"。例:

   to make contact with someone

   to be in contact with someone

3. such being the case:因此; 既然是這樣。

4. in this matter:關於這件事。

No.21 對於提供徵信報告的謝函

Dear Sirs:

We have received your letter of May 24 giving us a report on the firm we inquired about. We wish to say that your information will certainly be of great assistance to us in deciding the course we have to take.

> We express our best thanks for the trouble you have taken, and you may rest assured that your information will be retained in our files in a confidential manner.
>
> Respectfully yours,

【註】

1. the course we have to take:我們應採取的行動方針。

第三節　有關徵信的有用例句

一、請求提供信用資料

1. $\left\{\begin{array}{l}\text{Please}\\\text{Kindly}\end{array}\right\}$ $\left\{\begin{array}{l}\text{supply}\\\text{furnish}\\\text{provide}\end{array}\right\}$ us with $\left\{\begin{array}{l}\text{your opinion}\\\text{a detailed report}\end{array}\right\}$ $\left\{\begin{array}{l}\text{on}\\\text{regarding}\end{array}\right\}$ →

→ the $\left\{\begin{array}{l}\text{financial status and responsibility}\\\text{financial standing and reputation}\\\text{financial responsibility}\\\text{business, reputation, means and credit standing}\end{array}\right\}$ →

→ of the $\left\{\begin{array}{l}\text{above}\\\text{above-mentioned}\\\text{following}\end{array}\right\}$ firm.

請就 $\left\{\begin{array}{l}\text{上列}\\\text{上述}\\\text{下列}\end{array}\right\}$ 商號的 $\left\{\begin{array}{l}\text{財務情況及付款能力}\\\text{財務狀況及聲譽}\\\text{財力}\\\text{業務、聲譽、資力、信用狀況}\end{array}\right\}$ $\left\{\begin{array}{l}\text{提供}\\\text{惠示}\end{array}\right\}$ 卓見。

2. Please get information for us on the financial standing and reputation of...Co., Ltd.

* get information:蒐集情報；取得情報。

二、將予保密及結尾句

1. $\begin{Bmatrix} \text{Any information you may} \\ \text{You can be sure that any information you may} \end{Bmatrix}$ $\begin{Bmatrix} \text{supply} \\ \text{give us} \end{Bmatrix}$ →

→ will be kept $\begin{cases} \text{in strict confidence.} \\ \text{confidential.} \\ \text{strictly confidential.} \end{cases}$

2. $\begin{Bmatrix} \text{We will be very glad} \\ \text{We assure you of our willingness} \end{Bmatrix}$ to $\begin{Bmatrix} \text{reciprocate} \\ \text{return your courtesy} \end{Bmatrix}$ →

→ $\begin{cases} \text{at any time.} \\ \text{when a similar opportunity arises.} \\ \text{if an occasion arises.} \\ \text{if an opportunity arises.} \end{cases}$

* at any time:隨時。

* when a similar opportunity arises:有類似機會時。

* if an occasion arises:如有機會。

3. You are assured of our discretion and we will welcome an opportunity to reciprocate your courtesy.

* assured of our discretion:保證嚴守機密。

三、情況良好的徵信報告

1. The $\begin{Bmatrix} \text{firm} \\ \text{company} \end{Bmatrix}$ you inquired about in your letter of…(date)…→

→ $\begin{cases} \text{enjoyed a good reputation in the business circles here.} \\ \text{is considered to have an excellent business reputation here.} \end{cases}$

2. In our opinion, they $\begin{cases} \text{carry on their business satisfactorily.} \\ \text{are considered good for normal engagements.} \end{cases}$

$$(我們認爲\begin{cases}他們經營得很順利。\\可與他們從事正常的交易。\end{cases})$$

3. They are quite punctual in meeting their obligations.

   (履行債務很準時。)

4. We have a favorable regard for the company and its management.

   (對該公司及其管理階層表示讚許。)

5. We have every confidence to recommend to you the firm you inquire about as one of the most reliable exporters in our district.

   (以充分的自信向你推薦…)

6. Messrs. Smith & Co. is A1 in every respect. They always display sound judgement in the conduct of their business, which we suppose, is based on their many years' experience.

   (各方面都是第一流的，在業務方面的判斷總是表現得很穩當…)

7. Our records show that they have never failed to meet our bills since they opened an account with us. The monthly limit of credit we feel we may safely grant them is approximately $…In addition, their sincere attitude toward trade and their extensive business activities merit high esteem.

   * never failed…bills:從不欠帳。* monthly limit of credit:每月信用額度。

   * grant:給與。* sincere attitude:誠實的交易態度。* merit high esteem: 值得尊敬。

四、情況令人起疑或欠佳的徵信報告

1. According to the talks in our district, they have recently commenced rather risky speculative business for an amount beyond their capacity.

   (據本地傳說，他們最近開始從事超出他們能力所及的金額的相當危險的投機交易。)

2. The reports in circulation indicate that they are in an awkward situation for meeting their obligations.

（傳聞該公司窮于應付債務的履行。）

3. We advise you to proceed with every possible caution in dealing with the firm in question.

（忠告宜特別謹慎。）

五、無法提供資料

1. We are sorry we are $\left\{\begin{array}{l}\text{not able}\\\text{unable}\end{array}\right\}$ to give a precise information you ask.

2. We are sorry that we $\left\{\begin{array}{l}\text{cannot}\\\text{are unable to}\end{array}\right\}$ $\left\{\begin{array}{l}\text{furnish}\\\text{provide}\end{array}\right\}$ you with the information you desire.

3. This firm does not maintain an account relationship with us. For this reason, we are not able to comment on the financial responsibility and reputation of the firm.

4. We regret we are unable to give you positive answers about the said concern.

　　* positive answers:肯定的答覆。

六、保密、不負責的聲明及結尾

1. $\left\{\begin{array}{l}\text{This}\\\text{The above}\end{array}\right\}$ information is $\left\{\begin{array}{l}\text{given}\\\text{furnished}\end{array}\right\}$ $\left\{\begin{array}{l}\text{confidentially}\\\text{in strict confidence}\end{array}\right\}$ →

→ and without responsibility on $\left\{\begin{array}{l}\text{our part}\\\text{the part of this bank}\end{array}\right\}$ and or →

→ its officers.

2. We hope that $\left\{\begin{array}{l}\text{this}\\\text{the above}\end{array}\right\}$ information will be of assistance to you.

3. The information is $\begin{Bmatrix} \text{furnished} \\ \text{given} \end{Bmatrix}$ $\begin{Bmatrix} \text{confidentially} \\ \text{in strict confidence} \end{Bmatrix}$ at your →

→ request without any guarantee or responsibility on our part

→ and with the understanding that its sources will not be disclosed
  in anyway.

＊ its sources will not be disclosed in anyway:將不以任何方式洩漏該項
資料的來源。

習　　題

一、將下列中文譯成英文。

　1. 煩請將下列商號的業務、聲譽、資力以及信用狀況惠示卓見爲荷。

　2. 該公司創立於 1960 年, 從事於塑膠靴(plastic shoes)之製造與外銷, 公司管理
　　階層(management)能幹而可靠, 似乎不至於承諾無法履行的債務(not likely
　　to commit the company to any engagement it could not fulfil)。

　3. 傳聞該公司窮于應付債務的履行(in awkward situation for meeting their
　　obligations)。

　4. 您提供給我們的任何消息, 將嚴加保密。

　5. 歡迎賜予我們回報的機會。

二、將本書 No.14 的英文信譯成中文。

三、試用下列片語造句。

　1. if an occasion arises

　2. such being the case

　3. carry on

　4. much as

第十章　一般交易條件協議書

（Agreement on General Terms and Conditions of Business）

第一節　一般交易條件協議書的意義

　　謹慎的進出口商與對方取得連繫後，第一步為調查對方信用，如調查結果認為對方信用良好，值得往來，第二步就與對方洽訂一般交易條件，議訂雙方權益，以作為日後實際交易的基準。尤其對對方的信用調查結果未能十分滿意時，更應訂定相當嚴密的交易條件，約束對方，俾免日後遭受意外的不利或不必要的損失。這種一般交易條件是就雙方所開示的希望交易條件，經往返磋商最後達成時，通常均須作成書面協議書，雙方簽署後，各執一份。協議書簽立後，雙方即可以電報、Telex 或書信進行交易。這種電報、Telex 或書信，僅記載個別交易的主要條件(品名、品質、數量、價格、交貨期) 而已。換言之，憑這種個別交易所簽定的個別契約需與預先簽立的協議書合併起來才構成一完整的契約。所以在雙方簽有協議書的情形下，買賣雙方進行交易，毋需每次都詳述一般交易條件，這對雙方都省事省錢，頗為方便。

　　上述協議書通稱為「一般交易條件協議書」(Agreement on General Terms and Conditions of Business)或稱為協議書(Memorandum of Agreement)或 Mutual Understanding 或 Basic Agreement。其內容隨交易對方的身分、信用、買賣貨物種類以及市場習慣等而有若干的差異，但一般而言，不外包括下列各事項：

1. 約定交易的性質：確定交易雙方身分。即契約雙方爲賣方與買方的關係。
2. 約定有關報價與接受事項。
3. 約定每筆交易訂貨事項：約定憑電報（或 Telex）成交後，以書面確認，且一經確認，非經同意不得取消。
4. 約定品質、包裝、裝運、保險、付款事項。
5. 約定索賠及不可抗力事項。
6. 其他事項的約定。

No.22　徵信結果，同意建立業務關係

Gentlemen:

Following our letter of May 17, we are pleased to say that our credit files have now been completed with favorable information from your Bankers, and therefore we are quite willing to accept your proposal made in your letter of May 10 and start business with you in electronic products.

Samples and Prices of Calculators. As requested, we are airmailing you samples of our calculators under Nos. $\frac{1}{3}$, particulars of which are given on the Price List No.50 enclosed. We ask you to note that each of the prices quoted is based on 1,000 sets, which is the minimum quantity of an order we can book. The prices are, of course, without engagement, and we are able to make you a firm offer by cable on receipt of your definite inquiry. We wish you will closely examine the samples and are convinced of our goods being found superfine and prices competitive. We are in a position to supply you with lower qualities, which, however, will be unsuitable for your trade.

Our electronic products cover various descriptions. They are all in brisk demand in the U.S. markets. We shall be glad to learn whether you have an opening for TV sets, TV game, Electronic Blocks, etc., in which we have been doing a good business with New York customers. Samples and prices will be forwarded immediately on receiving your inquiries.

<u>Terms and Conditions.</u> We agree to your terms, i.e. draft at 60 d/s under an Irrevocable L/C to be opened simultaneously with the placing of an order. In order to preclude any possible misunderstanding which may arise in our transactions, we are enclosing an Agreement on General Terms and Conditions, on which all our future business will be based, for your approval.

We look forward to your further communications and hope that the relations now being established between us will last long and become mutually profitable.

Very truly yours,

【註】

1. without engagement:不受約束，即價格可變更之意。

2. superfine:極精緻的。

3. brisk demand: 不斷的需要。Trade is brisk: 生意興隆。

4. have an opening for=to be open for=be in the market for

5. preclude:排除。

第二節　一般交易條件協議書例示

No.23　一般交易條件協議書

AGREEMENT ON GENERAL TERMS AND CONDITIONS OF BUSINESS AS PRINCIPAL TO PRINCIPAL

THIS AGREEMENT entered into between ABC Co., Ltd., 111 Lin-Sheng N. Road, Taipei, Taiwan, hereinafter referred to as SELLER, and XYZ Co., Ltd., 222 Broadway Street, New York, N.Y., U.S.A., hereinafter referred to as BUYER, witnesses as follows:

1) Business:Both SELLER and BUYER act as principals and not as agents.

2) Commodities:Commodities in business and their unit to be quoted, are as stated in the attached list.

3) Quotations and Offers:Unless otherwise specified in cables or letters, all quotations and offers submitted by either party to this Agreement shall be in U.S. dollars on C.I.F. New York basis.

4) Firm Offers:All firm offers shall be subject to a reply within the period stated in respective cables. When "immediate reply" is used, it means that a reply is to be received within three days from and including the day of

the despatch of a firm offer. In either case, however, Sundays and all official Holidays are excepted.

5) Orders:Any business concluded by cable shall be confirmed in writing without delay, and orders thus confirmed shall not be cancelled unless by mutual consent.

6) Payment:Payment to be effected by BUYER by usual negotiable and irrevocable letter of credit, to be opened 30 days before shipment in favor of SELLER, providing for payment of 100% of the invoice value against a full set of shipping documents.

7) Shipment:All commodities sold in accordance with this Agreement shall be shipped within the stipulated time. The date of Bill of Lading is taken as conclusive proof the day of shipment. Unless expressly agreed to, the port of shipment is at SELLER's option.

8) Marine Insurance:All shipments shall be covered All Risks for a sum equal to the amount of the invoice plus 10 percent, if no other conditions are particularly agreed to, all policies shall be made out in U.S. currency and payable in New York.

9) Quality:Quality to be guaranteed equal to description and/or samples, as the case may be.

10) Inspection:Commodities will be inspected in accordance with normal practice of supplier, but if BUYER desires special inspections, all additional charges shall be borne by BUYER.

11) Damage in Transit:SELLER shall ship all commodities in good condition and BUYER shall assume all risks of damage, deterioration or breakage during transportation.

12) Exchange Risks:The price offered in U.S. dollars is based on the prevailing official exchange rate in Taiwan between the U.S. dollar and the New Taiwan dollar. Any devaluatin of the U.S. dollar to the New Taiwan dollar at the time of negotiating draft shall be for BUYER's risks and account.

13) Change in Freight and Insurance Rate:Any change in marine freight rate and marine insurance rate is for BUYER's account.

14) Shipment Samples:In case shipment samples be required, SELLER shall forward them to BUYER prior to shipment under the contract of sale.

15) Cable Expenses:Expenses relating to cabling shall be borne by the respective senders.

16) Claims: Claims, if any, shall be submitted by cable within fourteen (14) days after arrival of commodities at destination. Certificates by recognized surveyors shall be sent by mail without delay. All claims which cannot be amicably settled between SELLER and BUYER shall be submitted to arbitration in New York, the arbitration board to consist of two members, one to be nominated by SELLER and one by BUYER, and should they be unable to agree, the decision of an umpire selected by

the arbitrators shall be final, and the losing party shall bear the expenses thereto.

17) Force Majeure:SELLER shall not be responsible for the delay of shipment in all cases of force majeure, including mobilization, war, riots, civil commotions, hostilities, blockade, requisition of vessel, prohibition of export, fires, floods, earthquakes, tempests, and any other contingencies which prevent shipment within the stipulated period. In the event of any of the aforesaid causes arising, documents proving its occurrence or existence shall be sent by SELLER to BUYER without delay.

18) Delayed Shipment:In all cases of force majeure provided in the Article No.17, the period of shipment stipulated shall be extended for a period of twenty one (21) days. In case shipment within the extended period should still be prevented by a continuance of the causes mentioned in the Article No.17 or the consequences of any of them, it shall be at BUYER's option either to allow the shipment of late goods or to cancel the order by giving SELLER the notice of cancellation by cable.

19) Shippping Notice:Shipment effected under the contract of sale shall be immediately cabled.

20) Packing & Marking:All shipments shall be packed for export and be marked XYZ in Diamond.

In witness whereof, ABC Co., Ltd. have hereunto set

their hand on the 1st day of June, 1990, and XYZ Co., Ltd. have hereunto set their hand on the 20th day of June, 1990. This Agreement shall be valid on and from the 1st day of July, 1990, and any of the articles in this Agreement shall not be changed and modified unless by mutual written consent.

SELLER   BUYER

ABC Co., Ltd.   XYZ Co., Ltd.

General Manager   General Manager

〔中譯〕

貨主間一般交易條件協議書

ABC 公司(以下稱爲賣方)，地址：臺灣臺北市林森北路111號，與 XYZ 公司(以下稱爲買方)，地址：美國紐約市百老滙街222號，茲訂立本協議書，(約定)證明下列各事項：

1. 交易形態：當事人雙方均爲法律上的本人，而非代理人。

2. 買賣貨物：買賣的貨物以及報價的單位，如附表。

3. 報價：除非電報或書信中另有規定，本協議書任何一方提出的報價，均按美金計算，而且均以"C.I.F. NEW YORK"爲條件。

4. 穩固報價：所有穩固報價，均須在個別電報所載的期限內回覆。在使用「立即回覆」字樣時，應解爲：回覆須於穩固報價發出之日起(包括發出之日)三日內被收到。但，星期日及公休日除外不計。

5. 訂貨：憑電報成交的買賣，應迅速用書面確認，訂貨經確認後，除非經雙方同意，不得取消。

6.付款方式：買方應以通常的、可讓購、不可撤銷信用狀付款；此項信用狀應以賣方為受益人，於裝運前三十天開出；規定憑全套貨運單證支付發票金額全額。

7.裝貨：所有憑本協議書售出的貨物，都必須在約定期間內裝出。提單日期，應視為裝貨日的決定性證據。除非有明確約定，裝貨港由賣方選擇。

8.海上保險：如未特別約定其他條件，所有裝出的貨物，均應投保全險，保險金額等於發票金額加一成。所有保險單應載明投保美金，在紐約理賠。

9.貨物品質：賣方應保證，所裝貨物在品質及狀況方面，與說明及／或樣品相符。

10.檢驗：貨物將依供應商通常方式實施檢驗，但如買方欲實施特別檢驗時，所有額外費用皆歸買方負擔。

11.運輸途中的損壞：賣方應裝出情況良好的貨物，買方則須負擔貨物在運輸中損壞、變質或破損的危險。

12.滙兌風險：以美金報價的價格，乃以現行臺灣美金對新臺幣的官定原價為準，在押滙時，如美金對新臺幣有任何貶值，則此項風險及損益歸買方負責。

13.運價、保險費率的變動：任何海運費率及海上保險費率的變動，歸買方負擔。

14.裝貨樣品：買方要求裝貨樣品時，賣方應於依約裝貨前，將樣品寄給買方。

15.電報費：拍發電報的費用，應由各發報人自行負擔。

16.索賠：如須索賠，應於貨物到達目的地十四日內，用電報提出。認可的公證行所簽發的證明書，應速即郵寄。買賣雙方不能友好解決的所有索賠事件，應在紐約交付仲裁，仲裁庭包括兩人，其中一個由賣方指定，另一人由買方指定，如兩人的意見不能一致，則以仲裁人所選評判人的決定為準。而敗方須負擔費用。

17.不可抗力：在所有不可抗力情形下，賣方對裝運遲延不負責，包括動員、戰爭、騷擾、民變、軍事衝突、封鎖、徵用船隻、禁止出口、火災、洪水、地震、風暴以及在約定期間阻礙裝貨的其他意外事故。萬一上述事故發生，證明事故發生或存在的文件，應由賣方迅速寄給買方。

18.遲延裝運：在第17條所列不可抗力情形下，約定的裝運期限應延長廿一日。如延長期間的裝運，仍因第17條所列事故或其影響繼續存在而受到阻礙，則應由買方選擇，或接受後來遲裝的貨物，或用電報通知賣方取消訂貨。

19.裝運通知：賣方依買賣契約裝出貨物時，應立即發出電報通知。

20.包裝及刷嘜：所有貨載須施以出口包裝並刷上 xyz 的嘜頭。為證明上述約定，ABC 公司於 1990 年 6 月 1 日在本協議書簽字，XYZ 公司於 1990 年 6 月 20 日在本協議書簽字，本協議書自 1990 年 7 月 1 日起生效，除非經雙方書面同意，本協議書中任何條款不得變更或修改。

賣方　　　　　　　　　買方
ABC Co., Ltd.　　　　　XYZ Co., Ltd.
總經理　　　　　　　　總經理

習　　題

一、試述訂立一般交易條件協議書的用意。

二、試解釋及區別下列用語：

1. Principal 與 Agent

2. Quotation 與 Offer

3. General terms and conditions 與 Basic terms and conditions

4. Trade terms 與 Terms and conditions of sale

5. Shipment 與 Delivery

6. Force majeure 與 Act of God

7. Arbitrator 與 Umpire

三、試用下列片語造句：

1. as requested

2. to be convinced of

3. in brisk demand

4. for buyer's risks and account

四、將下列中文譯成英文：

1. ABC 公司（以下稱為賣方）與 XYZ 公司（以下稱為買方），茲訂立本協議書，約定下列各事項。

2. 本協議書自 1990 年 7 月 1 日起生效，除非經雙方同意，本協議書中任何條款不得變更或修改。

3. 賣方於裝出貨物時，應立即發出電報通知買方。

4. 在押滙時，如美金對新臺幣有任何貶值，此項風險應由買賣雙方平均負擔。

5. 任何海運費率及保險費率的變動，均歸賣方負擔，賣方不得以任何理由向買方要求補償。

第十一章　交易條件

(Terms and Conditions)

　　國際貿易一如國內買賣，一筆交易的成立，買賣雙方必須就其交易的內容有所約定，雙方才能遵照履行。買賣雙方所約定的內容稱爲交易條件 (Terms and Conditions of the Transactions)。交易條件的詳略，視貨物種類、買賣習慣以及事實需要而定。但國際買賣每一筆交易通常至少應就商品名稱、品質、價格、數量、包裝、交貨、保險及付款等條件有所規定。這八條件，在報價時，爲報價的基本內容，爲構成有效報價的基本要素。報價一經有效接受，買賣契約卽告成立，而這八條件卽轉而成爲買賣契約的內容，以下分節說明。

第一節　商品與商品目錄

一、商品名稱

　　買賣的標的物通常稱爲商品，但有時也稱爲產品、貨品、貨物、物資、物料或器材等，在英文則有 commodity, merchandise, goods, product, produce, ware, line , materials, supplies, article, item 等等稱呼，在保險界及航運界則又稱爲 cargo。

　　交易上所使用商品名稱，應爲國際市場上一般通行者，如以地方性的名稱做爲交易商品名稱，在各方面都不方便。

　　茲將各種有關貨物的名詞列舉如下：

GOODS（貨物、商品、貨品、物資、原料）

Air-borne goods	空運物資	Half-finished goods	半製品
Bargain goods	特價品	Heavy (light) goods	重(輕)量貨
Canned goods	罐頭品	Household goods	家庭用品
Tinned goods	罐頭品	Luxury goods	奢侈品
Capital goods	資本財	Inflammable goods	易燃品
Coarse(crude) goods	粗製品	Low-priced goods	廉價品
Clearance goods	出清存貨	Manufactured goods	製成品
Consumer goods	消費品	Measurement goods	體積貨
Consumption goods	消費品	Quality goods	高級品
Contraband goods	走私貨、違禁品	Perishable goods	易腐品
Cotton goods	綿製品	Processed goods	加工品
Customable goods	應課稅品	Piece-goods	布疋
Damaged goods	損壞品	Sporting goods	運動用品
Durable goods	耐用品	Seasonable goods	季節性貨品
Dangerous goods	危險品	Staple goods	重要物資
Earthen-ware goods	瓦器	Strategic goods	戰略物資
Dry goods	布疋(美)	Substitute goods	代替品
Imported goods	進口貨	Sundry goods	雜貨
Fancy goods	精巧品	Miscellaneous goods	雜貨
High-quality goods	高級品	General goods	雜貨
Finished goods	完成品	Woolen goods	毛織品
First-rate goods	一級品	Wet goods	酒類(美)
Second(third)class goods	二(三)級品		

MERCHANDISE（商品，集合名詞）

General merchandise	雜貨	Standard merchandise	標準品
Returned merchandise	退貨	Unclaimed merchandise	貨主不明貨物

COMMODITY（商品、物品、物資）

Daily commodity	日用品	Perishable commodity	易腐品
Essential commodity	基本物資	Staple commodity	重要物資
Marketable commodity	適銷品	Vital commodity	生活用物資

PRODUCT（產品、製品）

Agricultural products	農產物	Intellectual products	智慧產物
Fishery products	漁產物	Foreign products	外國產品
Forestry products	林產物	Marine products	海產物
Industrial products	工業製品	Staple products	重要產品

WARES（手工藝品、物品、製品）

Bamboo-ware	竹器	Iron-ware	鐵製品，鐵器
Brass-ware	銅器	Lacquered ware	漆品
Earthen-ware	瓦器	Luxury ware	奢侈品
Enamel-ware	油漆品	Tableware	餐具
Glass-ware	玻璃製品	Wooden-ware	木器
Ceramic-ware	陶瓷器	Silver-ware	銀器
Hardware	金屬器具	Aluminum-ware	鋁器

CARGO（貨物，保險界航運界用語）

General cargo	雜貨	Corrosive cargo	腐蝕性貨物
Fine cargo	精良貨物	Perishable cargo	易腐貨物
Clean cargo	精良貨物	Refrigerating cargo	冷藏貨物
Rough cargo	粗貨	Chilled cargo	冷凍貨物
Dirty cargo	不潔貨	Valuable cargo	貴重物
Liquid cargo	液體貨	Heavy cargo	笨重貨(超重貨)
Dangerous cargo	危險性貨物	Bulky cargo	笨大貨(超大貨)
Explosive cargo	爆炸性貨物	Lengthy cargo	超長貨
Poisonous cargo	有毒性貨物		

二、商品目錄

商品目錄又稱爲型錄(catalog)，是廠商爲便於推銷商品，以文字、圖片等說明其所經營商品性能、規格、形狀、重量、尺碼、顏色、包裝方法的印刷物。商品目錄是一種沉默的推銷員(silent salesman)，因此廠商在編印時力求內容精彩，外表美觀，以求引人入勝。

商品目錄編印目的既爲拓展市場，但國外市場區域廣大，各國所用文字不盡相同。因此欲使各國讀者便於閱覽，商品目錄自宜採用各國文字，使閱覽者有親切感。惟英文已成爲國際商業語言，所以目前各廠商所編印的商品目錄多以英文爲主。

商品目錄的種類有：

Illustrated catalog （有插圖的目錄）	Export catalog	（出口貨品目錄）
Latest catalog	Complete catalog	（完整目錄）
Revised catalog （修訂目錄）	Supplemental catalog	（增補目錄）

New catalog

Recent catalog

Catalog of sporting goods

General catalog　　　（總目錄）

Descriptive catalog　（有說明的目錄）

Spring
Summer
Fall
Winter
New Year's
} catalog

Mail order catalog　（郵購用目錄）

第二節　品質條件

一、約定品質的方法

在洽談買賣時，首先須確定品質。貿易糾紛以品質糾紛爲最多，所以對於品質的約定方法應特別小心，約定品質的方法有：

1.以樣品爲準(Sale by sample)

【例】Quality: Same as sample submitted by seller on May 10,

1990.

樣品的種類

①依寄送實務分：

Original sample 　　　　　　　　　　　　　正份樣品(寄給對方的)

Duplicate sample(file sample, keep sample)　副份樣品(自己保留的)

②依樣品功能分：

Selling sample　　推銷用樣品

Approval sample　核准用樣品

Sample for test　試驗用樣品

Claim sample　　　索賠用樣品

Umpire sample　　仲裁用樣品

③依提示樣品的人分：

Seller's sample　　賣方樣品

Buyer's sample　　買方樣品

Counter sample　相對樣品

④依代表部分區分：

Quality sample　　　　　品質樣品

Color sample　　　　　　色彩樣品

Pattern(design)sample　花樣（圖樣）樣品

⑤依取樣時間分

Advance sample　先行樣品

Shipping sample　裝船樣品

Outturn sample　卸貨樣品

2.以標準物爲準(Sale by standard)

　　各業公會製有標準樣品(Standard sample)者，可約定以標準樣品爲品質標準，主要適用於棉花、黃豆、玉米等。

【例】如擬以美國二級黃豆爲品質標準，可約定：

Quality: U. S. Grade No.2 yellow soybeans.

至於甚麼是 U. S. Grade No.2 可不必列明，因爲 U. S. Grade No.2 的標準規格爲：

Bushel Wt.: 54 lbs. min., Moisture: 14% max, Splits: 20% max.

Damaged total(including heat damaged): 3% max.

Foreign material: 2% max.

Brown, black/bicolored: 2% max.

3.以規格爲準(Sale by grade)

有些商品如水泥、鋼板、鋼筋等已由政府或產業團體或學會制定有關品質的標準規格(Standard Specifications)，這種商品的買賣，其品質可約定以某種規格碼爲準。

【例】Quality: conforming to ASTM description C-150-88 requirement for Portland Cement type.

4.以平均品質或適銷品質爲準(Sale on FAQ 或 GMQ)

FAQ 爲 Fair Average Quality 的略語。FAQ 條件就是指所交商品品質以「裝運時裝運地該季所運出商品的中等平均品質爲準」之意。(Fair average quality of the season's shipment at time and place of shipment.)

GMQ 爲 Good Merchantable Quality 的略語，GMQ 條件乃謂保證商品品質在某種商業用途上良好可銷的品質條件。

【例】按 FAQ 約定時：

Quality: Brazilian soybeans, 1990 new crop, FAQ

按 GMQ 約定時：

The quality of the goods to be of GMQ.

（品質須適合商銷）

5.以牌記爲準(Sale by brand or trade mark)

即憑牌名或商標約定品質。

【例】Quality: Toyota truck, Model: kp 366, Year: 1990, Standard equipment.

6.以說明爲準(Sale by specification of catalog)

即以說明書或型錄說明商品規格、構造、材料、形狀、尺寸、性能等。

【例】⑴Specifications as per maker's catalog No. 123.

⑵Specifications: Nitrogen: 46%min, Uncoated, In granules （粒狀）.

二、有關品質的用語

1.有關樣品條件的表現法

Quality to be
{
as per
up to
exactly the same as
identical to
exact to
a match to
in conformity with
conforming to
in accordance with
according to
correspondent with
similar to
about equal to
}
the sample sent to you on...(date)

Sample(s)
- which is (are) the nearest to you required. （與你要求的最接近）
- which is (are) most likely to suit your market. （似乎最適合你的市場需要）
- which has (have) been specially prepared for export. （專為外銷而製的）
- representing the bulk. （代表正貨）
- closely resembling to what you want. （與你所需者近似）

The
- quality
- specifications
- color
- size
- material

must be
- strictly same as
- equal to
- up to
- conforming to

→

→
- sample
- pattern
- swatch

submitted.

The goods must
- agree
- comply

in every respect with our
- specifications.
- samples.
- patterns.

2. 有關樣品的動詞

- to maintain the standard
- to keep up to the standard

保持標準

- to be of inferior quality
- to be inferior to the sample
- to be below the sample

- 劣等品質
- 比樣品差

- to examine
- to inspect

a sample

- 檢查
- 檢驗

樣品

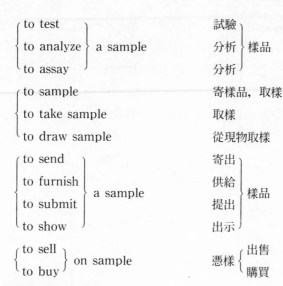

to test	試驗	
to analyze } a sample	分析 } 樣品	
to assay	分析	

to sample	寄樣品，取樣
to take sample	取樣
to draw sample	從現物取樣

to send	寄出	
to furnish	供給	
to submit } a sample	提出 } 樣品	
to show	出示	

to sell } on sample	憑樣 { 出售	
to buy	購買	

3. 有關品質的形容詞

(1)表示高級品質

Al, O. K., Extra O. K., Extra best, Extra fine, Very best, Very superior, Superior, Superfine, Prime, Fine, Good

(2)表示中等品質　　　　(3)表示劣等品質

second class	
good fair average	
middling	
medium } goods	
common	
usual	
ordinary	

inferior	
low grade	
bad } { goods
poor	{ article
third class	

well-conditioned		情況良好的		
perfect		完整的		
standard		標準的		
sound		良好的		
defective		有瑕疵的		
damaged		損壞了的		
deteriorated	goods	變質的	貨品	
faded		褪了色的		
imitated		仿造的		
spurious		偽造的		
meretricious		俗氣的		
shock-proof		防震的		
water-proof		防水的		

三、有關品質的有用例句

1. If you have something new in this style, we would thank you to transmit us some patterns.

2. It is difficult to judge of a cloth by a narrow cutting, and therefore we expect you to send us fair-sized patterns, with colorings, of your next season goods.

   * fair-sized patterns: 大小合適的花樣, * colorings: 配色

3. Your offer looks very promising, and we should like to have a complete set of your samples as we cannot do anything without samples.

   * promising: 很有希望

4. In a few days we will send you some patterns of quality which we consider suitable for your market.

5. We are enclosing a copy of our recent catalog with a few samples which may possibly interest you, and shall be glad to hear from you at any time.

6. We have made such a selection as will suit your market, and sent you per sample post.

   * sample post: 樣品郵件

7. The goods must be in strict conformity with the shades and patterns supplied.

8. The quality must be $\left\{ \begin{array}{l} \text{quite equal to} \\ \text{same as} \end{array} \right\}$ the sample submitted.

9. Should you be unable to find an exact match, kindly send us samples you have to offer so as to enable us to submit them to our customers for approval.

10. Our article is the fruit of nearly twenty years' earnest, conscientious work and has had the test of time and wear; this fact, we believe, is the best guarantee of quality and make.

    * fruit: 成果, * earnest: 熱心, * test of time and wear: 時間的考驗, * make: 牌子, 式樣

11. If it is rather a question of price than of quality, the way out of the difficulty is comparatively easy. We suggest that you stock a lower grade of quite a similar appearance.

    * the way out of the difficulty: 克服困難

12. We admit there are, as you say, some other lines much lower-priced than ours, but we feel confident that a com-

parison will reveal to you a wide difference in quality.

* other lines:別家的貨品, * reveal: 顯出

13. Our quality has never been sacrified to meet price; we would rather suffer a blame for high price than produce a low-priced article to the detriment of the quality.

* to the detriment of the quality: 有損品質

14. We guarantee our machines to wear five years. New machines will be furnished gratis, if they fail to sustain the guarantee.

* gratis: 免費, * fail to sustain the guarantee: 與保證不符

15. We give our full guarantee that these are all wool and fast-dyed.

* fast-dyed 不褪色

16. One-year's guarantee goes with our goods, and if they do not come up to our representations, we will replace them free or refund the money.

* do not come up to our representations: 不符我們所言

第三節　數量條件

商品品質一經約定，跟著就要約定買賣數量。

一、數量單位

　　1. Weight（重量）

　　　　Avoirdupois ounce(oz)＝28.35 grams　　　　　　　（盎司）

　　　　Troy ounce＝31.104 grams

　　　　Pound(lb.)＝16 oz＝454 grams　　　　　　　　　（磅）

Kilogram(kg.)=1,000 grams=2.2046 lbs.=35.274 oz　（公斤）

Ton $\begin{cases} \text{long ton(L/T)=2240 lbs.=British ton=gross ton} \\ \text{short ton(S/T)=2000 lbs.=American ton=net ton} \\ \text{metric ton(M/T)=2204.6 lbs.=1000 kgs=French ton} \end{cases}$

Hundredweight(cwt) $\begin{cases} 英制=112\ \text{lbs.=long cwt} \\ 美制=100\ \text{lbs.=short cwt} \end{cases}$ （匈威特）

2. Number（個數）

Piece(pc., pcs.)	（個，件）	Bale	（包）
Set	（套，組，臺，部）	Unit	（部）
Dozen(doz.)	（12 個，多數不加 s）	Each (ea.)	（件，條）
Gross	（籮，12 打）	Bundle	（束）
Great Gross	（大籮，12 籮）	Sheet	（張）
Roll	（捲，matting 用）	Deca	（10 個）
Coil	（捲，wire 用）	Bag	（包）
Reel	（捲，線捲）	Case	（箱）
Ream	（令，紙張用）	Carton	（箱）
Pair	（雙）	Pack	（包）
Dozen pair	（打雙）	Head	（頭，多數不加 s）

3. Area（面積）

Square feet(sq. ft.)平方呎

Square yard(sq. yd.)平方碼

Square meter(sq. m.)平方公尺

4. Length（長度）

Yard(yd.)=3 feet=91.4 cm.

Foot(ft.)=12 inches

Centimeter(cm.)

Meter (m)=39.37 inches=3.28 ft.=1.09 yds.

5. Capacity（容積）

Gallon $\begin{cases} 英制：277.42 \text{ cubic inches...imperial(British)gallon} \\ 美制：231 \text{ cubic inches...U. S. gallon} \end{cases}$

Liter=0.264 gallon

Cubic centimeter(c.c.)

6. Volume

Cubic feet(cft., cu. ft.)

Cubic yard(cu. yd.)

Measurement ton $\begin{cases} 40 \text{ cft.} \\ 1 \text{ CBM} \end{cases}$

Cubic meter=CBM=35.3 cft.

二、有關數量條件用語

1. Gross weight(G. W.); Tare weight; Net weight(N. W.); Net Net Weight(N.N.W.); Legal weight: "Gross weight"爲毛重，即包括包裝材料在內的重量。"Tare weight"爲包裝材料的重量。"Net weight": Gross weight 減去 Tare weight 爲 Net weight（淨重）。"Net Net Weight"：純淨重,爲毛重除去包裝材料後的重量，也即商品本身的實際重量，所以又稱爲 Actual net weight。在很多場合 Net weight 即爲 Net Net weight。"Legal weight"：法定重量，爲商品重量包含裝飾包裝的重量。

2. Shipped quantity terms 與 Landed quantity terms: "Shipped quantity terms"爲「裝運數量條件」，即賣方交給買方的數量以裝運時的數量爲準者。"Landed quantity terms"爲「卸貨數量條件」，即賣方交給買方的數量以卸貨時的數量爲準者。

【例】(1) The shipped weight and/or count at the time and

place of loading port shall be final.

⑵ Landed net weight at port of destination shall be final.

⑶ 500 metric tons, G. S. W.(gross shipped weight)

3. More or Less, Plus or Minus, Increase or Decrease:有些貨物因性質關係，技術上不易按約定數量準確交貨，因此常約定可多交或少交若干或要求多交或少交若干，規定這種條件的條件稱爲過與不足條款(more or less clause)，例：

Sellers have the option of shipping 5% more or less of the contracted quantity.

"more or less"又可以"plus or minus"或"increase or decrease"或"±"代替。

三、有關數量的有用例句

1. For any special coloring process, we set a minimun quantity for order. The smallest (minimum) for No.123 is 25,000 yards.

2. The smallest (minimum) order we can fill for this quality is 3,000 yds. If your requirement is below that, we suggest that you review the enclosed color cards and choose the shade most nearly like it.

   ＊ quality: 貨色, ＊ shade: 色度

3. The above price is based on a minimum order of 1,000 doz.

4. Quantity, unless otherwise arranged, shall be subject to a variation of 5% plus or minus at seller's(buyer's, ship's) option.

5. Any shortage or excess within one percent of B/L weights shall not be taken into consideration.

＊shall not be taken into consideration: 不予考慮，不計

6. As to weight and/or measurement, an accredited surveyor's certificate shall be final at loading port.

＊accredited:可信賴的

7. Public surveyor's certificate of weight at loading port to be final.

8. All landed weight shall be considered final and conclusive.

9. Weighing to be done within six days after delivery into consignee's craft.

10. Weighing at the seller's works or at the place of despatch shall govern.

第四節　　價格條件

買賣雙方來往折衝，多是為了價錢的問題，商人所以熱衷於逐什一之利，也無非想從交易中獲得差價，賺取利潤。

一、交易所使用貨幣種類

在國際貿易，用以表示價格的貨幣不外本國貨幣，對方國貨幣及第三國貨幣，我國目前使用較多的是美金(U.S. Dollars)。

二、價格基礎

在國際貿易做為價格基礎的貿易條件(Trade terms)以 FOB, C&F, 或 CIF 使用最多，但有時也偶爾以其他條件做為價格基礎。茲據 Incoterms 將其所規定者列舉於下：

1. 以出口地為交貨地的貿易條件

　(1) Ex Works(ex factory, ex mills...)　　　　工廠交貨價

　(2) FOR(Free on Rail), FOT(Free on Truck)　　鐵路交貨價

(3) FAS (Free Alongside Ship) 　　　　　　　　船邊交貨價

(4) FOB (Free on Board) 　　　　　　　　船上交貨價，離岸價格

(5) C&F (Cost and Freight) 　　　　　　　　運費在內價

(6) CIF (Cost, Insurance, Freight) 　　運費、保費在內價，到岸價格

(7) FRC (Free Carrier) 　　　　　　　　向運送人交貨價

(8) FOA (FOB Airport) 　　　　　　　　機場交貨價

(9) DAF (Delivered at Frontier) 　　　　　　　　邊境交貨價

(10) DCP (Freight or Carriage Paid to) 　　　　　　　　運費付訖價

(11) CIP (Freight/Carriage and Insurance Paid to) 　　運保費付訖價

2. 以進口地爲交貨地的貿易條件

(1) Ex ship 　　　　　　　　目的港船上交貨價

(2) Ex Quay (duty paid) 　　　　目的港碼頭交貨價(關稅付訖)

(3) DDP (Delivered Duty Paid) 　　　　　　　　稅訖交貨價

【例】

DM 34 per case Ex Factory Taipei, delivery during
　　August.
　　　　（每箱 34 馬克，臺北工廠交貨價，八月份交貨）

DM 35 per case FOR Hwa—shan Station, shipment during
　　August.
　　　　（每箱 35 馬克，樺山車站交貨價，八月份交運）

DM 38 per case FAS Keelung, shipment during August.
　　　　（每箱 38 馬克，基隆船邊交貨價，八月份交運）

DM 39 per case FOB Keelung, shipment during August.
　　　　（每箱 39 馬克，基隆船上交貨價，八月份交運）

DM 44 per case C&F Hamburg, shipment during August.
　　　　（每箱 44 馬克，至漢堡運費在內價，八月份交運）

DM 45 per case CIF Hamburg,shipment during August.

　　　　(每箱45馬克，至漢堡運費、保費在內價，八月份交運)

DM 46 per case Ex Ship Hamburg, delivery during August.

　　　　(每箱46馬克，漢堡船上交貨價，八月份交貨)

DM 60 per case Ex Quay (duty paid) Hamburg, delivery during August.

　　　　(每箱60馬克，漢堡碼頭交貨價〔關稅付訖〕，八月份交貨)

　　貿易條件一方面可表示貨物在運輸中風險移轉時、地，他方面又可表示買賣雙方如何分擔運費、保險費，及其他構成價格因素的費用，所以有的人將貿易條件狹義地稱為價格條件。

　　國際上解釋貿易條件的規則，除上述 Incoterms 之外，尚有美國對外貿易定義(Revised American Foreign Trade Definitions,1941)；因為其解釋與 Incoterms 略有出入，買賣雙方為避免誤會，宜在報價單、訂單或契約中訂明應以何種規則為準，例如約定：

Unless otherwise agreed upon, the trade terms in this contract shall be governed and constructed under by the provisions of INCOTERMS 1980.

三、價格變動條款

　　在國際買賣，自訂約至交貨付款完畢，通常有一段相當長的時間，在此期間如遇到成本、滙率等劇烈變動，其中一方難免將遭受嚴重損失，因此，在物價、滙率變動幅度較大的時期，買賣雙方常約定價格變動條款。

　　1. 原料成本變動條款

　　Seller reserves the right to adjust the contracted price, if prior to delivery, there is any substaintial variation in the cost of labor or raw materials.

　　2. 運費變動條款

Ocean freight is calculated at the prevailing rate, and increase in the freight rate at the time of shipment shall be for buyer's account.

3. 滙率變動條款

Exchange risks, if any, for buyer's account.

四、關於價格的用語

1.各種價格

Actual price	實　價	Unit price	單　價
Fixed(set, settled)price	標　價	Asked price	要　價
Import price	進口價格	Bid price	開　價
Export price	出口價格	Blanket (lump) price	總括價格
List price	定　價	Reduced price	減　價
Market price	市　價	Opening price	開盤價
Tag price	掛牌價	Closing price	收盤價
Factory price	出廠價	Ruling price	市　價
Buying price	購　價	Current price	時　價
Purchasing price	購　價	Prevailing price	時　價
Selling (sale) price	售　價	Base price	基　價
Net price	淨　價	Fair price	平實價格
Gross price	總　價	Firm price	確定價格
Prime cost	成　本	High(low) price	高(低)價
Marked-down price	削碼價	Moderate price	廉　價
Cash price	付現價	Official price	公定價格
Credit price	賒帳價	Black market price	黑市價格
Contract price	契約價格	Quoted price	報　價
Average price	平均價格	Retail price	零售價格
Flat price	統一價格	Wholesale price	批發價格

Unit price	單　價	Price terms	價格條件
Trade price	同業價格	CIFC3 price	包括佣金3%的起岸價格
Upset price	開拍價格	Price fluctuation	價格變動
Price current	行　情	Price limit	限　價
Price index	物價指數	FOB price	離岸價格

2.表示價格高、好

High price		高　價
Extravagant Excessive Exorbitant Unreasonable } price		過高價格
Favorable Handsome } price Fine(good)		好 價 格
Maximum(highest) price		最高價格
Ridiculous Fabulous } price Absurd		荒唐的價格
Fancy price		高昂價格

3.表示價格正常

Moderate (reasonable) price	中庸 (合理) 價格
Satisfactory price	令人滿意的價格
Usual price	通常價格
Remunerative price	合算價格
Competitive price	競爭性價格

4.表示價格低廉、便宜、差

| Low price | 低價 | losing price | 賠錢價 |
| Minimum (lowest) price | 最低價 | Rock-bottom price | 最低價 |

Floor price	底價	Bed-rock price	最低價
Unfavorable (bad) price	價錢差	Half price	半價
Special price	特價	Attractive price	誘人的價格
		Popular price	廉價

【註】price 的高低用"high"或"low"形容，不能用"dear"或"cheap"形容，例：
The prices rule high. The goods command a high price. "Dear"爲
"high in price"之意，"Cheap"爲"low in price"之意，因此我們說：This
is dear, but that is cheap. These goods are high in price. The
price is too high for me.

5.與動詞連用

(1)一般性的

To quote on the goods	就該貨報價
To quote a price for the goods	就該貨報價
To offer a price	報　　價
To bid a price	要　　價
To overcharge	索價過高
To knockdown a price	報　　價
To raise (advance, increase) a price	漲　　價
To lower (lessen, bring down) a price	降　　價
To fix a price	決定價格
To work out a price	算出價格
To make up a CIF price	算出 CIF 價格
To reduce a price	減　　價

To { get / make / allow } a discount off a price　　折　　價

To shade a price	減一點價錢

(2)表示上漲

$$\text{Price(s)} \begin{cases} \text{is (are) up} \\ \text{advance(s)} \\ \text{show(s) strength} \\ \text{tend(s) upward} \\ \text{is (are) on the advance} \\ \text{show(s) an advancing tendency} \end{cases}$$

$$\text{Goods} \begin{cases} \text{jump high} \\ \text{fluctuate} \\ \text{run up} \\ \text{soar} \end{cases} \text{in price}$$

(3)表示下跌

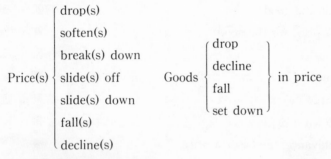

$$\text{Price(s)} \begin{cases} \text{drop(s)} \\ \text{soften(s)} \\ \text{break(s) down} \\ \text{slide(s) off} \\ \text{slide(s) down} \\ \text{fall(s)} \\ \text{decline(s)} \end{cases} \quad \text{Goods} \begin{cases} \text{drop} \\ \text{decline} \\ \text{fall} \\ \text{set down} \end{cases} \text{in price}$$

五、有關價格條件的有用例句

1. The price is net price, without any commission.

2. The price includes 5% commission on FOB basis.

3. US$20 per $\begin{Bmatrix} \text{dozen} \\ \text{set} \end{Bmatrix}$ CIF New York, including your commision 5% on FOB basis.

4. The price of this offer is calculated at the rate of US$1 to NT$26, when there is any surplus or deficiency in NT$ proceeds at the time of negotiating a bill of exchange, such

difference shall be adjusted. Any surplus shall be refunded to buyer, and any deficiency shall be compensated by buyer.

（本報價係按 US$1＝NT$26 滙率計算，押滙時如臺幣收入有多或不足，則應予調整，多餘時應退還賣方，不足時應由買方補足。）

5. Bunker surcharge imposed by the steamship company, if any, shall be for buyer's account.

＊ bunker surcharge imposed by the steamship company: 船公司所徵收的燃料附加費

6. Ocean freight is calculated at US$ 20 per metric ton, any increase in freight rate at the time of shipment shall be for the buyer's risk and account.

7. The goods are sold at US$ 103 per M/T CIFC3 Kuwait.

8. In view of the present state of the market, it is evident that prices will rule high for some considerable period.

9. Values all round have an upward tendency, and we believe that you would be well advised to lay in stock sufficient for your requirements during the next season.

10. It is believed that prices will again advance, perhaps higher than the point reached in the last advance.

11. Under the present conditions, when the cost of labor and raw materials may go up any day, it is impossible for us to guarantee the prices for any definite period.

12. Stocks are low, the Japanese crop is late, and there are disturbances in Korea, all which causes may have the effect of bringing about an improvement in prices.

13. You are no doubt aware that the wool market is very unsettled, and we are therefore obliged to allow ourselves reasonable margin in view of possible disturbances in the present prices.

   * very unsettled: 很不穩定, * reasonable margin: 合理的利用, * disturbances: 不正常

14. This price is quite impossible under present conditions being far above the market here and also the above quotations received from other sources.

15. We are surprised that some other exporters in your area are underselling us, so much so because we are the originators of the machine which is protected by patent.

   * underselling us: 售價低於我們, * originators: 創作人; 發明人

16. If you can $\left\{ \begin{array}{l} \text{make the prices a little easier} \\ \text{shade the prices a little} \end{array} \right\}$, we shall probably be able to see our way to place an order with you.

17. Your competitors, who made us an offer the other day, charge for the same class of goods, the quality being exactly the same as yours, only US$9 per dozen.

18. We see no possibility of competing against rival makes unless you can reduce your quotation.

   * rival makes: 競爭品 (牌子)

19. As there is keen competition for this article, your price will put us out of the market.

* put us out of the market: 將我們從市場驅逐出去

20. We are of the opinion that the present quotation US$10 per set FOB Taiwan is a rock-bottom price and a most profitable investment.

21. We are able to make this low prices, becanse the manufacturers have lowered their prices to us, and rather than keep the difference as extra profit for ourselves, we are going to give our customers the entire benefit of this cut in the prices.

22. Owing to the continual fall in prices, buyers are really afraid to place orders for forward shipment, as prices which appear to be attractive one day, are found within a week or so to be high.

23. Your quotation seems to be exorbitant, in view of the prices prevailing in this market, which stand at something like US$ 12 per dozen.

第五節　包裝及刷嘜條件

一、包裝種類

進出口貨物因大多需經長途的轉輾運輸，所以其包裝必需適當，才不致在運輸途中遭到損壞。至於何種包裝才適當，因貨物的性質、種類、搬運方法、經過路程、轉運與否、目的地碼頭設備等等情形而異。茲將常見的包裝類別及適用貨品列表於次頁。

| 外形 | 包　裝　名　稱　略　字 | | | | 適　用　貨　物 |
	英　　　　　文	中　文	單數	多數	
箱　狀	WOODEN CASE	木　　箱	C/	C/S	雜貨、儀器、機件
	WOODEN BOX	木　　箱	Bx	Bxs	雜貨、儀器、機件
	CHEST	茶　　箱	Cst	Csts	茶葉
	CRATE	板　條　箱	Crt	Crts	玻璃板、機器
	CARTON	紙　　箱	Ctn	Ctns	雜貨、塑膠品、奶粉
	SKELETON CASE	漏　孔　箱	—	—	陶瓷、蔬菜、青果
捆　狀	BALE	捆　　包	B/	B/S	棉花、布疋
袋　狀	SACK	布　　袋	Sx	Sxs	麵粉
	BAG	袋	Bg	Bgs	肥料、水泥等
	GUNNY BAG	麻　　袋			玉米、小麥、雜糧
	PAPER BAG	紙　　袋			水泥、肥料
	STRAW BAG	草　　袋			鹽
桶　狀	BARREL	鼓　形　桶	Brl	Brls	酒、醬油
	HOGSHED	菸　葉　桶	Hghd	Hghds	菸葉
	CASK	樽	Csk	Csks	染料
	KEG	小　　樽	Kg	Kgs	鐵釘
	TUB	木　　桶	—	—	醬油
	DRUM	鐵　　桶	Dum	Dums	油、染料
罐　狀	CAN	罐			罐頭食品
	TIN	聽			罐頭食品
圓筒狀	CYLINDER	鋼　　桶			液體瓦斯、氧氣
	BOMB	鋼　　桶			液體瓦斯、氧氣

瓶　裝	BOTTLE	瓶	Bot	Bots	酒類
	DEMIJOHN	大　罐			酸性化學品
	CARBOY	酸　瓶			酸性化學品
	JAR	甕			酒類
	FLASK	細頭鋼瓶			水銀
簍　狀	BASKET	簍			蔬果

二、嘜類

刷嘜(marking)是指包裝容器上用油墨、油漆或以模板(stencil)加印嘜頭或標誌(mark)，用以有別於其他船貨。所謂嘜頭，廣義的說，是指包裝上所標示的圖形、文字、數字、字母及件號的總稱。

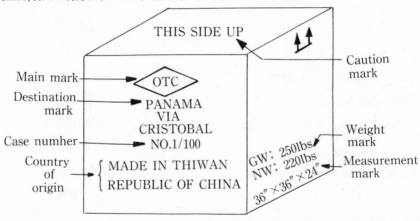

茲將常見的注意標誌列出若干於下：

This side up or This end up	此端向上
Handle with care or With care or Care handle	小心搬運
Use no hook or No hook	請勿用鈎
Keep in cool place or Keep cool	放置冷處(保持低溫)

Keep dry	保持乾燥
Keep away from boiler or Stow Away from boiler	遠離鍋爐
Inflammable	易燃貨物
Fragile	當心破碎
Explosives	易炸貨物
Glass with care	小心玻璃
Poison	小心中毒
Heave here	此處舉起
Open here	此處開啓
Sling here	此處懸索
Do not drop	小心掉落
To be kept upright	豎立安放
Keep away from heat	隔離熱氣
Perishables	易壞貨物
Guard against wet(damp)	勿使受潮
No smoking	嚴禁煙火
Keep flat(Stow level)	注意平放
Not to be thrown down	不可拋擲

三、有關包裝、刷嘜條件的有用例句

1. To be packed in
   - strong wooden case suitable for export
   - strong wooden case with iron hooped　　　（用鐵條箍住）
   - zinc-lined case,... doz each　　　　　　　（鋅板襯裏箱）
   - wooden kegs lined with pitch paper　　　　（瀝青紙）
   - hardwood iron bound barrels
   - cardboard box　　　　　　　　　　　　　（硬紙板箱）
   - water-proof canvas　　　　　　　　　　　（防水帆布）
   - three-ply kraft paper bags　　　　　　　　（牛皮紙袋）
   - seaworthy packing　　　　　　　　　　　　（耐航包裝）
   - customary export packing　　　　　　　　（習慣出口包裝）
   - regular export packing
   - conventional export packing　　　　　　　（慣例出口包裝）

2. One piece in a polybag, 12 polybags in a paper box, then 60 boxes packed in an export carton, its measurement about 26 cft.

3. Each piece must be wrapped up in paper and packed in a zinc-lined case.

4. Marking: Every package shall be marked with "OTC"in diamond and the package number.

5. The following shipping marks must be clearly stenciled on each package, and stated on invoices and packing list.

* stencil: 以模板鏤花

6. Unless otherwise requested and instructed by the buyer in time, the seller will decide the marking at their discretion.

　* at their discretion: 任由他們自由

第六節　保險條件

一、保險投保人

在一筆國際買賣中，貨物的保險究應由買方或賣方負責付保，端視其貿易條件而定。如以 C&I 或 CIF 條件成交，則賣方有義務購買保險，但裝船後如發生危險事故，則由買方負責向保險公司索賠。正因如此，在 C&I 或 CIF 交易時，買賣雙方應就保險條件作適當的約定。

二、保險的種類

1. 基本險類
   - ICC (A)　　　　　　　　　　　　　　　　　　A 款險
   - ICC (B)　　　　　　　　　　　　　　　　　　B 款險
   - ICC (C)　　　　　　　　　　　　　　　　　　C 款險

2. 附加險
   - TPND (Theft, pilferage and non-delivery)　竊盜、遺失險
   - RFW/D (Rain and fresh water damage)　雨水淡水險
   - COOC (Contact with oil and/or other cargoes)　接觸險
   - JWOB (Jettison and/or washing overboard)　投棄浪沖險
   - Heat & Sweat damage　發熱或汗濕險
   - Breakage　破損險
   - Hook hole　鈎損險
   - Shortage　短失險

上述 ICC (A)、ICC (B)、ICC (C)均不包括兵險(war risks)或罷工暴動險(SR&CC)。因此，如須獲得這種危險的保障，須另加保。

三、保險金額

根據 Incoterms 應按 110% of CIF Value 付保。

根據信用狀統一慣例至少應按 110% of CIF Value 付保。

四、有關保險條件的有用例句

1. We attend to insurance, so please inform us of the name of vessel and the date of sailing, loading port, quantity shipped, value, etc., at the same time of loading by telex.

2. Please advise Messrs. Hall & Co., agents for Lloyd's, of shipment, declaring invoice value of goods, with 10% added thereto.

3. Insurance to be covered by buyer, and any kind of possible loss/damage, such as breakage, shortage, theft, pilferage, etc., after loading shall be covered by insurance by buyer at buyer's option or risks.

4. Insurance to be covered by seller against ICC(B) including TPND, If not arranged otherwise, war risk insurance to be covered by buyer.

5. Premium of war risks is calculated at 0.1%, and increase in premium of war risk subsequent to conclusion of the contract shall be for buyer's account.

   Therefore, L/C must include following clause:

   "If war risk premium is higher than 0.1%, beneficiary is authorized to draw the difference in excess of L/C amount."

6. Marine insurance covering ICC(A), War Risk, and SR&CC for full CIF value plus 10% shall be effected by the seller and should be covered up to buyer's warehouse in Taipei.

第七節　交貨條件

所謂「交貨」(delivery)卽賣方將貨物交給買方之意。有關交貨條件的文字，可分爲下列各種:

一、交貨時間

　1. 即期交貨

　　⑴ Immediate shipment(delivery)　　　　隨卽裝運(交貨) ⎫
　　⑵ Prompt shipment(delivery)　　　　　卽期裝運(交貨) ⎬ 四星期內
　　⑶ Shipment as soon as possible　　　　儘速裝運 ⎭

　　⑷ Shipment by first available steamer(ship, boat, vessel)　有船卽裝

　　⑸ Shipment by first opportunity　優先裝運

　2. 定期交貨

　　⑴ Shipment $\begin{Bmatrix} \text{in} \\ \text{during} \end{Bmatrix}$ July

　　⑵ Shipment $\begin{Bmatrix} \text{by} \\ \text{on or before} \end{Bmatrix}$ August 20　　　　　八月廿日之前裝運

　　⑶ Shipment on or about August 20　八月十五日至八月廿五日之間裝運

　　⑷ Shipment in early November　　　　　　　十一月初旬裝運

　　⑸ Shipment within 30 days $\begin{Bmatrix} \text{after} \\ \text{on} \\ \text{of} \end{Bmatrix}$ receipt of remittance.　收到滙款後卅天內裝運

　　⑹ Shipment must be effected $\begin{cases} \text{on or before...} \\ \text{by end of...} \\ \text{during...} \\ \text{during April and May.} \\ \text{within...days after receipt of L/C.} \\ \text{early in December.} \end{cases}$

　3. 交貨時間附帶條件

⑴ Shipment during August subject to shipping space available.

⑵ Shipment by April 30 subject to L/C reaching seller on or before March 15.

二、交貨方法

　　1. 可否分批裝運

$$
\begin{array}{l}
\text{(1)} \quad
\begin{array}{l}
\text{Partial shipments(to be)} \\
\text{Transhipment(to be)}
\end{array}
\left\{
\begin{array}{l}
\text{allowed.} \\
\text{permitted.} \\
\text{not allowed.} \\
\text{unallowed.} \\
\text{prohibited.} \\
\text{forbidden.}
\end{array}
\right.
\end{array}
$$

⑵ Shipment in two approximately equal monthly instalments,→

$$
\rightarrow
\left\{
\begin{array}{l}
\text{beginning} \\
\text{commencing}
\end{array}
\right\}
\left\{
\begin{array}{l}
\text{with} \\
\text{in}
\end{array}
\right\}
\text{March.}
$$

⑶ Shipment to be spread equally over three months →

$$
\rightarrow
\left\{
\begin{array}{l}
\text{beginning} \\
\text{commencing}
\end{array}
\right\}
\left\{
\begin{array}{l}
\text{in} \\
\text{with}
\end{array}
\right\}
\text{May.}
$$

⑷ Shipment to be effected between June 15 and September, 1990, in four equal shipment with interval at least 15 days apart.

　　2. 運輸工具

$$
\text{Shipment to be}
\left\{
\begin{array}{l}
\text{effected} \\
\text{made}
\end{array}
\right\}
\text{per APL's steamer.}
$$

三、有關交貨條件的有用例句

　1. We can generally make delivery of fairly large quantities not later than prompt, but this must not be taken as a promise in so far as the time we shall require for delivery would naturally depend upon the size of your order and

the conditions at our mills at the time your order is received.

* not later than prompt: 3 星期以內

2. According to our contract with our clients we have to deliver it by semi-monthly instalments of 1,000 M/T, failing which we are under penalty of a heavy fine.

* failing which...a heavy fine: 否則我們將被罰鉅額違約金

3. Owing to great pressure at the mills, we are afraid we cannot guarantee delivery within less than three months from this date.

4. We have a large stock of this year's crop. and this, together with our unsurpassed shipping facilities, places us in a position to give delivery at any time on receipt of your order.

* unsurpassed: 最佳的

5. Please let us know per return whether you can undertake delivery within 14 days from receipt of order.

* undertake delivery: 交貨

第八節 付款條件

在一筆買賣中，賣方的首要義務為將約定貨物交給買方，而買方的首要義務為依約將貨款付給賣方。由於國際貿易的買賣雙方遠隔異地，交貨與付款不能同時履行，所以對于貨款的清償方式，須視雙方的各種情況而異。

一、付款方式

1.預付貨款(Payment in Advance)：屬於這種方式的有：

(1) CWO(Cash With Order)：訂貨時付現，也稱爲 CIA(Cash in Advance)。

(2) Anticipatory Credit(預支信用狀)。

2.裝貨付款(Payment on Shipment)：屬於這種方式的有：

(1) CAD(Cash against Documents)：通常指在出口地交單證時，進口商或其在出口地的代理人卽須付款，但也有解釋爲後述的 D/P。

(2) Sight L/C（卽期信用狀）。

3.延期付款(Deferred Payment)：屬於這種方式的有：

(1) COD(Cash on Delivery)：買方收到貨物時付款，美國則稱爲 "Collect on Delivery"。

(2) D/P(Documents against Payment)：付款交單。卽賣方將貨物交運後備齊貨運單證，經由銀行向買方收款，買方付款之同時取得單證。

(3) D/A(Documents against Acceptance)：承兌交單。與 D/P 不同之點在於交單條件，在 D/A 條件下，買方一經承兌賣方所簽發滙票便可取得單證，至於貨款則俟滙票到期時才支付。

(4) On Consignment (寄售)。

(5) Open Account(專戶記帳)：卽賣方先將買方所訂購的貨物陸續運交買方，貨款則暫時記入專戶帳，於一定時日後結帳滙付賣方，實際上卽爲賒帳。

(6) Instalment （分期付款）。

二、付款工具

1. Remittance （滙付）：係由買方將貨款滙交賣方的方法。

(1) T/T, T. T. (Telegraphic transfer) 電滙, 又稱 "Cable transfer"

(2) D/D, D. D. (Demand draft)　　　票滙

(3) M/T, M. T. (Mail transfer)　　　信滙

(4) Personal Check　　　　　　　私人支票

2. Drawing (發票): 卽由賣方向買方簽發滙票收取貨款的方法。

(1) Clean Bill(C/B)　　光票 (不附單證)

(2) Documentary Bill　跟單滙票 "Bill" 爲 Bill of Exchange
之簡稱。

三、有關付款條件的有用例句

1. Payment shall be made Cash With Order by means of
T/T or M/T.

2. Ten percent of the contract value shall be paid in advance
by cash, and ninety percent by sight draft drawn under an
irrevocable L/C.

3. Terms of payment: Net cash against documents payable in
New York.

4. Payment shall be made by a prime banker's irrevocable &
transferable L/C in favor of seller, available by draft at
sight for 100% invoice value.

5. L/C must be opened within 10 days after conclusion of
contract, otherwise this contract shall be cancelled uncon-
ditionally.

6. Payment shall be made by draft drawn under L/C payable
180 days after presentation of documents to the drawee
bank, together with an interest of eight percent for buyer's

account.

7. Payment to be made by draft drawn on buyer payable at sight, D/P.

8. Payment shall be made by 180 days' sight bill, D/A.

9. For payment, we shall arrange with the Bank of Taiwan, Taipei for an irrevocable L/C in your favor for the amount of US$10,000.

習　　題

一、請回答下列問題:

　1.商品品質的決定方法有那幾種?

　2.噸(ton)的種類有幾種?

　3.為什麼保險金額通常係按 CIF 的 110%投保?

　4.貿易貨款的清償方法有那幾種?

二、將下列中文譯成英文:

　1.貨品必須各方面都與樣品一致。

　2.在六月間裝運一千打，一個月後再運其餘數量。

　3.必須在收到信用狀後三十天內裝運。

　4.自五月份起分三個月裝運。

　5.一件裝一塑膠袋，一打裝一紙盒，六十盒裝一出口紙箱，紙箱體積約廿六立方呎。

　6.保險由賣方按發票金額的 110%投保 B 款險、兵險及罷工暴動險。

　7.數量 100 公噸，賣方有多交或少交 5%的選擇權(option)。

第十二章 詢　　價

(Trade Inquiry)

第一節　詢價信的寫法

買賣雙方取得聯繫或獲悉對方之後，出口商卽向對方進口商推銷其貨品，進口商則向出口商寫信探詢有關貨品的種種問題。由進口商主動寫信向賣方提出有關貨品的詢問時，這種詢問稱爲詢價(trade inquiry)或業務詢問信(business inquiry)。詢價又稱探詢，實際上就是買方對某種貨品的查詢。買方的查詢固然多屬於價格方面，但並不以此爲限。諸如索取目錄、樣品、往來條件、貨品的有無、種類、數量、交貨期等也在查詢之列。這種 Trade Inquiry，有時將其希望購買貨品名稱、品質、數量、交貨期等具體記載，已略具 Buying Offer 的雛型。再者，如採行單刀直入的做法，則這種 Trade Inquiry 與前述 Trade Proposals，在事實上往往很難區別。

寫詢價信內容約有三：

1. 開頭先將所要詢問的問題提出，如問題簡單，一句話卽可，如問題複雜，應逐項清楚列出，不要嚕囌陳述許多無謂的話。

2. 其次陳述詢問目的：如採單刀直入做法者，應先自我介紹。

3. 有禮貌的結束。但宜避免"Thank you in advance"的濫調。

No:24　查詢卡式錄音機價錢，並索取目錄樣品

Dear Sirs,

Cassette Tape Recorder

Thank you very much for your extensive consideration in establishing business relations with our company. With the prospects of great success we wish to start off with an initial order for 500 sets of your most popular cassette tape recorder, Model CRC-137.

As the demand for inexpensive cassette tape recorders is high, we may expect a successful sale depending on the cost and quality of your machines.

Incidentally we shall be much obliged if you will send us your latest catalog listing your tape recorders and price with samples.

Your prompt reply will be much appreciated.

Yours very truly,

【註】

1. your extensive consideration:深思熟慮；深入的考慮。

2. start off:開始。

3. with the prospect of great success:預期未來往來關係可獲致很大的成就。

4. initial order: 初次訂單。cf. trial order:嘗試訂單。

5. successful sale:銷售良好；暢銷。

6. the demand for...is high(or active, bullish, brisk):對…需求殷切。

7. incidentally:附帶地，又可以"meanwhile", "in the meantime"代替。

8. depending on:決定於。

9. latest catalog:最新的貨品目錄，不可用"newest catalog"。

No.25　請寄府綢樣品及報價

Gentlemen:

We appreciate the information you have so kindly furnished us in your letter of January 30.

We deeply regret that business between us has been suspended for some time owing to the slackness of trade. Business in general, however, seems to be picking up, and demands will revive. We feel confident, therefore, that we shall be able to resume dealings in your Silk Piece Goods. As to Tussore Silk, in which you wish to do business, prospects are also hopeful as we have a fairly large outlet for it. We ask you to send us samples of Tussore Silk and quote best rates. On receipt of your samples we shall place them before our prospects, and if it is possible for us to do business at your figures, we shall cable you an offer in the usual way.

We appreciate your keeping us well informed as to the trend of your market and welcome any suggestions you may offer for our import operations.

Very sincerely yours,

【註】

1. slackness＝dullness; depression; stagnancy; stagnation:不景氣;蕭條。

2. to be picking up ＝to be improving:轉好。

3. revive:復蘇。　　　　　4. tussore silk:府綢。

5. outlet＝market:銷路，出路 cf. Bombay is an excellent outlet for gar-ments.

6. prospects＝prospective buyers(or customers):可能的顧客，英國人常用 "likely buyers"一詞。

7. figure＝price

8. to keep one well informed:使某人消息靈通。informed＝posted

9. trend＝condition; state; tendency:趨勢；情勢；狀態。

No.26　詢購壓克力毛線衫

Acrylic Sweaters

Gentlemen,

　　We have recently received many inquires from retailing shops in Hamburg area about the subject articles and sure that there would be very brisk demands therefor at our end. We, therefore, are writing to you for your quoting us your most competitive prices on a CIF Hamburg basis for the following:

　　Commodity: Acrylic Sweaters for men and ladies in different color/pattern assortments.

　　Quantity: 1,000 doz.

　　Size assortments: $\dfrac{S}{3}$ $\dfrac{M}{6}$ $\dfrac{L}{3}$

　　Packing: to be wrapped in polybags and packed in standard export cardboard cartons.

Since this inquiry is an urgent one, please indicate in your quotation the earliest shipment you are able to make for delivery. Competition of these articles is very keen here, therefore, not only your prices should be most competitive but also the quality should be the best.

Meanwhile, please also send us by air parcel one sample each of these garments for our evaluation.

Should the prices you quoted be acceptable and the quality of your samples meets with our approval, we will place orders with you forthwith.

Your prompt response is requested.

Faithfully yours,

【註】

1. acrylic sweaters: 壓克力毛線衫。

2. retailing shops: 零售店。

3. color/pattern assortments: 顏色及式樣的搭配。

4. size assortments:尺寸的搭配。

5. cardboard:硬紙板。

6. evaluation:評估。

7. forthwith: 立卽。

8. response: 答覆。

No.27　詢購特級白砂糖

Re: SWC Sugar

Dear Sirs,

We have just received an enquiry from one of our Singapore customers who needs 5,000 M/T of subject sugar and shall appreciate your quoting us your best price immediately.

For your information, the quality should be Superior White Crystal(SWC)sugar packed in new gunny bags of 100 kgs.net each. In the meantime, the sugar should be inspected by an independent public surveyor as their quality and weight at the time of loading.

The buyers in Singapore will arrange shipping and insurance, therefore, the price to be quoted on an FOB Stowed Kaohsiung basis.

As there is critical shortage of sugar in Singapore, the goods should be ready for shipment as soon as possible. Please be assured that if your price is acceptable, we will place order with you right away.

Your prompt reply is requested.

Yours very truly,

【註】

1. SWC: 爲 Superior White Crystal 的略語，指特級白砂糖。

2. gunny bags: 蔴袋。

3. independent public surveyor: 獨立公證行。

4. at the time of loading: 裝船時。

5. FOB stowed:船上交貨包括艙內堆積費在內條件，在大宗物質的交易，如係袋裝者（例如水泥等）在出口港船上的堆積費往往約定由賣方負擔。

6. critical＝serious 嚴重。

7. ready for shipment:儘速（早）裝運；馬上裝船。

8. if your price is acceptable: 如果價格相宜。

9. right away: 馬上。

No.28　詢購柳安合板

Lauan Plywood

Dear Sirs,

As we are in need of one milllion sq. ft. of the subject commodity for prompt delivery, would you please quote us your best price therefore as early as practicable.　For your reference, we are giving below the details of this enquiry.

Specification: Lauan Plywood, Rotary Cut,

Type Ⅲ, 3-ply

Size: 1/8″ x 3′ x 6′

Quantity: One million sq. ft.

Price: Either FOB Taiwan or CIF Bangkok

Shipment: Prompt delivery

Payment: By Irrevocable Sight L/C.

It will be appreciated if you will let us have your quotation before the end of this month.

Yours very truly,

【註】

1. sq.　ft.＝square　feet 平方呎，爲合板的數量單位，至於單價數量單位通常以 MSF（＝1,000 sq. ft.）表達。

〔例〕US$68 per MSF FOB Taiwan port。

2. specification: 規格，簡稱爲"Spec."

3. lauan plywood: 柳安合板，"3-ply lauan plywood"爲三夾板。

4. rotary cut:圓切。

5. size: 1/8″×3′×6′: 尺寸:厚度 $\frac{1}{8}$ 吋，寬3呎，長6呎。

6. sight L/C: 即期信用狀，與遠期信用狀(usance L/C)相對稱。

No.29 詢購魚油脂

Gentlemen:

Your very helpful letter of April 28 and the samples of Sardine Oil Nos. 1/3 have been received with thanks.

We are specially interested in this line, and having heard that you are one of the most reliable refiners and exporters of Fish Oils, we have no doubt that we could not do better than avail ourselves of your valuable services.

We have closely examined the samples which you were good enough to send us and found them suitable for our trade. We, therefore, have the pleasure of sending you an inquiry as detailed below:

1,000 cases Sardine Oil No.1.

60 lbs. net to be contained in a can; 2 cans to be packed in a case.

Quote your finest price C. I. F. San Francisco per ton (2,000 lbs.) of net shipped weight.

Loss in weight exceeding 2% to be allowed for by you.

Your terms and conditions are agreeable to us, and should business result, we shall arrange a credit in your favor. As we are likely to place large orders regularly, we must make it clear from the very beginning that a competitive price is essential.

Yours earliest reply with full particulars will be appreciated.

Very truly yours,

【註】

1. sardine oil:沙丁魚油脂。

2. Nos. 1/3 為 No.1 至 No.3 共三種（樣品）。

3. refiner:煉製廠。

4. could not do better than:最好。

5. avail ourselves of your valuable services:利用你寶貴的服務。

    cf. avail yourselves of this good opportunity:利用這個好機會。

6. net:這裡指"net weight"

7. finest＝lowest＝best

8. shipped weight＝intake weight:裝船重量，與此相對的是"landed weight"
    （卸貨重量），油類在運輸中易發生漏損，所以特別言明按裝船重量交易。

9. loss in...by you:「重量減少2%時，超過部分由你負擔。」to be allowed for by
    you＝for your account 即"for seller's account."

10. should business result:假如生意有結果，即假如成交之意。

11. in your favor:以你為受益人。

12. credit＝letter of credit."arrange a credit"為「按排信用狀」也即開發信用狀
    之意。

13. make it clear:說明。說明清楚。

14. from the very beginning:一開始就。"very"是一種加強語氣的字眼。

    cf. that's the very thing!:就是那個！

    He is the very man I saw yesterday:他就是我昨天看到的那個人。

15. full particulars＝full details:全部詳情。

No.30 請報美棉價

American Raw Cotton

Gentlemen,

Would you please quote us by cable 1200 bales of the subject cotton with the following particulars:

Grade: Strict Low Middling

Staple: 1-1/16″

Pressley: 90,000

Micronaire: 3.8 NCL

Please quote the price either on FAS Gulf ports basis or on CIF Keelung basis. We are in urgent need of the cotton and should like these 1200 bales to be delivered at 300 bales each monthly from July through October, 1989.

We are expecting to receive your earliest reply to this enquiry.

Yours faithfully,

【註】

1. raw cotton:原棉。

2. particulars:細節，這裡可譯成「規格」。

3. Strict Low Middling:原棉等級的一種，簡稱爲"SLM"。美國將美國產白棉 (white cotton)分爲九級，即:

Middling fair	Strict low middling
Strict good middling	Low middling
Good middling	Strict good ordinary
Middling	Good ordinary

4. staple:指「棉花纖維長度」，1-1/16″，一又十六分之一吋。

5. pressley:拉力（以 78,000 磅以上為好棉）。

6. micronaire:纖度

7. NCL: 為"no control limits"（無上下限）的縮寫。

8. Gulf port:指位于墨西哥灣的美國港口。

9. in urgent need of:急需。

10. bale: 美國原棉每包約 500 lbs.

（以下 No.31-32 是單刀直入法的例子）

No.31 請寄脚踏車目錄及通知價格

Dear Sirs,

It has come to our attention through our Chinese friends in Taiwan that you are one of the foremost manufacturers and exporters of Bicycles in Taiwan.

Being in the market for Bicycles and Acces ories, we shall be greatly obliged if you will send us a copy of your illustrated catalogue, informing us of your best terms and lowest prices C. I. F. Singapore. As we are in a position to handle large quantities, we hope you will make an effort to submit us really competitive prices.

For any information as to our financial standing, we refer you to:

The Hongkong & Shanghai Banking Corporation Singapore

We recommend this matter to your prompt attention.

Yours faithfully,

【註】

1. in the market for＝on the lookout for; in want（＝need）of...:擬購買…。

   cf. to be on the market: 推出市場；上市。to put(place)on the market:上市。

   to be out of the market:停止出售。

   關於 market 的用例:

   The market is active(brisk):市場很活躍；呈現活氣。

   The market is dull(weak, heavy, slack, depressed):市場陷入低潮；沒有活力。

   The market is firm(steady):市場堅挺（堅穩）。

   The market is rising(in the upward tendency):行情看漲中。

   The market is falling(in the downward tendency):行情看跌中。

   The market remains unchanged(staionary):行情堅守原盤。

   The market is strong(bullish):行情看漲。

   The market is weak(bearish):行情看跌。

   The market is sagging:行情平抑。

   The market soars up(advances rapidly):行情猛漲。

   The market falls sharply (slumps):行情猛跌。

2. accessories:附件。

3. illustrated catalogue:插圖貨品目錄。

4. recommend:建議。

No.32　詢購罐頭蘆筍

Canned Asparagus

Dear Sirs,

　　We are much indebted to CETRA for the name and address of your company and are pleased to learn that you are one of the leading producers of the subject item in Taiwan.

We are now in the market for the subject product and shall appreciate your quoting us therefor either on FOB Taiwan port or on CIF Hamburg basis at your earliest convenience. For the moment, we are in need of approximately 300 cartons each of Tips & Cuts, Center Cuts, and End Cuts to be packed in export cartons of 48 cans each. Since consumers at our end are very demanding about quality, you are requested to quote us of Al quality.

As there are many other competing brand scrambling for the consumers' dollars in our market, it is, therefore, imperative that the prices quoted have to be very competitive.

We are looking forward to receiving your quotation very soon.

Yours faithfully,

【註】

1. leading producers:主要生產廠商。

   cf. main producers

   　　reputable producers

   　　reliable producers

2. quote us therefor:quote us for that product.

3. tips & cuts:尖端及切片。

4. center cuts:中段切片。

5. end cuts:後段切片。

6. to be packed...cans each:用外銷紙箱包裝，每箱 48 罐。

7. at our end＝on our side:在此地。

8. very demading about＝very critical about:對…很苛求，對…很挑剔。

9. imperative:必須。

10. competing brand:競爭牌子。

11. scrambling for:爭取。

第二節　有關詢價的有用例句

一、開頭句

1. $\left\{\begin{array}{l}\left.\begin{array}{l}\text{Will you please}\\\text{Please}\end{array}\right\}\left\{\begin{array}{l}\text{send us}\\\text{let us have}\end{array}\right\}\\\text{We}\left\{\begin{array}{l}\text{will}\\\text{would}\end{array}\right\}\text{appreciate}\left\{\begin{array}{l}\text{receiving}\\\text{your sending us}\end{array}\right\}\\\text{We shall be}\left\{\begin{array}{l}\text{pleased}\\\text{glad}\\\text{happy}\end{array}\right\}\text{to}\left\{\begin{array}{l}\text{receive}\\\text{have}\end{array}\right\}\end{array}\right\}\rightarrow$

$\rightarrow\left\{\begin{array}{l}\text{your}\\\text{recent}\\\text{latest}\end{array}\right\}\left\{\begin{array}{l}\text{catalog}\\\text{samples}\end{array}\right\}\rightarrow\left\{\begin{array}{l}\text{full details}\\\text{particular}\end{array}\right\}\text{of your...}\rightarrow$

$\rightarrow\left\{\begin{array}{l}\text{together with}\\\text{with their}\end{array}\right\}\rightarrow\left\{\begin{array}{l}\text{detailed offer.}\\\text{best prices.}\\\text{your rockbottom price.}\end{array}\right.$

2.
$$\left\{ \begin{array}{l} \text{We are interested in...→} \\ \text{We are in the market for...→} \\ \text{We are in urgent need for...→} \\ \text{We} \left\{ \begin{array}{l} \text{have} \\ \text{had} \\ \text{received} \end{array} \right\} \left\{ \begin{array}{l} \text{an inquiry} \\ \text{inquiries} \\ \text{many inquiries} \end{array} \right\} \text{from our} \left\{ \begin{array}{l} \text{customer} \\ \text{trade connections} \end{array} \right\} \\ \left\{ \begin{array}{l} \text{here} \\ \text{in...} \end{array} \right\} \begin{array}{l} \text{for your} \\ \text{...→} \end{array} \\ \text{We have} \left\{ \begin{array}{l} \text{read} \\ \text{seen} \end{array} \right\} \text{your ad in} \left\{ \begin{array}{l} \text{May issue} \\ \text{a recent issue} \end{array} \right\} \text{of Taiwan Export →} \\ \text{Your ad in today's China News interests us →} \end{array} \right.$$

$$\rightarrow \left\{ \begin{array}{l} \text{and we will be} \left\{ \begin{array}{l} \text{glad} \\ \text{pleased} \end{array} \right\} \text{to} \left\{ \begin{array}{l} \text{receive} \\ \text{have} \end{array} \right\} \rightarrow \\ \text{and we would like to} \left\{ \begin{array}{l} \text{receive} \\ \text{have} \end{array} \right\} \rightarrow \\ \text{and we would be} \left\{ \begin{array}{l} \text{grateful} \\ \text{obliged} \end{array} \right\} \text{if you would send us →} \\ \text{and we would} \left\{ \begin{array}{l} \text{welcome} \\ \text{appreciate} \end{array} \right\} \rightarrow \end{array} \right.$$

$$\rightarrow \left\{ \begin{array}{l} \text{your} \left\{ \begin{array}{l} \text{price list.} \\ \text{your samples.} \end{array} \right. \\ \text{samples with your} \left\{ \begin{array}{l} \text{best prices.} \\ \text{best terms.} \\ \text{rockbottom prices.} \\ \text{lowest prices.} \\ \text{most competitive quotations.} \end{array} \right. \end{array} \right.$$

二、條件

1. Will you please inform us (as soon as possible) of the

$\left\{\begin{array}{l}\text{prices at which} \\ \text{terms on which}\end{array}\right\} \rightarrow$ you can supply?

2. $\left\{\begin{array}{l}\text{Please} \\ \text{Will you please}\end{array}\right\}$ quote us your competitive prices $\left\{\begin{array}{l}\text{for} \\ \text{on}\end{array}\right\} \rightarrow$

$\rightarrow$ the following $\left\{\begin{array}{l}\text{goods} \\ \text{articles.} \\ \text{commodities.}\end{array}\right.$

3. $\left\{\begin{array}{l}\text{We would appreciate your sending us} \\ \text{Please send us} \\ \text{Please let us have}\end{array}\right\} \rightarrow$

$\rightarrow \left\{\begin{array}{l}\text{samples} \\ \text{patterns} \\ \text{a full range of sample}\end{array}\right\} \rightarrow$

$\rightarrow$ of...which $\left\{\begin{array}{l}\text{you can supply} \\ \text{you can deliver} \\ \text{can be shipped}\end{array}\right\} \rightarrow$

$\rightarrow \left\{\begin{array}{l}\text{promptly.} \\ \text{immediately.} \\ \text{during May.} \\ \text{in(within)...after receipt of our order.}\end{array}\right.$

4. Please $\left\{\begin{array}{l}\text{inform us} \\ \text{let us know}\end{array}\right\}$ $\left\{\begin{array}{l}\text{how soon you can }\left\{\begin{array}{l}\text{ship} \\ \text{deliver}\end{array}\right\}\text{ them.} \\ \text{how long it will take you to }\left\{\begin{array}{l}\text{fill} \\ \text{ship} \\ \text{execute}\end{array}\right\}\text{ an order.}\end{array}\right.$

＊ fill an order: 配發一批訂貨

＊ ship an order: 交運一批訂貨

＊ execute an order: 配交一批訂貨

If you can supply us with goods of high quality at competitive prices →

If your $\left\{\begin{array}{l}\text{prices}\\\text{terms}\end{array}\right\}$ are $\left\{\begin{array}{l}\text{competitive,}\\\text{attractive,}\\\text{satisfactory,}\\\text{reasonable,}\end{array}\right\}$ →

* attractive: 引人感興趣

If your quality is $\left\{\begin{array}{l}\text{good}\\\text{excellent}\\\text{superior}\end{array}\right\}$ and the price

5. $\left\{\begin{array}{l}\text{is competitive,}\\\text{is suitable for our market,}\end{array}\right\}$ →

* right: 合適

If $\left\{\begin{array}{l}\text{your}\\\text{the}\end{array}\right\}$ goods $\left\{\begin{array}{l}\text{are equal to the samples,}\\\text{meet with our approval,}\\\text{are of superior quality,}\end{array}\right\}$ →

* meet with our approval: 合我們的要求

$\left.\begin{array}{l}\text{If}\\\text{Provided}\\\text{Suppose}\\\text{So long as}\end{array}\right\}$ you can $\left\{\begin{array}{l}\text{guarantee regular supplies,}\\\text{promise prompt delivery,}\end{array}\right\}$ →

* guarantee regular supplies: 保證經常供應

we may place a $\begin{cases} \text{big} \\ \text{substantial} \\ \text{large} \\ \text{trial} \end{cases}$ order with you.

　　　* trial order: 試驗性訂單; 嘗試訂單

we may place $\begin{cases} \text{big} \\ \text{substantial} \\ \text{large} \\ \text{bulk (龐大的)} \\ \text{regular} \end{cases}$ orders with you.

we would consider to sign $\begin{cases} \text{an agency} \\ \text{a sole agency} \\ \text{a long-term} \end{cases}$ contract with you.

your goods should $\begin{cases} \text{sell well} \\ \text{sell readily} \\ \text{find a ready sale} \end{cases}$ in our market.

　　* sell well: 暢銷, * sell readily: 容易出售

　　* find a ready sale: 立卽找到銷路

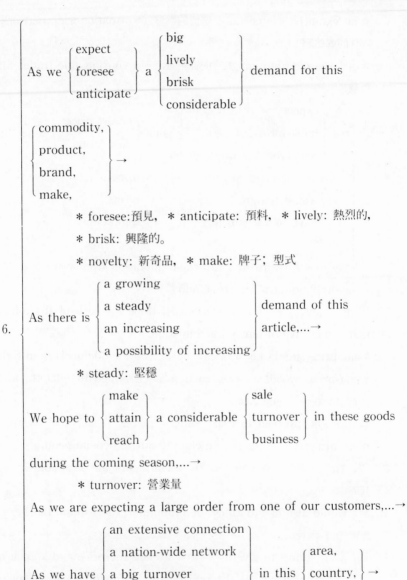

As we { expect / foresee / anticipate } a { big / lively / brisk / considerable } demand for this { commodity, / product, / brand, / make, } →

    ＊ foresee:預見，＊ anticipate: 預料，＊ lively: 熱烈的,

    ＊ brisk: 興隆的。

    ＊ novelty: 新奇品；＊ make: 牌子；型式

6. As there is { a growing / a steady / an increasing / a possibility of increasing } demand of this article,...→

    ＊ steady: 堅穩

We hope to { make / attain / reach } a considerable { sale / turnover / business } in these goods during the coming season,...→

    ＊ turnover: 營業量

As we are expecting a large order from one of our customers,...→

As we have { an extensive connection / a nation-wide network / a big turnover / a large distribution / a considerable trade } in this { area, / country, / district, } →

* an extensive connection: 廣泛的聯繫，* a nation-wide network: 全國性的聯絡網

* a large distribution: 大量的經銷量，* a considerable trade: 相當的生意

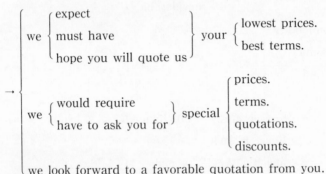

* favorable quotation: 有利的報價

7. Please quote your lowest prices CIF Kobe for each of the following items, inclusive of our 5% commission.

8. Your price should be on a CIF basis, and include packing in tinlined water-proof wooden cases, each piece wrapped in oilcloth, and 30 pcs. packed in one case.

9. Before you proceed to make sample pieces, please give us approximate prices CIF Singapore including our 5% commission.

10. We should like to know if these garments are available in nylon tricot.

11. We would also like to know the minimum export quantities per color and per design.

12. Do your utmost to give us a firm offer at the price and keep it open as long as possible, as the buyer would probably increase his offer considerably.

三、結尾用語

1. We $\left\{\begin{array}{l}\text{are expecting}\\\text{look forward}\\\text{are looking forward to}\\\text{await}\\\text{are awaiting}\\\text{wait for}\\\text{are waiting for}\end{array}\right\}$ $\left\{\begin{array}{l}\text{your}\\\text{an early}\end{array}\right\}$ $\left\{\begin{array}{l}\text{reply.}\\\text{answer.}\\\text{response.}\end{array}\right.$

2. We hope to $\left\{\begin{array}{l}\text{receive your reply}\\\text{hear from you}\end{array}\right\}$ $\left\{\begin{array}{l}\text{soon.}\\\text{by air.}\\\text{by telex.}\\\text{by cable.}\\\text{at your earliest convenience.}\end{array}\right.$

3. Your $\left\{\begin{array}{l}\text{prompt}\\\text{immediate}\end{array}\right\}$ attention to this matter $\left\{\begin{array}{l}\text{would}\\\text{will}\end{array}\right\}$ be appreciated.

4. $\left\{\begin{array}{l}\text{An}\\\text{Your}\end{array}\right\}$ early reply $\left\{\begin{array}{l}\text{by telex}\\\text{by cable}\end{array}\right\}$ $\left\{\begin{array}{l}\text{will}\\\text{would}\end{array}\right\}$ be appreciated.

5. Your $\left\{\begin{array}{l}\text{cooperation}\\\text{kind consideration}\end{array}\right\}$ in this matter $\left\{\begin{array}{l}\text{will}\\\text{would}\end{array}\right\}$ be appreciated.

習　題

一、將下列中文譯成英文。

1. 敬請惠賜貴公司有關洋傘(umbrellas)的目錄及價目表為荷。

2. 請儘速告訴我們貴公司所能供應的價格。

3. 倘貴公司能以合理價格(at reasonable prices)供應優等品質(superior quality)的貨物，我們將向貴公司簽發巨額的(substantial)訂單。

4. 倘貴公司的產品都符合我們的要求(meet with our approval)，貴公司產品會在此地暢銷(sell well)。

二、試用下列片語造句。

1. right away　　　2. in need of　　　3. in demand for

4. make it clear　　5. on the market

三、請將本書 No.30 的英文信譯成中文。

第十三章　答覆詢價

(Reply to Inquiry)

第一節　答覆詢價信的寫法

賣方收到詢價信之後，應儘速答覆，答覆的內容要正確不含糊，並設法抓住買方的買意，以免貽誤商機。

答覆詢價信時，其內容應注意下列各點：

1. 對來信表示謝意。

2. 針對所詢問題作確切的答覆。

3. 檢送所索資料，諸如商品目錄、價目表及樣品等，或說明另行寄送。

4. 適當地說明市況，激發對方早日訂購的意願。

5. 如果無對方擬購的貨品，可藉機推薦代替品，或介紹其他貨品。

6. 以期待能收到訂單做結束。

No.33　對卡式錄音機詢價的答覆

Dear Sirs,

Cassette Tape Recorder

Thank you very much for your inquiry of July 26 concerning your purchase of 500 sets of our popular cassette tape recorder, Model CRT-137.

In compliance with your request, we have enclosed

our latest tape recorder catalog with our price list No. 250. Among our best selling products, we recommend our Model CRT-508 because it has won the Chinese Government Good Design Award, and has been enjoying excellent sales ever since.

If satisfactory, we will send 400 sets from the production line. Owing to a rush of orders, this will be our maximum quantity for the present.

Together with the printed materials requested, we have sent samples, including several other models, by special delivery.

We are very pleased to have concluded business relations with you and hope that our appreciation will continue for a long time to come.

　　　　　　　　　　　Yours very faithfully,

【註】

1. inquiry of July 26 concerning: 7 月 26 日有關…的詢價。

2. in compliance...request:「遵從」,「依照」之意, 新派人士認為這是 overformal 的措詞, 貿易書信中不必如此拘謹, 可以"as requested"代替。

3. Chinese Government Good Design Award: 中華民國政府優良設計獎。

4. ever since: 從那時候起到現在; 此後一直。

5. from the production line: 從生產線。

6. rush of orders: 訂單湧入。

7. together with: 可以"along with"代替。

8. printed materials: 印刷品，包括目錄、技術資料、說明書等。

9. special delivery: 限時專送。

10. conclude business relations: 建立業務關係。

11. to come: 未來的；將來的。"for a long time to come"爲「未來長時間」，即「今後長久」之意。

No.34　對府綢詢價的答覆

Gentlemen:

Tussore Silk

Your letter No. 100 of February 18 has given us much pleasure as it serves as a means of stimulating business between us.

Samples and Prices. We have sent you by airmail a pattern book of our Tussore Silk. The book contains all the patterns that are typical of the goods, which, we trust, are best adaptable for your selling purposes. From the Price List enclosed you will observe that our prices are exceptionally low. This sacrifice is entirely due to our recognition of the necessity for price cutting in order to counter competition.

Market. The market here is recovering somewhat from the long period of depression under which it has labored, and there is a decided improvement in the general tone. Inquiries, too, are more frequent and more often lead to business than has of late been the case. Manufacturers in most branches are getting busier, and we may look forward before long to an upward tendency in prices all round. We believe, therefore, that the present is a very favorable opportunity to make bull purchases.

We wish to assure you that we always adhere to our policy of providing high-grade products at competitive prices. Perhaps you would be good enough to entrust us with a trial order, which will assuredly redound to your advantage.

Your inquiry will be promptly answered, and more detailed information furnished upon request.

Very sincerely yours,

【註】

1. a pattern book＝a book of patterns: 樣品簿 (款式簿)。

2. sacrifice＝sacrifice price: 犧牲價, 即 low price 之意。

3. to counter competition: 應付競爭。

4. recovering: (從不景氣) 恢復, 又稱爲 "rally", 如從上漲跌回到稱爲 "react"。

5. depression: 不景氣。

6. labored: 受…不利的影響; 爲…所支配, 與 under 連結。

7. decided improvement: 決定性的改善。指市況轉好, 價格上漲。

8. more frequent: 更頻繁。

9. the general tone: 一般景況。這是市場用語之一種。

10. lead to business: 成交。

11. of late＝recently。

12. branches＝trades 行業。

13. getting busier: 越來越忙。生意好，自然忙。

14. all round: 一齊，全盤。

    cf.

    ⒜ Prices in silks are soaring *all round with* the approach of the new season's demands.

    ⒝ There being a good deal of business doing in silks, prices *all round* are sure to appreciate very soon.

    ⒞ An *all round* decline was registered in the Yokohama silk market.

15. bull purchases＝to buy in expectation of a rise. 預期價格會上漲而買進。

16. adhere to our policy: 固守我們的政策。

17. entrust us with a trial order: 惠賜試驗性訂單。

18. redound to your advantage: 增加你的利潤。

19. upon request: 一經要求。

No.35 對詢購油漆及樣品的答覆

Dear Sirs,

We are obliged for your letter of the 3rd May and are glad to learn that you are particularly interested in our "Lion"Brand Paint and intend placing a standing contract should our product and prices be found satisfactory.

We are pleased to say that the standing contract you are going to make deserves our special consideration, and we are accordingly prepared to allow you a special discount of 5 per cent off our catalogue prices, in addition to the usual quantity discount of 3 per cent. This special discount will involve our selling practically at cost, and we trust you will recognize that this concession must be regarded as exceptional.

You will readily admit that this concession is made on the stipulation that you bind yourselves to place with unfailing regularity an order monthly, running for one year, for 500 drums of our "Lion" Brand Paint in assorted colours, or the special concession is to be revoked if your monthly order falls below this figure, i.e., you are to pay the full prices for the quantity you do take. This condition does not, of course, prejudice your right to reject any goods that are not up to samples, or are unsuitable or shipped later than the time stipulated in each contract.

As requested, we have sent our sample tins for your test. We also enclose our Catalogue No. 51, in which you will find that our regular prices for "Lion" Brand Paint are the same as described in Catalogue No. 50.

We trust that the result of your test of our product and our special concession in the prices will induce you to enter into the contract.

Yours faithfully,

【註】

1. deserve our special consideration: 值得我們的特別考慮。

2. discount of 5 per cent off our catalogue price: 按目錄價格折扣 5 %。

3. this special discount will involve...cost: 這個特別折扣難免使我們按成本出售。"involve"為…「難免」,「使…必需」之意。

4. concession＝discount

5. readily＝promptly

6. on the stipulation that: 以…為條件。

7. bind yourselves: 拘束你; 使你負有義務。

8. with unfailing regularity: 正確無誤地。

9. runnng for one year＝continuous for one year

10. in assorted colors: 以各種顏色。

11. revoke: 取消 (約定); 解除 (契約)。

12. falls below...: 低於…。

13. full prices＝regular prices: 正規的價格。此外又有「可獲充分利益的價格」之意。

14. do take: "do"為加強語氣的助詞。

15. prejudice your right: 損及貴方權利。

16. not up to samples: 達不到樣品的標準。

17. sample tins: 樣品罐。

18. induce＝cause, 促使 (某人做某事, 後接不定詞)。

    cf. What induced you to do such a thing? 什麼誘導你做這樣的事?

19. enter into contract: 締約。

20. 這封信雖然寫得相當長, 但是寫得很具體、詳細, 使收信人覺得寫信人的認真、精細、週到。

No.36 覆告無法供貨

Dear Sirs,

PVC Resins for Wire Coating

We thank you for your letter of May 10 concerning PVC Resins for Wire Coating.

We appreciate your interest in our products, but regret to inform you that due to the current critical shortage we are not in a position to offer this material.

The present outlook for PVC supplies in this country is not good and it appears that this shortage will continue throughout 1990.

Towards the end of 1991, we expect to have a new large PVC plant on stream and hope at that time we will be in a position to serve our foreign customers.

We thank you again for your interest in our products.

Yours faithfully,

【註】

1. concerning: 同義詞有"in relation to", "relating to", "regarding", "in reference to"

2. PVC resins for wire coating: 做電線護皮用塑膠樹脂。

3. due to current critical shortage: 由於目前嚴重缺貨。

4. present outlook: 目前的展望。

5. not in a position to offer: 無法供應。

6. a new large PVC plant on stream: 一座新的大塑膠工廠開始生產。

7. will be in a position to serve our customers: 將能對顧客服務（意指可供貨）。

No.37　對於詢購貨品無法供應，另推薦代替品

Gentlemen:

We refer to your letter of August 15 with regard to "Sunshine" brand recording tape.

Unfortunately, Sunshine Industries, Ltd., has already been tied up with an agent on recording tapes on your side, and consequently we are unable to offer this line. We have, however, made a contact with another supplier, Moonshine Industrial Co., Ltd., who manufactures a practically complete line of recording tapes, and the quality of their product is kept up with, even better than, that made by Sunshine Industries, Ltd.

The enclosed price list will give you our best CIF Singapore prices which we believe are highly competitive.

We realized that it might be pretty difficult for you to push an unknown brand, but if you think there is a possibility of promoting this line, we shall be happy to send you sample.

Yours faithfully,

【註】

1. be tied up with: 與提携，名詞為"tie-up"。"be tied up"也用於船運方面，指因碼頭工人罷工被套牢無法進退。

2. push an unknown brand＝promote the sale of unknown brand: 拓銷未

出名牌子的東西。

第二節　有關答覆詢價的有用例句

一、開頭句

1. $\left\{\begin{array}{l}\text{Many thanks}\\ \text{Thank you}\\ \text{We thank you}\end{array}\right\}$ for your $\left\{\begin{array}{l}\text{enquiry}\\ \text{letter}\end{array}\right\}$ of (date) $\left\{\begin{array}{l}\text{for}\\ \text{about}\end{array}\right\}$ ...

2. $\left\{\begin{array}{l}\text{We acknowledge with thanks}\\ \text{We appreciate}\end{array}\right\}$ your $\left\{\begin{array}{l}\text{letter}\\ \text{inquiry}\end{array}\right\}$ of (date)

   $\left\{\begin{array}{l}\text{for}\\ \text{about}\end{array}\right\}$ ...。

3. $\left\{\begin{array}{l}\text{As requested }\left\{\begin{array}{l}\text{by you}\\ \text{in your letter of...}\end{array}\right\}\\ \\ \text{In answer}\\ \text{In reply}\\ \text{With reference}\end{array}\right.$ to your $\left\{\begin{array}{l}\text{letter}\\ \text{enquiry}\end{array}\right\}$ of... $\left.\begin{array}{l}\\ \\ \\ \\ \end{array}\right\}$, we

   $\left\{\begin{array}{l}\text{are glad to enclose}\\ \text{are happy to enclose}\\ \text{enclose}\\ \text{are enclosing}\\ \text{are pleased to enclose}\\ \text{are sending (you)}\end{array}\right\}$ →

   → a copy of our $\left\{\begin{array}{l}\text{illustrated}\\ \text{revised}\\ \text{spring}\\ \text{new}\end{array}\right\}$ catalog →

...in which you will find a detailed description of our line of product.

...in which you will find $\begin{Bmatrix} \text{many} \\ \text{a number of} \end{Bmatrix}$ items that will interest you.

...which we recommend to your *careful perusal*. (詳核)

→ ...contaning quantities for large orders taken from our *existing stock*. (庫存)

...of sporting goods, on page 24, you will find marked a machine that may interest you.

...with a few samples which may possibly interest you, and shall be glad to hear from you at any time.

...and price list, in the hope that you may find something to suit you.

二、條件

1. All $\begin{Bmatrix} \text{details} \\ \text{particulars} \end{Bmatrix}$ are $\begin{Bmatrix} \text{indicated} \\ \text{given} \\ \text{shown} \end{Bmatrix}$ in our $\begin{Bmatrix} \text{catalog.} \\ \text{price list.} \end{Bmatrix}$

2. We $\begin{Bmatrix} \text{can} \\ \text{are ready to} \end{Bmatrix}$ $\begin{Bmatrix} \text{despatch} \\ \text{deliver} \\ \text{supply} \\ \text{ship} \end{Bmatrix}$ any quantity of...→

　＊ are ready to: 隨時都能

→ $\begin{Bmatrix} \text{from stock.} \\ \text{immediately.} \\ \text{within...} \begin{Bmatrix} \text{days} \\ \text{months} \end{Bmatrix} \begin{Bmatrix} \text{of} \\ \text{from} \\ \text{after} \end{Bmatrix} \text{receipt of order.} \end{Bmatrix}$

　＊ from stock: 由現貨中

3. We are $\left\{\begin{array}{l}\text{pleased}\\\text{happy}\\\text{glad}\end{array}\right\}$ to quote you as follows and can promise

$\rightarrow\left\{\begin{array}{l}\text{dispatch}\\\text{delivery}\\\text{shipment}\end{array}\right\}\rightarrow$ within 30 days on receipt of your L/C.

4. $\left\{\begin{array}{l}\text{Our prices}\\\text{All prices}\end{array}\right\}$ are subject to $\left\{\begin{array}{l}\text{alternation.}\\\text{change without notice.}\\\text{market fluctuations.}\end{array}\right.$

* subject to change without notice: 隨時變更，恕不另行通知

* subject to market fluctuations: 隨市況波動

5. $\left\{\begin{array}{l}\text{For quantity of}\\\text{For order of}\\\text{In case of an order for}\end{array}\right\}$ 10,000 sets and $\left\{\begin{array}{l}\text{more}\\\text{over}\end{array}\right\}\rightarrow$

$\rightarrow$ we $\left\{\begin{array}{l}\text{will }\left\{\begin{array}{l}\text{allow}\\\text{give}\end{array}\right\}\text{ you a special discount of 2\%.}\\\text{can offer a discount of 10\% on }\left\{\begin{array}{l}\text{list}\\\text{catalog}\end{array}\right\}\text{ prices.}\end{array}\right.$

6. $\left\{\begin{array}{l}\text{Regarding}\\\text{As regard}\end{array}\right\}$ payment, you are requested to $\left\{\begin{array}{l}\text{open}\\\text{issue}\\\text{establish}\end{array}\right\}$ through$\rightarrow$

* as regards 中的"regards"是 transitive verb, 那"s"不可省去, 後面不可加"to"

$\rightarrow\left\{\begin{array}{l}\text{first class bank}\\\text{prime bank}\end{array}\right\}$ an irrevocable L/C in our favor for the full

invoice $\rightarrow$ value within 30 days after confirmation of sale.

7. Our payment terms are
$$\left\{\begin{array}{l}\text{cash with order (CWO).}\\ \text{by sight L/C.}\\ \text{by 90 d/s bill, D/A.}\\ \text{by sight bill, D/P.}\end{array}\right.$$

8. The catalog incolsed gives you a small knowledge of the vast range of the articles we are now handling. If you have requirement for any other sorts of new style, please give us an opportunity to demonstrate our ability of making a creative design.

9. We believe our new model illustrated in our catalog will make outmoded other products of the similar type.

三、結尾句

1.
$$\left\{\begin{array}{l}\text{As our } \left\{\begin{array}{l}\text{stock is}\\ \text{stocks are}\end{array}\right\} \text{ running } \left\{\begin{array}{l}\text{low}\\ \text{out}\\ \text{short}\end{array}\right\} \rightarrow\\ \text{* running low: 逐漸減少}\\ \text{* running out: 即將售完, * running short: 逐漸短缺}\\ \text{As prices are rising,}\\ \text{As we are booking heavy orders every day,}\\ \quad\text{(每天接受大量訂單)}\\ \text{In view of the } \left\{\begin{array}{l}\text{heavy}\\ \text{great}\end{array}\right\} \text{ demand for this article,}\\ \text{As we fill all orders in strict rotation,}\\ \quad\text{(嚴格依序發貨)}\end{array}\right\} \rightarrow$$

$\rightarrow$ we would advise your $\left\{\begin{array}{l}\text{order}\\ \text{place an order}\end{array}\right\} \rightarrow$

$\rightarrow \left\{\begin{array}{l}\text{soon.}\\ \text{without delay.   (不可延誤)}\\ \text{without loss of time.   (不可猶豫)}\end{array}\right.$

2. We hope that you will find the article(s) you want in our catalog and are looking forward to receiving your order.

3. Please let us know if our $\begin{Bmatrix} \text{offer} \\ \text{quotation} \end{Bmatrix}$ does not contain what →

→ you $\begin{Bmatrix} \text{want} \\ \text{equire} \end{Bmatrix}$ in order to send you further samples.

4. May we $\begin{Bmatrix} \text{hear from you soon?} \\ \text{have the pleasure of hearing from you soon?} \end{Bmatrix}$

5. We look forward to your answer and a pleasant business $\begin{Bmatrix} \text{cooperation.} \\ \text{association.} \\ \text{relationship.} \end{Bmatrix}$

   * association: 提攜

6. We stand ready to be at your service and await your order, which shall have our best and quickest attention.

7. Any order that you may place with us will have our prompt and careful attention.

8. We welcome you as a customer and hope that this will lead →

   to $\begin{Bmatrix} \text{permanent} \\ \text{lasting} \end{Bmatrix}$ business $\begin{Bmatrix} \text{relations} \\ \text{cooperation} \end{Bmatrix}$ of $\begin{Bmatrix} \text{mutual advantage.} \\ \text{with mutual benefits.} \end{Bmatrix}$

   $\begin{Bmatrix} \text{永久性的} \\ \text{永恒的} \end{Bmatrix}$　　　　　　$\begin{Bmatrix} \text{互利的} \\ \text{互惠的} \end{Bmatrix}$

9. We $\begin{Bmatrix} \text{hope} \\ \text{are sure} \end{Bmatrix}$ that these samples will $\begin{Bmatrix} \text{prove satisfactory to you} \\ \text{meet with your approval} \\ \text{meet your requirement} \end{Bmatrix}$ →

   → and $\begin{Bmatrix} \text{hope to receive your order.} \\ \text{that we may be favored with your order.} \end{Bmatrix}$

   * will prove satisfactory to you: 會使你滿意,

   * will meet with your approval: 能獲得你的贊許,

＊ will meet your requirement: 符合你的需要

10. We strongly advise you to $\left\{\begin{array}{l}\text{take advantage} \\ \text{avail yourselves}\end{array}\right\}$ of →

→ this exceptional opportunity.

＊ take advantage of: 利用,　　＊ avail yourselves of: 掌握; 利用

11. We $\left\{\begin{array}{l}\text{look forward to} \\ \text{hope to have}\end{array}\right\}$ the $\left\{\begin{array}{l}\text{pleasure} \\ \text{opportunity}\end{array}\right\}$ of $\left\{\begin{array}{l}\text{serving you.} \\ \text{receiving your order.}\end{array}\right.$

習　　題

一、將下列中文譯成英文。

1. 隨函寄上有關人造絲、特多龍等縫線(sewing threads, tetron, etc.)的樣品各一種，價格標明於各種樣品之上(The prices are marked on each sample)。

2. 竭誠爲貴公司服務(at your service)。

3. 貴公司所需的物品(the articles you require)暫時售罄(temporarily out of stock)甚抱歉，茲寄上最相近物品的樣品(samples of the nearest)，希能適合貴公司的需要(meet your requirements)。

4. 訂購數量在 1000 打以上時，我們願給予 1％的特別折扣(special discount)。

二、將本書 No.35 的英文信譯成中文。

三、用下列片語造句。

1. together with

2. all round

3. on the stipulation that

4. on your side

第十四章 報 價

(Offer)

第一節 報價信的寫法

當賣方答覆買方有關商品的詢價時，其答覆往往就是一封報價函電。報價的一方稱爲「報價人」(Offeror)，被報價的對象爲被報價人(Offeree)。報價必須經被報價人的接受(Acceptance)，契約才能成立。

一、報價的種類

1. Selling Offer 與 Buying Offer：所謂 Selling Offer 乃賣方將擬售商品名稱、品質、數量、價格、裝運及付款等條件以函電等通知買方表示願按這些條件將商品賣給對方的意思表示。所謂 Buying Offer（購方報價）則係買方將希望購買的商品名稱、品質、數量、價格、裝運及付款等條件等以函電等通知賣方，表示願依這些條件向對方購進商品的意思表示。

2. Firm Offer 與 Non-firm Offer：Firm Offer（穩固報價）即報價中載明接受期限，在期限內不變更所報各項條件，Offeree 只要在期限內接受，契約即告成立。Non-firm Offer（非穩固報價）則指報價中未載明接受期限的報價，且不受拘束的報價。Non-firm Offer 又可分爲：

   ⑴ Offer without engagement(不受約束報價)：即報價人可隨時改變條件的報價，這種報價實際上只是 Invitation-to-offer 的一種。

   ⑵ Offer subject to confirmation（確認後有效報價）：即 Offeree

的接受報價，必須經原報人的確認，契約才能成立。

(3) Offer subject to·prior sale
Offer subject to being unsold ⎫⎬ （有權先售報價）：即 Offeree

的接受報價，以商品未售出才算有效的報價。

3. Counter Offer（還價）：如 Offeree 就 Offeror 所報價的條件，加以
變更接受，則等於 Offeree 就 Offeror 的報價提出還價。Firm Offer
一經 Offeree 的 Counter Offer，即失去效力。

二、報價實例

報價可以信函、報價單（Offer sheet）或電話傳眞（FAX）方式進行，也
可以電報或 Telex 進行。在後者的情形，除了報價時效很短者外，往往於
拍發電報或 Telex 後，立即再以信函確認。撰寫確認報價的函件，應先說
明拍發電報或 Telex 的日期和內容，然後將有關各項再予說明或補充。

No.38 對於砂糖詢價的 Firm Offer

Dear Sirs,

SWC Sugar

We have received your letter of June 6, 1990 asking us to offer 5,000 metric tons of the subject sugar for shipment to Singapore and appreciate very much your interest in our product.

In compliance with your request, we hereby offer you as follows:

1. Commodity & Description: Taiwan Superior White Crystal Sugar

2. Quantity: Five Thousand (5,000) metric tons.

3. Price: US$250 per metric ton FOB Stowed Kaoh-

siung.

4. Packing: To be packed in new gunny bag of 100 kgs. net each.

5. Insurance: Buyer's care.

6. Shipment: Within 30 days after receipt of L/C from Kaohsiung to Singapore, by vessel.

   Partial shipments to be permitted.

7. Payment: By irrevocable L/C in our favor through a prime bank available by draft(s) at sight.

8. Validity: This offer is good until June 30, 1990, our time.

Your attention is drawn to the fact that we have not much ready stock on hand. Therefore, it is imperative that, in order to enable us to effect early shipment, your L/C should be opened within 10 days after our offer is accepted by you.

We look forward to your prompt reply.

Faithfully yours,

【註】

1. in compliance with your request＝to comply with your request＝as requested

2. buyer's care:由買方照顧, 即由買方負責投保之意。

3. to be good until:有效到…。

4. validity＝expiry:有效期限。

5. our time:以我方時間為準。

6. your attention is drawn to:請注意…。

7. ready stock on hand:手頭存貨。

No.39　對於柳安合板詢價的 Firm Offer

Re: Lauan Plywood

Dear Sirs:

　　We thank you for your letter of May 30, 1990 enquiring about the captioned building material and take pleasure to enclose our Offer No. TT-110 for your consideration.

　　May we draw your attention to the facts that the price we quoted is on C&I Bangkok basis instead of on either FOB Taiwan or CIF Bangkok basis and that our offer will be valid until June 30, 1990. As the demands for plywood have been very heavy recently, your early decision and reply are requested.

　　We are waiting for your prompt and favorable reply.

　　　　　　　　　　　　　　　　　　Faithfully yours,

【註】

1. re＝Reference:案由。新派的人認爲不宜用此字。

2. captioned:上開。新派人士認爲這種字不該用, 而應改爲"subject"或"above"或 "above-mentioned"等字樣, 其實這是迷信。

3. C&I:保險費在內價, 即等於 FOB 加上保險費的價格。當運價變動不定時, 常用 此條件代替 CIF。

No.40　報價單

OFFER SHEET

To: Bangkok Trading Co., Ltd.　　　　　　June 5, 1990

    P.O. Box 123

    Bangkok

NO. TT-110

Dear Sirs:

    We take pleasure in offering you the following commodity at the price and on the terms and conditions set forth below:

    Payment:Against 100% confirmed, irrevocable and transferable Letter of Credit in our favor.

    Insurance: All Risks plus war for 110% of invoice value.

    Shipment: During August, 1990 subject to your L／C reaches us by end of July, 1990.

    Packing: Export Standard Packing.

    Validity: June 30, 1990, Our Time.

    Remark: Mill's inspection to be final.

Item	Commodity Description & Specifications	Quantity	Unit Price	Total Amount
TT-12	Lauan Plywood, Rotary cut, Type III, 1/8″, 3-ply, Grading	1,000,000 sq.ft.	C&I Bangkok US$60／	US$60,000.00

as per JPIC Standard. Size: 1/8″ x3′ x 6′	MSF	
		Yours very truly,

【註】

1. our time:以我們的時間為準，意指買方的接受報價函電須在有效期間內到達賣方才有效。

2. mill's inspection to be final:品質以工廠檢驗為準。

3. JPIC＝Japan Plywood Inspection Council:日本合板檢驗協會。

No.41 對詢購府綢的 Cable Confirmation of Firm Offer

Gentlemen:

We thank you for your telegram of February 28 and now confirm our cable dispatched to you to offer firm the undermentioned goods, subject to your acceptance in our hands by March 7:

Goods: 300pcs. No. 450 Tussore Silk Crepe, 28″×40″ yds. per pc.

Price: @$1.20 per yd. C.I.F. New York.

Shipment:Per steamer during April via Panama.

Packing:20 pcs. each in a tin-lined case.

Terms: As usual.

For your information we may say that it would be advantageous for you to buy the goods during the present prevalence of this low price, as large orders for Silk Goods are expected to come in at this time of the year.

We have no doubt that these orders will affect the market to a considerable extent and that any delay in purchasing will make you pay higher prices.

Your specific inquiries for any of our lines would be appreciated. We assure you that we will make every effort to meet your requirements.

Sincerely yours,

【註】

1. $1.20 讀成"one dollar twenty"但$0.50 則讀成"fifty cents"。

2. per steamer...via Panama:於四月份由經由巴拿馬運河的船運出。

3. a tin-lined case:錫箔襯裏箱。

　　cf. a zinc-lined case:鋅箔襯裏箱。

　　an iron-hooped case:加箍鐵皮箱。

　　a wire-hooped case:加箍鐵線箱。

4. terms:指付款條件而言。

5. it would be...low price＝it would be to your interests to buy the goods while the price is so greatly in your favor (while the price rules low).

6. will effect... extent: will have a great effect on the market.

No.42　對於詢購毛線衣的 Offer Subject to Prior Sale

Dear Sirs,

　　We acknowledge with thanks the receipt of your letter of May 12 and as requested we are pleased to offer you the following items subject to their being unsold upon receipt of your reply.

Commodity: Acrylic Sweaters for men and ladies in different color pattern assortments.

Quality: as per samples submitted to you on May 18 by air parcel.

Quantity: 1,000 doz., Size assortments: $\dfrac{S}{3}$ $\dfrac{M}{6}$ $\dfrac{L}{3}$

Price: US$14 per doz. CIF Hamburg

Packing: To be wrapped in polybags and packed in standard export cardboard cartons.

Insurance: To cover ICC(A) plus war for 110% of invoice value.

Shipment: September／OCtober, 1990

Paymemt: by Al bankers irrevocable and without recourse L／C available by draft(s) at sight. Such L／C must reach us one month prior to shipment.

This low price can be quoted because we have employed a new mass production system since last Spring. However, we are now receiving many inquiries from other areas we have recently contacted. The price will surely rise when the present stock is exhausted with the approach of the season. Therefore, it is advisable that you take advantage of this rare opportunity by accepting this offer as soon as possible.

We look forward to your early acceptance soon.

Very truly yours,

【註】

1. $\dfrac{S}{3}$ $\dfrac{M}{6}$ $\dfrac{L}{3}$:意指每一打的尺碼搭配爲小號(S) 3 件，中號(M) 6 件，大號(L) 3 件。

2. without recourse L／C:無追索權信用狀。與 with recourse L／C 相對。
   憑有追索權信用狀開出的滙票，萬一遭到拒付(unpay)時，持票人可向背書人請求償還票款。反之，憑無追索權信用狀開出的滙票，如遭到拒付時，持票人不能向背書人請求償還票款。換言之，前者有追索權，後者則無追索權。

3. employ:啓用。

4. new mass production system:新式大量生產系統。

5. when...is exhausted:當目前存貨售罄時。

6. take advantage of this rare opportunity:把握此一難得的機會。

No.43 對詢購原棉的 Firm Offer

Dear Sirs,

Subj: American Raw Cotton, SLM

Thank you for your letter of June 1, 1990 informing us that you are in urgent need of the subject cotton. To meet your requirements, we confirm we have cabled you today to offer the following goods, subject to your acceptance arrives here by the end of June, 1990.

(decoded)

Commodity: American Raw Cotton; Type: TASK; Grade: Strict & Low Middling; Staple: 1 -1／16″; Pressley: 90,000

Spec: Micronaire: 3.8 NCL

Quantity: 1200 bales of approximately 500 lbs. each.

Price: US ¢ 52 per lb. FAS Gulf ports.

Shipment: August, 1990 through November, 1990 at 300 bales each monthly.

Payment: By irrevocable L／C in our favor available by draft(s) at sight one month prior to each monthly shipment.

We have to inform you that we are unable to comply with wish so far as shipment is concerned because the 1990 new crop would not be available until August and trust that you would agree to our proposed shipping schedules. Meantime, we are awaiting your early acceptance in order to catch up with our shipping schedules.

Yours sincerely,

【註】

1. Subj:爲 Subject 的略字, 即案由之意。

2. subject to...here:以你的接受（函電）在（某日以前）到達此地才有效的條件, 因規定了接受期限, 所以本報價爲穩固報價。

3. decoded:指將電碼譯成普通文字之意。原來的報價係使用電碼, 在本電報證實書, 則將其譯成普通文字。

4. new crop:新作物, "crop"主要指土地生產物而言。

5. proposed shipping schedules:所提議的裝運預定日期。

6. catch up with:趕上。

No.44 對詢購罐頭蘆筍的 Firm Offer

Dear Sirs,

Canned Asparagus, Al Grade

We thank you very much for your letter, Ref. INQ／90／123, of May 3, 1990 inquiring about the subject item and take pleasure in quoting you as follows:

ITEM	QUANTITY (carton)	FOBC₃ in DM	CIFC₂ Hamburg in DM
		(Unit price per cans)	carton of 48
Tips & Cuts	300 48 cans× 12 oz.nw	30.00	35.00
Center Cuts	300 48 cans× 12 oz.nw	27.50	32.50
End Cuts	300 48 cans× 12 oz.nw	24.00	29.00

N.B.1. Packing:Standard export cardboard cartons.

2. Insurance:ICC(B) plus TPND and war risk for 110% of CIF value.

3. Shipment:To be shipped in one lot during August, 1990.

4. Payment:By a prime banker's irrevocable L／C to be opened in our favor 30 days before shipment.

5. Validity:15 days from the date hereof.

Please note that the prices we quoted above are the rock-bottom ones and that our products would compete favorably with other brands in your market because of their superior quality.

As our stocks have been running low and the demands therefor are brisk right now, we advise you to make decision as soon as possible.

Very truly yours,

【註】

1. FOBC₃:船上交貨價含佣金百分之三。"C"代表 commission 而且此項佣金係指回佣(return commission)，即將來需退還給對方的佣金。

2. DM＝Deutsche Mark:德國馬克。

3. carton of 48 cans:每箱裝 48 罐。

4. N.B.＝Nota Bene (拉丁語)＝Note well (注意)。

5. cardboard cartons: (硬) 紙板箱。

6. in one lot:以一批 (裝出)。

7. 15 days...hereof:hereof 指"of this offer"。

第二節　有關報價的有用例句

一、開頭句

1. $\left\{\begin{array}{l}\text{Thank you}\\\text{Many thanks}\\\text{We thank you}\end{array}\right\}$ for your $\left\{\begin{array}{l}\text{inquiry}\\\text{letter}\end{array}\right\}$ of (date) $\left\{\begin{array}{l}\text{about}\\\text{on}\\\text{concerning}\end{array}\right\}$ ...→

→ and (we) are $\begin{cases} \text{pleased} \\ \text{happy} \\ \text{glad} \end{cases}$ to offer them as follows:

2. $\begin{cases} \text{In response} \\ \text{In reply} \\ \text{With reference} \end{cases}$ to your inquiry of (date) $\begin{cases} \text{about} \\ \text{on} \\ \text{concerning} \end{cases}$ ... →

→ we have the pleasure of offering you on the following terms and conditions.

3. In compliance with your request of..., we are pleased to →

→ $\begin{cases} \text{submit} \\ \text{make} \end{cases}$ our offer as follows:

4. Thank you for your inquiry of...concerning...As requested,

we are $\begin{cases} \text{pleased} \\ \text{glad} \\ \text{happy} \end{cases}$ to offer them as follows:

二、條件

1. $\begin{cases} \text{This} \\ \text{The} \\ \text{The above} \end{cases}$ offer is $\begin{cases} \text{firm} \\ \text{valid} \\ \text{good} \\ \text{effective} \\ \text{open} \\ \text{in force} \\ \text{available} \\ \text{to remain in force} \end{cases}$ $\begin{cases} \text{for...days.} \\ \text{until Nov. 10.} \\ \text{till Nov. 10.} \\ \text{until further notice.} \end{cases}$

2. All $\begin{cases} \text{offers} \\ \text{quotations} \end{cases}$ and sales are subject to the terms and →

→ conditions printed $\begin{cases} \text{on the reverse side hereof.} \\ \text{on the back hereof.} \end{cases}$

3. This is a combined offer on all or none basis.

(此爲聯合報價，所列貨品必需全部接受或全部不接受。)

4. We renew our offer of June 20 →

→ { on the same terms and conditions.
    subject to the following modifications.

(我們 { 基於原來條件
        依照下列變更 } 更新 6 月 20 日的報價。)

5. { This
    Our
    The } { offer
            quotation } is subject to →

→ { prior sale.
    being unsold on receipt of your acceptance (reply)

→ your reply { received
              reaching } here { by...
                                not later than...
                                on or befor... }

此報價 { 有權先售。
        收到接受時未售才有效。
        以你的答覆 { 在……以前
                    不能晚於
                    在……或以前 } { 收到
                                    到達這裡 } 才有效。

6. { Due to
    Because of } the sensitive market situation here, we are →

→ { unable
    not in a position } to { submit our offer
                             make our offer
                             offer } that might be

outstanding for over 15 days.

({ 由於
    因爲 } 本地市場的敏感性，我們無法 { 提出
                                      報出 } 有效至 15 天以上的價。)

7. We are now pleased to submit to you a firm offer Taiwan jades as required by you some time ago, subject to your reply here by noon our time on Monday, June 25.

8. This offer is CIF any port on the Pacific Coast subject to our confirmation and for your reply here on June 30, our time.

9. We are busily working on your offer, in the hope that we can bring about business.

　　＊ working:處理,　＊ bring about:促成;完成;成交。

10. These goods being in great demand we cannot hold this offer open; we are therefore obliged to make the offer subject to prior sale.

11. As the goods now under offer are commanding a ready market, there is no knowing when they will be sold out. You will, therefore, understand that we hold this offer good only for acceptance by Friday, May 25, our time, and after that, it is subject to the goods being unsold.

　　＊ commanding a ready market:擁有現成的市場, 意指暢銷, 可隨出銷出。

習　　題

一、術語解釋

　1. Firm offer　　2. Non-firm offer　　3. Counter offer

二、將下列中文譯成英文

　1. 二月十一日大函收悉。很高興知道貴公司決定生產人造纖維混合紗及布(blended yarns／fabrics)。為覆貴公司詢價, 特報價如下:

　2. 本公司今日以 telex 報價, 詳見所附 telex 副本。

　3. 我們奉勸貴公司接受本項報價, 因為現已有跡象(as there are already indications)顯示我們的市場正轉為賣方市場(turning into seller's market)。

4. 由於料價及工資的上漲(rising cost of raw materials and wages)本公司不得不(reluctantly compelled to)將價格提高 5%。

5. 本公司確信我們的價格低得足(our price is low enough)與任何其他公司競爭(compete with)。

三、將本書 No.38 的英文信譯成中文

第十五章 接　　受

(Acceptance)

第一節　接受信的寫法

接到報價之後，應迅速檢討報價的內容，以便決定接受全部條件，或另提出條件，或因條件相差太遠根本不接受。無論那種情形，均應迅速答覆為原則，不可置之不理。因此對於報價的答覆可分為三種：

1. Complete Acceptance（全部接受）：即對於對方報價所開的全部條件無條件接受。如前所述，Acceptance 意指被報價人願依報價人所開條件成立契約的意思表示。接受的內容除接受的意思表示外，尚應將商品名稱、品質規格、數量、價格、裝運、包裝、保險及付款條件等內容予以重述。Acceptance 一如 Offer，往往以電報或 Telex 完成，所以發出電報或 Telex 後，通常尚須發出 Confirmation，並附上 Purchase Order 或 Order Sheet。

2. Conditional Acceptance（附條件的接受）：即將對方的報價條件加以變更或追加新條件而接受。這種接受是不完全接受(Imcomplete Acceptance)，所以實際上是「反報價」也即還價(Counter Offer)，還價本質上不僅為拒絕原報價，而且也是被報價人對原報價人的新報價。

3. Non-acceptance（不接受）：即拒絕對方的報價(Offer declined)之意。

拒絕報價時，其拒絕信應包括：

(1)謝謝其報價　　　　　　(2)表示不能接受的遺憾

(3)解釋不能接受的理由　　(4)暗示將來還有機會往來

No.45　覆告白砂糖報價已接受

Dear Sirs:

Re:SWC Sugar

We thank you very much for your letter of June 12, 1990 offering us 5,000 M／T of the subject sugar and wish to confirm this order with the following particulars:

Commodity& Specifications:Taiwan Superior White Crystal Sugar.

Quantity:　5,000M／T.

Price:　　　US$250.00 per M／T FOB Stowed Kaohsiung.

Packing:　　To be packed in new gunny bag of 100 kgs. net each.

Insurance: Our care.

For your information, we have already applied for the issuing of an L／C in your favor to Bank of Canton, Singapore and trust you would receive it in a few days. It will be appreciated if you will effect shipment of these 5,000 M／T's in about two equal lots by direct steamer as soon as you receive our L／C. In the meantime, although the price contracted is on an FOB Stowed basis, we shall appreciate your arranging with shipping companies on our behalf for delivery of the goods with ocean freight to collect.

Yours faithfully,

【註】

1. applied for...to Bank of Canton:向廣東銀行申請開發 L／C, "for"後面接申請的東西, "to"後面接「人、機構等」。

2. effect shipment:裝運（船），也可以"make shipment"代替。

3. in about two equal lots:分約二等批，意指分兩批裝，兩批數量大約相等。

4. arrange with shipping companies on our behalf...:代我們與船公司安排交貨事宜，指由賣方代洽船，"on our behalf"可以"on behalf of ourselves"或"in our behalf"代替。

5. freight to collect:運費到付。freight prepaid 爲運費預付。

No.46　接受柳安合板報價並發出購貨訂單

Dear Sirs:

Lauan Plywood

We acknowledge with thanks the receipt of your letter of June 5, 1990 together with one copy of your offer sheet No. TT-110 for one million sq. f. of the captioned commodity.

After careful perusal of the terms and conditions of your offer, we have found them quite acceptable, therefore, we are enclosing our Purchase order No. PO-123 in duplicate for your signature. Please sign and return one signed copy thereof for our files.

　　As we are in urgent need of this materials, you are requested to effect shipment during August, 1990 as promised in your offer. Meanwhile, we shall apply for the opening of an L/C in your favor within one week after we have received the copy of our purchase order duly signed by you.

　　We are looking forward to receiving your confirmation soon.

Very truly yours,

Encl: a/s

【註】

1. careful perusal:仔細的審閱；詳核。

2. one signed copy thereof: one signed copy of our purchase order

3. for our files:俾供我們存檔。

4. duly signed:簽妥字。

No.47　購貨訂單

PURCHASE ORDER

Taiwan Trading Co., Ltd.　　　　　　　　June 20, 1990

Taipei, Taiwan R.O.C.　　　　　　　Order No. PO-123

Dear Sirs,

We confirm having purchased from you the following commodity on the following terms and conditions:

1. Commodity:　　Lauan Plywood, Rotary Cut.

2. Specifications: Type III, 1/8″, 3-ply, Grading
　　　　　　　　　conforming to JPIC standard.

                    Size:1/8″ x 3′ x 6′

3. Quantity: 1,000,000 sq. ft.

4. Price: US$60 per MSF C&I Bangkok.

5. Packing: Export standard packing.

6. Insurance: ICC(A) plus war for 110% of Invoice
                    value.

7. Shipment: To be shipped in August, 1990 in one
                    lot.

8. Payment: By irrevocable and transferable L/C
                    in your favor to reach you by end
                    of July, 1990.

9. Inspection: Mill's inspection at mill to be final.

                              Bangkok Trading Co., Ltd.

【註】

1. transferable L/C:可轉讓信用狀，以前又稱爲"assignable L/C"，現在最好不要用。

2. Mill:工廠，木材廠，麵粉廠，紙廠等的工廠多用"mill"這個字。

No.48 接受壓克力毛線衫報價並發出訂單

Dear Sirs,

                    Acrylic Sweaters

    Your letter of May 25, 1990 offering the subject gar-
ments has been received, and we are pleased to inform
you that your terms and conditions are acceptable to us.

    By means of this letter, we are pleased to place our

order with you for the garments of the following particulars:

Acrylic Sweaters for men:

Quantity:　500 doz.

Colors and size assortments per dozen:

Size Color	S	M	L
Red	1	2	1
Green	1	2	1
Yellow	1	2	1

Acrylic Sweaters for ladies:

Quantity:　500 doz.

Colors and size assortments per dozen:

Size Color	S	M	L
Navy	1	2	1
Blue	1	2	1
Purple	1	2	1

As the season is coming, you are requsted to effect shipment the total quantity of this order in one lot not later than October 5. Meanwhile, please send us your

Proforma Invoice immediately and we will open our L/C for the total value in due course upon receipt of it.

Your prompt confirmation is awaited.

Yours faithfully,

【註】

1. by means of this letter:憑此信。

2. color and size assortment:顏色及尺寸的搭配。

3. in due course:在適當時期；不久以後。

No.49　買方訂購罐頭蘆筍函

Gentlemen:

Canned Asparagus, Al Grade

We have received your letter of May 15, 1990 submitting your offer for three different qualities of the subject item.

Since the prices you quoted are quite reasonable, we hereby place the following order with you:

Item	Quantity
Tips & Cuts	300 cartons, 48 cans x 12 oz., N.W./CTN
Center Cuts	ditto
End Cuts	ditto

So far as prices are concerned, we prefer CIFC2 Hamburg to FOBC3. It is understood that the prices for the former are DM35, DM32.50, and DM29 per carton for these three qualities respectively.

As requested in your letter under reply, we will issue

an L／C through our bankers, Dresdner Bank, Hamburg, in your favor around June 30, in order to enable you to make delivery during August, 1990.

Please let us have your confirmation at your earliest convenience.

Very truly yours,

【註】

1. quite reasonable: (價錢) 很公道。
2. ditto: 「同上」之意。
3. so far as...concerned:就……而論，也可以"as far as...concerned"代替。
4. prefer...to...:prefer 常與 to 連在一起，意指「喜歡……而不喜歡……」。例如: I prefer tea to coffee. （我喜歡茶而不喜歡咖啡。）
5. it is understood:諒解; 同意; 心照不宣。
6. around June 30: 6 月 30 日左右, "around"也可以"about"代替, 不過比較不確切。

No.50　賣方寄出罐頭蘆筍售貨確認書函

Dear Sirs,

Re:<u>Canned Asparagus, Al Grade</u>

Thank you very much for your letter of May 25, 1990 placing an order with us for 300 cartons each of three qualities of the captioned canned food and we take pleasure to inform you that we have already started processing this order. Please be assured that we will effect shipment of the total quantity during August provided your L／C is received prior to June 30, as indicated in your letter under reply.

When the goods are ready for shipment, we will telex you upon shipping space has been arranged. Enclosed please find our Sales Confirmation in duplicate for your signature and we shall appreciate your returning to us one duly signed for our file.

We thank you again for your interest in our product.

Yours faithfully,

【註】

1. take pleasure to inform＝take pleasure in informing

2. for our file:可以"for our records"代替。

No.5l 售貨確認書

TAIWAN TRADING COMPANY, LTD.

Taipei, Taiwan

Sales Confirmation

X.Y.Z. A.G.   June 1, 1990

P.O. Box 123   Sale No. 90/123

Hamburg

Dear Sirs,

We confirm that we have sold to you the following goods on the following terms and conditions:

1. Commodity:Canned Asparagus. Al Grade of Tips & Cuts, Center Cuts, and End Cuts.

2. Quantity: 300 cartons each of Tips & Cuts, Center

	Cuts, and End Cuts.　Each carton contains 48 cans×12 oz., N.W.
3. Price:	Tips & Cuts: DM35 per carton CIFC2 Hamburg.　Center Cuts: DM32.50 per carton CIFC2 Hamburg. End Cuts: DM29 per carton CIFC2 Hamburg.
4. Packing:	Standard export cardboard packing.
5. Insurance:	ICC(B) plus TPND and War for 110% of invoice value.
6. Shipment:	During August, 1990.
7. Payment:	By a prime bank's irrevocable L/C which must be opened in our favor 30 days prior to shipment.

Accepted on...(date)　　　　　　　　　Yours faithfully,

by:　　　　　　　　　　　Taiwan Trading Company, Ltd.

No.52　確認接受縐紗報價，發出購貨確認書

Dear Sirs,

　　We have pleasure in confirming our telex, accepting your firm offer of the 28th March on 200 pieces No. 450 Crepe de Chine, and now send you herewith our Purchase Note No. 150, which we trust you will find in order.

　　In order to cover the amount of this purchase, we have arranged with Lloyds Bank, Ltd., London, for a Confirmed Letter of Credit in your favour.　This credit will be telexed to your bank as usual. This business being very important on our part, we ask you to give it your best attention so as to be able to satisfy us in every respect.

Tussore Crepe No. 380. We have just been approached by our reliable friends in Liverpool to make them an offer on this commodity, and accordingly we have just telexed offer, as per copy enclosed, asking you to telex us a firm offer C.I.F. on 500 pieces for May shipment. We hope you will do everything you possibly can to make us your best offer, as we think it is a good chance to dispose of a large parcel of this commodity. We look forward to your telex offer, so that we may close the business to our mutual advantage.

Faithfully yours,

WM: FS   H. WHITEHALL & CO., LTD.

Enclos. 3 Copies of telexes   *W Meyer*

     1 Purchase Note.   President

P.S. Your telex offer of the 30th March on 200 pieces Pongee Silk No. 190 has just reached us. We regret that we find it unworkable at your figure, but we will make every effort to telex a feasible offer at an early date.

W.M.

【註】

1. crepe de chine:縐紗的一種。

2. purchase note＝bought note＝purchase confirmation.

3. in order...purchase＝in settlement for our purchase:為清償購貨價款。

4. parcel:貨物，包成一定數量的貨物。

5. to close:成交，締結（契約）。

6. unworkable:難於處理，意指價高無法接受。

7. feasible offer:可行的報價。feasible＝practicable 這裡的"offer"為"buying offer"。

No.53　對尼龍內衣報價的還價

Dear Sirs,

　　We thank you for your letter of July 20 and for the samples of nylon underwear you kindly sent us.

　　We appreciate the good quality of these garments, but unfortunately your prices appear to be on the high side for garments of this quality. To accept the prices you quoted would leave us little profit on our sales since this is an area in which the principal demand is for articles in the medium price range.

　　We like the quality of your goods and also the way in which you have handled our enquiry and would welcome the opportunity to do business with you. May we suggest that you could perhaps make some allowance on your quoted prices that would help to introduce your goods to our customers. If you cannot do so, then we must regretfully decline your offer as it stands.

　　　　　　　　　　　　　　　　　Yours faithfully,

【註】

1. the sample of nylon underwear you kindly sent us:承蒙惠寄的尼龍內衣樣品。本句中的關係代名詞因係受格並係限定使用法，故省略。請參閱下列：The offer（which）you made is too high. 貴公司所作的報價太高。These are the best terms（which）we can grant. 此乃本公司能給予的最優惠條件。

2. appear to be on the high side for garments of this quality:與同品質的成衣相較價格似嫌偏高。

3. to accept the prices you quoted would leave us little profit on our sales:如接受貴公司所開之價，將使本公司的銷售幾乎無利可剩。本句的 little 乃強調所剩無幾之意，如用 a little 則強調尚有幾分利潤。

4. articles in the medium price range:中等售價的貨品。

5. the way in which you have handled our enquiry:貴公司對本公司詢價的處理方式。

6. you could perhaps make some allowance on your quoted prices:可否對所作的報價略予讓步。此句中的 could 是一種請求的禮貌說法。例：Could you do me a favor? 另"allowance"也可譯為折扣。

7. then, I must regretfully decline your offer:否則祇能謝拒貴公司的報價了。"decline"比"refuse"要客氣。

8. as it stands:照現狀；照現在的報價。

No.54　接受尼龍內衣還價

Dear Sirs,

We are sorry to learn from your letter of August 5 that you find our prices too high. We do our best to keep prices as low as possible without sacrificing the quality and to this end, are constantly enquiring into new methods of manufacture.

Considering the quality of the goods we quoted, we do not feel that the prices are at all excessive, but bearing in mind the special character of your trade, we have decided to offer you a special discount of 3% on a first order for US$12,000. We make this allowance because we should like to do business with you if possible, but we must stress that it is the furthest we can go to help you. We hope this revised offer will now enable you to place an order.

Sincerely yours,

【註】

1. without sacrificing the quality:不犧牲品質水準。

2. to this end, are constantly enquiring into new methods of manufacture:爲達到此目標，不斷探求生產製造的新方法。

3. bearing in mind the special character of your trade:考慮到與貴公司交易的特別性質。

4. We must stress that it is the furthest we can go to help you:我們必須強調這是本公司能給予貴公司最大的優待。

5. this revised offer:此經過修正的報價。

No.55　不同意還價

Dear Sirs,

Many thanks for your letter of May 18, in which you ask us for a keener price for our Pattern 102.

Much as we should like to help you in the market you mention in your letter, we do not think there is room for a reduction in our quotation as we have already cut our price in anticipation of a substantial order. At 8 cent per yard this cloth competes well with any other product of its quality on the home or foreign markets.

We are willing, however, to offer you a discount of 5% on future orders of value $1,000 or over, and this may help you to develop your market.

Yours faithfully,

【註】

1. a keener price:更克己的價格。

2. there is no room:無餘地。

No.56 對還價同意減價若干

Dear Sirs,

We have given your letter of September 23 very careful consideration.

As we have done business with each other so pleasantly for many years, we should like to comply with your request for lower prices.

However, our own overhead has increased sharply in recent months and we cannot reduce price 15% without lowering our standard of quality...and that we are not prepared to do.

We suggest an overall reduction of 5%, through the line.

> We hope you will consider this satisfactory, and that we can continue our long and friendly association.
>
> 　　　　　　　　　　　　　　　　　　　Yours truly,

【註】

1. pleasantly:愉快地。

2. overhead:營業開支。

第二節　有關接受的有用例句

一、接受報價

1. $\left\{ \begin{array}{l} \text{In reply to your} \\ \text{With reference to your} \end{array} \right\} \left\{ \begin{array}{l} \text{cable offer} \\ \text{quotation} \\ \text{offer} \end{array} \right\}$ of...(date)...,

→ $\left\{ \begin{array}{l} \text{we accept on the terms and conditions as follows:} \\ \text{we accept on the terms and conditions quoted as follows:} \end{array} \right.$

2. We are pleased to confirm acceptance of your offer of...(date)...on 10,000 sets TV sets Model TR-123 at US$ 130 per set FOB Keelung, shipment to be effected within 60 days after receipt of L／C.

3. Thank you for your $\left\{ \begin{array}{l} \text{offer} \\ \text{quotation} \end{array} \right\}$ of...(date)..., which we →

→ accept subject to $\left\{ \begin{array}{l} \text{your confirmation reaches} \\ \text{your confirmation received by} \end{array} \right\}$ us →

$\left\{ \begin{array}{l} \text{→ before}\cdots\text{date}\cdots \\ \text{the approval of import licence.} \\ \text{modification of the following terms.} \end{array} \right.$

4. We accept your offer of July 10 on Garments but price do better, if possible.　(如可能，請酌減價錢。)

二、還價

1. In view of the prevailing prices in this market, your →

   → $\left\{\begin{array}{l}\text{offer}\\\text{quotation}\end{array}\right\}$ is a little expensive. We have just $\left\{\begin{array}{l}\text{cabled}\\\text{telexed}\end{array}\right\}$ →

   → you a counter offer $\left\{\begin{array}{l}\text{asking}\\\text{requesting}\end{array}\right\}$ for a large discount. Unless

   approved we regret that business agreement will not be acceptable.

2. $\left\{\begin{array}{l}\text{If}\\\text{In case}\end{array}\right\}$ you can reduce your price to US\$ 10 per dozen, →

   → it is possible for us to place an order for considerable quantity of this article.

3. We counter offer US\$10 per dozen subject to your confirmation

   $\left\{\begin{array}{l}\text{reaches us}\\\text{arrives here}\end{array}\right\}$ on or before July 10.

   (本公司還價每打 10 元，但以貴公司確認在 7 月 10 日以前達到我方爲條件。)

4. Please reduce your price as much as possible without any change of other terms.

   (在其他條件不變下，請儘量減價。)

5. Please confirm whether you can shorten the delivery time of your offer of...(date)...

三、對還價的答覆

1. We are glad to confirm acceptance of your counter offer of August 10 on our portable transistor radio Model BX-123.

2. The prevailing prices in this market are nearly 10% higher than yours and therefore we regret our inability to accept the counter offer mentioned in your telex.

3. Owing to rising cost of raw materials and wages, we are not in a

position to reduce our prices any more.

四、拒絕接受報價

1. Unfortunately we cannot accept your offer. Your prices are prohibitive. （太貴了）

2. We have the impression that your price is higher than that in the average market which is now showing a decline.
   (我們以爲你的價格比目前起趨跌的平均市價要高。)

3. Unfortunately we are not in a position to accept your offer because another supplier in your market offered us the similar article at a price 3% lower.

習　題

一、將下列中文譯成英文

1. 貴公司七月五日電報報價，謹確認同意接受 10,000 碼棉織襯衣布料(cotton shirtings)，每碼美金七角，基隆船上交貨。

2. 倘蒙寄下預期發票列報(a proforma invoice quoting) OK-102 型(model OK-102)電晶體收音機 1,000 臺新嘉坡到岸價格(CIF)則深爲感激。

3. 謝謝貴公司四月三日報價，謹依下列條件(on the terms quoted as follows)接受。

4. 本公司接受貴公司三月三日報價，如可能(if possible)，請酌減價格(do better)爲盼。

5. 以其他條件不變爲條件(without any change of other terms)，請盡量減價。

二、試將本書 No.50 英文信譯成中文。

第十六章　推銷與追查

(Sales Promotion and Follow-up)

第一節　推銷信的寫法

推銷信(Sales Letters)就是賣方向買方主動洽銷生意的信。廣義地說，賣方所發出的每一封信都可視爲推銷信，但這裡所要談的是指賣方向已有往來的客戶所發出的狹義的推銷信而言。出口商推銷貨品的方法，或發出推銷信或印發 Circular Letters (通函)、商品目錄、價目表、樣品、市況報告(Market Report)，藉以引起客戶的注意, 由而達成推銷的目的。

一般而言，一封完善的推銷信，其內容應以達成下列四項爲要：

1.引起注意(Attention)　　2.發生興趣(Interest)

3.喚起欲望(Desire)　　　　4.決定行動(Action)

上述四項可簡稱爲"AIDA"。茲略作說明於下。

一、引起注意: 推銷信的第一段必須寫得引人入勝(attract the reader's attention)。所以推銷信的第一段，必須寫得動人。

信文的第一段，在性質上，應以讀者的立場(即 your attitude)，用積極的語氣及直接的措詞。在寫法上，可用詢問式、命令式、說明式、假定式或故事式，並力求 Courteous 及 Impressive。

1. You attitude:就是要從讀者立場, 說明讀者可以有怎樣的利益, 例如:

"As an importer of novelties, you are certainly seeking for more new products to meet your customers' growing requirement. Where are you looking for?"

2. Positive tone（積極的語氣）：與其說：

"You wouldn't import any tropical fish food, would you?"

不如說：

"You want to import tropical fish food, don't you?"

3. Direct expression（直接的表示）：使收信人一看就知道信的內容，例如：

"We have recently developed some new styles of ladies' handbags made of plastics for our customers in West Germany. They may be of your intertest."

Sales letter 的第一句，如用詢問式，則所問的應該集中讀者的注意，並適合所說的貨物。例如：

"Do you encounter the problem of locating suitable machinery suppliers and/or mechanical man? If you do, why not let us solve your problem?"

說明式可以採用合適的格言或俗語。例如：

"A penny saved is a penny earned."

假定式，第一為使讀者相信一種好的標準；再說明你的貨品等於或勝於這個標準。例如：

"If you wish to import cars that will bring you profit, import the"Ford".

故事式是利用人類喜歡聽故事的心理，先講故事，再說明你要推銷的貨品。

二、發生興趣：客戶對於你所要推銷的貨品情況不明白，當然就無法發生興趣。所以推銷信在引起了收信人的注意之後，跟著就應敍述貨品的用處、品質、式樣、優點、價格、付款條件、交貨時間等。爲求敍述能引起其興趣，寫信的時候，必須從讀者的觀點出發，研究收信人的需要和環境，使信的內容能夠迎合客戶的心理。譬如收信人喜歡價廉的貨物，推銷信就應著重於價格的低廉；如喜歡 Stocklots（存貨），就要著重於 Stocklots 一事以引起他的興趣(arousing the reader's interest)。例如：

"We have a great number of inexpensive ladies garment of various styles, designs, and colors in our stock."

三、喚起欲望：有時收信人對於推銷的貨物，雖則發生興趣，但並無購買的欲望，要喚起收信人的購買欲望(Create buying desire)，可以用理由或例證，使他確信貨物易於出售、易於獲利、服務好，以及其他優待條件。例如：

"In addition to the trade discount stated, we would allow you 5% discount for order exceeding 1,000 doz."

四、決定行動：喚起了收信人欲望之後，應進一步促其立即決定購買。促使收信人決定購買的行動，可以用存貨不多，優待期有限，或價格有上漲趨勢等等詞句，而促其採取行動(induce action or urge him for taking action)。例如：

1. On June 1, the price of this item will be positively raised from US$...to...US$...per dozen, FOB Taiwan. To take advantage of the old price, you must send your order now.

2. Remember too, our supply in stock is limited. Send your order to us before...(date).

No.57 推銷原子筆

Dear Sirs,

Ball Pens

We have the pleasure of forwarding you, under separate cover, some samples of the ball pens made by us.

We have to draw your attention to our prices mentioned in our price list enclosed that we can offer our products at from 5% to 10% lower than those of other brands.

Notwithstanding the lower prices, you may rest assured that our products are of the first-class quality and of the best workmanship, as can be proved by our samples.

With these advantages you can develop market for our products without difficulty, and so we trust we shall be able to receive your orders soon.

Yours very truly,

【註】

1. ball pens:原子筆, 又寫成"ball point pens", "ball-point pens"或"ball point"。

2. at from 5% to 10% lower than those of other brands:比其他廠牌便宜5% 至10%, "those of"爲"the prices of"之意。

3. develop market: 開拓市場。

No.58　推銷華美新奇品

Gentlemen,

Fancy Novelties

As an importer of novelties, you are certainly seeking for more new products to meet your customers' requirements. Where are you going to look for?

Look for us please. Right here is the Fancy Enterprise Co., Ltd. You can get many newly-developed items of novelties that you may be of interest.

Being one of the leading manufacturers and exporters of novelties, we are always manufacture various fancy novelties to fulfil our customers' needs. We have done outstanding services for our regular customers, so we can do the same for you too.

For your reference, we are pleased to enclose a list of novelties that are manufactured by us. You can find from the list that we have wide range of products. Please don't hesitate to let us know if you find any specific item (s) interesting. We will, upon request, promptly send you detailed information including catalog, prices, and even samples for your evaluation.

We are ready to serve you, why not write us now!

Sincerely yours,

【註】

1. fancy novelties:華美新奇品。

2. newly-developed items:新開發的項目，即新產品。

3. you may be of interest:也許你感興趣。

4. wide range of products:產品種類多。

5. please don't hesitate to＝please feel free to:請不必客氣。

6. why not write us now! :請現在就寫信來吧!

No.59　推銷油漆

Gentlemen:

New "Sealex" Paint

You may be interested in the new "Sealex" paint we have just introduced to the trade. A sample has been sent to you today by parcel post.

"Sealex" is the result of many months of careful research. It is made from a special formula and owes its superiority over other exterior paints to its remarkable ability to allow for the movement of those paint-peeling cracks just visible to the naked eye. This quality to expand with the cracks comes from a very special combination of granite, mica and resin that provides a rich, thick coating twice the thickness of that of the average finish, thus giving long-term protection.

"Sealex" is available in twenty-one basic colors and, as you will see from the enclosed list, prices are surprisingly low. We are nevertheless allowing a special 5% discount to you if you place orders before the end of the current month.

> So why not take advantage of the opportunity now
> and send us an immediate order.
>
> 　　　　　　　　　　　　　　Yours faithfully,

【註】

1. have just introduced to the trade:甫推出上市。

2. the result of many months of careful research:經過很多月份（長久時間）精心研究的結果。

3. made from special formula:由特別處方製成。

4. superiority over:優於。

5. paint-peeling cracks:油漆脫落的痕跡（裂縫）。

6. just visible to the naked eye:肉眼即可看見。

7. to allow for the movement...the naked eye 意指:使油漆脫落的痕跡不致看得出來。

8. granite:花崗石。

9. mica:雲母。

10. resin:樹脂。

11. basic colors:基本色彩。

12. so why...opportunity:所以，為何不利用機會…。

No.60　推銷牛肉精

> Dear Sirs,
>
> 　　　　　　　New Beef Extract
>
> 　A sample of our new beef extract, Vimbeff, has been
> sent to you today by air parcel post, which we hope will
> reach you in perfect condition.
>
> 　You will find that it possesses many unique features
> which definitely place it ahead of its many competing

brands. It dissolves readily in water, leaving no trace of sediment. As a result, its digestion presents no difficulty, rendering it particularly valuable in cases of convalescence and general debility. Other points are stressed in the leaflet enclosed.

The many inquiries we have received prove that the public are fully aware of the merits of our new product. There will accordingly be no prolonged holding of stocks, with the loss and difficulties such a practice entails.

After paying due regard to the amount of turnovers, you will agree that the 15 percent trade discount allowed is decidedly generous, particularly as settlement is to be effected on 90 days' D/A basis.

We look forward to adding you to our other importers who are reaping substantial profits from the sale of Vimbeff.

Yours faithfully,

【註】

1. extract:濃縮物、精華。

2. in prefect condition:情況完好。

3. unique feature:獨特的優點。

4. place it ahead...brands:使之列於同類貨品的前茅。

5. dissolve:溶解。

6. leaving...of sediment:毫無渣滓; sediment 爲渣滓或沈澱物。

7. digestion:消化。

8. rendering:致…使…。

9. convalescence: (病) 復元。

10. debility:虛弱。

11. stress:強調; 表明。

12. no prolonged holding of stocks:存貨將不致留存很久。

13. with the loss...entails: (以致存放過久) 招致損失與困難。entail 爲引起、招致之意。

14. decidedly generous:絕對的優厚。

15. settlement...basis:按 90 天 D/A 結帳。

16. we look forward...importers:我們深盼你們將成爲經手我們貨品的進口商之一。

No.61　推銷耶誕禮品

Dear Sirs.

Christmas Gifts Items

Have you been looking for some new items, fresh and fancy, to add to your Christmas gifts items of this year? Here is just what could make your dream come true: the Hollow Glass Ornaments made by us after months of careful designing and repeated experimenting. They are so cute and loveable that they would surely win the hearts of children as well as adults.

The Hollow Glass Ornaments are made in the shape of either dolls or animals, adorned with wigs and hats made of bright colored flowers cloth. The dolls include Snow White, Cinderella, Petter Pan etc. and the animals poodles, hounds, puppies, rabbits, bears, elephants, etc.

The color of glass varies in different shades of crimson, yellow, brown, purple, green, blue, grey and white, all transparent. The adornments to each figure are made out under special attention and direction of our artist-designers.

The color illustrations on the enclosed catalogs will show you how eye-catching they are. You would feel that You could hardly wait to see them in real shape. Besides, we believe you will find the price very attractive and reasonable.

It is sincerely hoped that you will take advantage of this opportunity and favor us with your order without loss of time.

Yours very truly,

【註】

1. fresh and fancy:新鮮又華美的。

2. make your dream come true:使你的夢想得以實現。

3. hollow:空心的；中空的。

4. cute and loveable:玲瓏可愛的。

5. ornaments:裝飾。adorned with wigs and hats:戴上假髮和帽子。

6. win the hearts of:獲得（小孩及大人的）心。

7. Snow White:白雪公主；Cinderella:仙履公主；Petter Pan:潘彼得；poodles:鬈毛狗；hounds:獵犬；puppies:小狗；crimson:深紅色；each figure:每一具形態；artist-designers:藝術設計家。

8. eye-catching:引人注目的。

9. you could hardly wait...:你幾乎等不及去；覺得非馬上去（看個真東西）不可。

10.without loss of time:刻不容緩。

第二節　追查信的寫法

賣方向顧客(customers)發出推銷信或循買方的要求報了價，甚至寄出樣品後，自然期待能收到覆信或訂單。假如寄出了信或樣品之後，如石沈大海，毫無音訊時，賣方自宜發出追查信(follow-up letters)查問。追查信的內容大致如下：

1. 提醒顧客曾向其發過信、寄過目錄、價目表或樣品。

2. 重敍貨品的優點，價格的低廉，願意為其效勞。

3. 提出若干理由（例如存貨不多、特價期限快屆滿、價格將提高等），促其趕快採取行動。

4. 誠懇要求顧客答覆。

No.62　寄出樣品後無音訊再發出追查信

Dear Sirs,

It was way back in August last year when we wrote you and sent you a sample of our products under separate cover.　Now Spring is here.The trees are budding and the flowers are blooming, but still no words from you.

Did you ever write a man about a matter you felt sure would interest him and then wait and wonder why he didn't reply?

That is what we are doing now.

We presume that our letter could have escaped your

attention due to the pressure of your work. So, this time we are going to make it easy for you to answer.

Just look below. Take your pencil and write "yes" or "no" opposite the questions that apply in your particular case. Then return this letter in the enclosed postage-paid envelope.

Do this now. It will pay you, because each of our customers has been successful.

Yours sincerely,

The questions are:

　　(1) Are our prices in line?

　　(2) Are our products suitable?

　　(3) Are you in the market for these products?

【註】

1. It was way back in August last year:那是去年八月的事了。

2. The trees are budding:樹木皆青。

3. the flowers are blooming:百花爭艷。

4. still no word from you:還是沒有收到你的音訊。

5. Did you...reply? 閣下是否曾經自己充滿了信心寫一封信給人而老是在等著，在猜測對方爲什麼不答覆呢？

6. That is...now:我們現在的情形正是如此！

7. Our letter could...of your work:可能因爲閣下貴人事忙而給疏忽了。

8. Just look below:請看一下下面。

9. opposite the questions：針對問題（填上"yes" 或 "no"）。

10. postage-paid envelope:已貼妥回郵郵票的信封。（這信封，顯然是用於國內的 follow-up letter,國際貿易用的信，不流行也無法如此做。因爲出口商不會有進口國的郵票, 所以 "postage-paid" 可改用 "addressed")。

11. it will pay you:這對貴公司有利; 對貴公司划得來。

12. Are our prices in line?:我們的價錢對不對?

13. Are our products suitable?:我們的產品合適不合適?

14. Are you in the market for...:你們是否想買 (這些貨品?)。

出口商將樣品與函件寄給進口商後, 毫無音訊時, 自宜再接再厲, 繼續追詢。

No.63　寄出雨傘報價單後無反應, 發出追查信

Dear Sirs,

Now that our offer of January 15 for 1,000 dozen of ladies' nylon umbrellas expired yesterday, we wish to know the reason for its failure in interesting your customers so that we may instruct our manufacturers to make better offers.

Since we made offer, we Have heard that Hongkong sources are underselling us. But to convince our manufacturers, will you please write to us stating the fact specifically?

This is also a test case for our manufacturers. As you may have seen in the various catalogs we sent you, they are very good manufacturers. So, please let us have a second chance by pointing out how to improve our offer.

　　　　　　　　　　　　　　　　Sincerely yours,

【註】

1. offer...expired:指報價時效已逾期, 原報價是 firm offer 定有時效。

2. Hongkong sources:香港方面。

3. 第二段可改用下列句子:

In quality and delivery we have full confidence, but we are afraid that our prices were not attractive enough, possibly in the face of Hongkong competition. Will you, therefore, let us have your comments on our offer, especially about competition, if any, and the price level workable to you?

4. "So, please...improve our offer" 可以下面句子代替:

But they are not the only source of supply. We have connections with many other leading manufacturers. So, we can choose any of them upon hearing from you.

本來出口商與工廠應同心協力推銷貨品才對。假如改用上面的句子, 似乎有使人覺得出口商淪於 Broker 的地位, 同時好像原來的工廠不一定是好工廠。所以, 非不得已, 不用上面的句子爲宜。

No.64　寄出男用拖鞋報價單、樣品後無音訊, 再發出追查信

Dear Sirs,

Gents' slippers

Upon receipt of your request of November 3, we mailed you immediately a letter together with a price list and catalog on November 10, and forwarded you one pair sample each of GM-123, GM-234, and GM-345 for your evaluation by air parcel post on November 12. We think you might have already received them.

Do our products meet your requirements? Do you find our prices competitive? Do you have good news for us?

We are keenly interested in assisting your continued

business development and we trust that you will never hesitate to contact us to explore areas of mutual interest.

　　As we mentioned to you in the last mail, we have a very warm feeling towards your firm and sincerely hope that we can stimulate a more active relationship.

　　We wish you every good fortune in the coming Christmas season.

Sincerely yours,

【註】

1. gents' slippers:男用的拖靴。gents'爲 gentlemen's 的簡寫。
2. explore:探索；開拓。
3. mutual interest:互利；cf. mutual benefit:互惠。
4. wish you every good fortune:祝您好運!
5. in the coming Christmas season:即將來臨的耶誕節。

No.65　寄出電晶體收音機樣品後無反應，再發出追查信

Transistor Radios

Dear Sirs:

　　In response to your inquiry of April 15, we sent you our price list together with two sample transistor radios per air parcel post. Since then we have heard nothing from your side and are beginning to wonder whether you have received these samples or not. Please let us know by return mail, because, if you did not receive them we will send again.

　　Perhaps the package has escaped your attention due

to the pressure of your work. If this is true, we wish you would do us a favor and spend about ten minutes right now, examing the quality of these unique radio sets and the price list. Then you will find the radio sets we have selected for you will be just the right models for your market, and the high quality of these items with the best price will convince you to place an order with us.

With the rapid growing demand for our quality products, our factory has a very tight schedule of production now. However, as one of our best regular customers, we would like to give you the first priority of executing your order and we promise that your order will be shipped within 30 days upon receipt of your L/C.

If you need our service, would you care to place your order now? A cable order will be more helpful.

Yours faithfully,

【註】

1. pressure of work:工作的壓力。

2. right now＝right away＝right off:立刻; 馬上。

3. convince you:使你確信; 使你信服。

4. quality products:高品質的產品。

5. tight schedule: (生產) 檔期緊湊。

6. first priority:第一優先。

7. cable order:電報訂貨。

No.66　寄出樣品後無回音，再發出追查信

Dear Sirs:

On April 15 we received a request from you for some prices and samples.

We replied to your letter on April 17, and when we did not hear from you, we wrote again to you on May 20.

At the same time we got the necessary information, after which we again wrote you on May 30, asking that you acknowledge receipt of this information, and let us know if you were interested.

To date, we have not had the courtesy of hearing from you.

Don't you think that when you ask for certain information, and a firm goes to the trouble of getting it for you, you should at least repay them with a reply even to say that you are not interested?

Yours very truly,

【註】

1. to date:到今天爲止…。

2. we have not had the courtesy of...:我們還沒有得到你們的賜示。

3. repay:報答；付還。

cf. to repay a kindness:報答恩惠。

It repays your labor well:你的力氣不會白費。

He repaid the money he had borrowed:他把借的錢還掉了。

4. 最後一段的譯文:你想當你向人家要某項資料, 人家花了很多功夫才找到寄給你, 你是不是至少該給人一個答覆, 即使說你不感興趣?

第三節　有關推銷的有用例句

一、開頭句

1. $\left\{\begin{array}{l}\text{We take the liberty of sending}\\ \text{We are very pleased to send}\end{array}\right\}$ you a copy of our latest catalogue
→ and pricelist.

2. $\left\{\begin{array}{l}\text{As you have placed many orders with us in the past}\\ \text{As you are one of our oldest and most regular customers}\end{array}\right\}$,
we → have decided to make you a special offer of....

3. We feel you will be interested in the new...which we are shortly to
→ $\left\{\begin{array}{l}\text{place on the market.}\\ \text{introduce to the trade.}\end{array}\right.$

4. We are sorry to note that we have been without an order from you for over five months....

5. With reference to the offer we made you on...,we regret that we have not yet had an order from you.

6. Looking through our records we note with regret that we have not had the pleasure of an order from you since last December.

二、結尾句

1. We hope you will take full advantage of this exceptional offer.

2. We are most anxious to serve you and hope to hear from you very soon.

3. We are offering you an article of the highest quality at very reasonable price and hope you will take the opportunity to try it.

4. We feel sure you will find a ready sale for this excellent material and that your customers will be well satisfied with it.

　＊ to find a ready sale:發覺好銷。

5. We should be pleased to welcome you to our showroom at any time to give you a demonstration.

6. We are allowing a special 5% discount to you if you place orders before the end of May and look forward to receiving one from you.

7. We look forward to the pleasure of your renewed custom.

　＊ renewed custom:再惠顧。

習　　題

一、一封完善的推銷信應包括那四項內容?

二、將下列中文譯成英文:

1. 頃接貴公司 3 月 6 日大函敬悉。承詢敝公司所產電視機之價目及折扣, 不勝感激。

2. 如承賜知足下需要何種成衣, 當再詳細奉告。

3. 不能效勞之處, 殊深抱歉。

4. 我們深信(feel sure)貴公司對於即將推出市場(shortly to be placed on the market)的新式電熱器(new electric heater)感到興趣。

5. 薄利多銷。

6. 捷足先登。

三、將本書 No.60 的英文信譯成中文。

第十七章 訂 貨

(Orders)

第一節 訂貨信的寫法

在國際貿易，交易的成立方式有二：

其一爲由出口商發出 Offer, 經過討價還價, 終於被 Accept 後, 由買方發出訂單(Order), 或由賣方發出售貨確認書(Sales Confirmation)。

其二爲由進口商根據出口商先前寄出的樣品、商品目錄、價目表、估價單, 主動向出口商發出訂單, 並經出口商接受(Acknowledge)或確認(Confirm)。關於第一種情形已在第十四、十五兩章中介紹。茲將第二種情形予以介紹。

進口商根據出口商原先寄出的樣品、商品目錄或價目表、估價單等主動向出口商發出訂單時, 如該訂單經出口商接受(Acknowledge)或確認, 買賣契約即告成立。

進口商下訂單時或用事先印妥的訂單(Order Sheet)附以簡單的伴書(Covering Letter), 或以書信方式下單一購貨信(Order Letter)。不論用那一種形式, 訂單的內容, 與前述報價單(Offer Sheet)相似, 應詳載交易的條件。

由進口商主動發出的訂單, 本質上是 Buying Offer, 訂單一經出口商接受, 契約即告成立, 所以內容應力求詳細完整。

No.67　初次訂購(Initial Order)

Dear Sirs,

　　Thank you for your samples and price list of 10th September.　We are pleased to find that your materials appear to be of fine quality.

　　As a trial order we are delighted to give you a small order for 100 pcs., white shirting S/235.　Please note that the goods are to be supplied in accordance with your samples.

　　The particulars are detailed in the enclosed Order Sheet No. 123.　We are in a hurry to obtain the goods, so please cable your acceptance. Upon receipt of which we will open an irrevocable L/C through Bank of Taiwan.

　　If this initial order turns out satisfactory, we shall be able to give you a large order in the near future.

Encl:a/s　　　　　　　　　　　　　　　Yours faithfully,

【註】

1. trial order:試驗訂貨; 嘗試性訂貨。

2. small order:少量訂貨。trial order 一般而言，均是 small order.

3. in accordance with:與…一致。

4. the particulars are detailed in...:細節詳載於…。

5. initial order:初次訂貨。

6. large order:大量訂貨。注意: a large order 又有「棘手之事」之意。

No.68　訂貨單

ORDER SHEET

Dear Sirs,　　　　　　　　　　23 rd September,19···

　We have the pleasure of placing the following order with you.

Quantity	Description	Unit Price	Amount
100 pcs. (2 cases)	Cotton White Shirting Sample No. 235 Exactly as shown in the sample.	CIF Osaka in US Dollars ⓐ 12.00/pc.	US$1,200.00

Packing:　　10 pcs.in Hessian bale.

　　　　　　5 balesin one wooden case.

Marks:　　　　　　　　with number 1 and 2 under port

　OSAKA　　　mark stating the country of origin.
　C/# 1-2

Insurance:　ICC (A) for full invoice amount plus 10%.

Shipment:　During October,19—.

Terms:　　　Draft at 30 d/s under an irrevocable L/C.

Remarks:　　Certificate of quality inspection & shipment

　　　　　　sample to be sent by airmail prior to ship-

　　　　　　ment.

　　　　　　　　　　　Osaka Trading Co., Ltd.

【註】

1. exactly as shown in the sample:與樣品完全一致。這種品質條件，猶如
　"exactly same as sample"或"exactly identical to sample"的條件，對出口
　商很不利，尤其紡織品的交易不宜以這種嚴格的條件做爲品質條件，否則很容易

　　引起糾紛。

2. ⓐ 12.00/pc:意指每疋 12 元。

3. shipment samples:裝船樣品。

No.69　訂購罐頭洋菇

Dear Sirs,

Canned Mushroom

We are pleased to receive your price list of January 1, 19…on the subject goods and hereby place our order with you as per enclosed Purchase Order No.123 in duplicate.

To cover this order, we will request our bankers to open an L/C in your favor immediately upon receipt of your confirmation of this order. Please arrange to ship the first two items by the first available vessel sailing to Melbourne direct after receipt of the L/C.

We shall appreciate your forwarding us the sample can each of these qualities by airfreight immediately for our evaluation of the quality and meantime, confirm your acceptance of this order by signing and returning the duplicate copy of our Purchase Order.

Yours very truly,

【註】

1. hereby:藉此；特此；茲。

2. to cover this order:為支付本訂貨價款。

3. sample can:樣品罐。

No.70　購貨單

PURCHASE ORDER

　　　　　　　　　No.123　　　January 30,19···

United Export Corp., Taipei

Dear Sirs,

　　We are pleased to place an order with you for the following goods on the terms/conditions set forth below:

Commodity

　　&

Description:Canned Mushroom of the following four

　　　　　　qualities:

　　　　　　A:6×68oz. Button　　C:24×8oz. Slice

　　　　　　B:24×16oz. Whole　　D:24×4oz.

　　　　　　Pieces & Stem

Quantity:　A:500 cartons;　　C:400 cartons

　　　　　　B:ditto;　　　　　　D:ditto

　　　　　　FOB Taiwan per carton in US Dollars

Price:　　A:16.20　　　　　　C:8.25

　　　　　　B:14.70　　　　　　D:5.00

Packing:　to be packed in export standard fiber

　　　　　　cartons.

Insurance:　Buyer's care.

Shipment:　For item A&B:ASAP; For items C&D:

　　　　　　within one month after receipe of L/C.

Payment:　At sight draft under Irrevocable L/C,

assignable and divisible.

Remark:　　　Sample cans of each quality to be

airfreighted to us for examination.

Yours truly,

Accepted by:

【註】

1. button:環狀。

2. whole:整片。

3. slice:切片。

4. ASAP:as soon as possible 的縮寫，大多用於 telexese.

No.71　訂購電晶體收音機電報確認書

Dear Sirs,

We are glad to confirm our cable order dispatched
this morning as follows:

LT

TAITRA

ARRANGE SHIPMENT BX 157 600 TRC 418 800

BOTH DURING AUGUST PV 123 BY SEPTEMBER

ELECTROLEADER

Article	Model	Quality	Shipment
Transister Radio	BX-157	600 sets	August
Tape Recorder	TRC-418	800 sets	August

| Portable TV | PV-123 | 1200 sets | September |

In your stock news we found these items available by the date mentioned above. Please arrange shipment accordingly.

We will cable the shipping instructions and the credit immediately upon receiving your acceptance by cable.

We hope that you will promptly accept our order and make your usual careful execution.

Yours faithfully,

【註】

1. cable order dispatched this morning:今晨以電報發出的訂單。

2. stock news:庫存（指 maker 對經銷商、代理商所發出的可供銷售庫存表）。

3. these items available by the date:到今日為止可以購到的項目。

4. arrange shipment:安排船運。

5. shipping instructions:裝船指示。"instructions"解做「指示」時，必須多數。

6. usual careful execution:如往例一樣，注意處理訂單，即指裝船之意。

No.72　確認訂購海灘用橡皮球

Gentlemen:

Enclosed you will find copies of cables, yours Nos. 14 and 15 and ours Nos. 11 and 12.

From the exchange of cables as well as your letters Nos. 25 and 26 and ours Nos. 200 and 212, it is our understanding that we have purchased from you, as a sample order, 100 dozen Rubber Beachball, 10″ and three-colored, at $2.30 per dozen F.O.B. Keelung.

We have today instructed the Chemical Bank & Trust Co., New York City, to cable a letter of credit in the amount of $270.00 in your favor.

We have to express our thanks for your having accepted this small order. However, we hope you realize that we are doing this in order to show our prospects the quality of the merchandise. We shall, therefore, be able to place our orders for larger quantities once this initial shipment has been made and the basis of operating established.

Please book space by the fastest route. We will pay the freight and insure the cargo at this end.

We believe that we can do a very considerable business in this item. We note from your letter that you are willing to accept our proposal and appoint us as exclusive agents for New York, provided that we can place orders with you for fairly large quantities. We would like to go into details regarding the agency and look forward to your further information.

However, the most important thing at this time is to get this sample order on the water. We trust you will do everything possible to expedite shipment.

Cordially yours,

【註】

1. sample order:樣品訂單, 即指做樣品用的小量訂貨, 與此相對的是"bulk order"(大量訂單)。

2. rubber beachball:海水浴場等遊戲用的海灘用橡皮球。

3. once this...has been established:「一旦這初次貨裝出, 且活動(業務)基礎確立」, 意指一旦這初次訂貨裝來, 且確立了銷售基礎的話。once＝when once

4. book space:洽訂艙位。

5. at this end＝at our end

6. get...on the water:把…運出。

7. to expedite＝to hasten; to quicken:加速。

第二節　有關訂貨的有用例句

一、開頭句

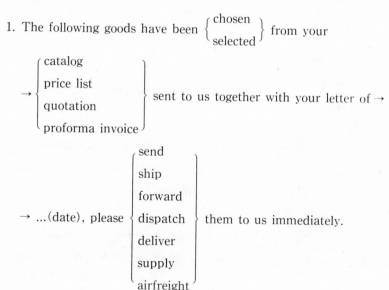

1. The following goods have been $\begin{Bmatrix} \text{chosen} \\ \text{selected} \end{Bmatrix}$ from your

→ $\begin{Bmatrix} \text{catalog} \\ \text{price list} \\ \text{quotation} \\ \text{proforma invoice} \end{Bmatrix}$ sent to us together with your letter of →

→ ...(date), please $\begin{Bmatrix} \text{send} \\ \text{ship} \\ \text{forward} \\ \text{dispatch} \\ \text{deliver} \\ \text{supply} \\ \text{airfreight} \end{Bmatrix}$ them to us immediately.

2. $\left.\begin{array}{l}\text{Enclosed is}\\\text{We enclose}\\\text{We are pleased to enclose}\\\text{We have the pleasure of enclosing}\end{array}\right\}$ our $\left\{\begin{array}{l}\text{order}\\\text{purchase order}\\\text{purchase note}\\\text{indent}\end{array}\right\} \rightarrow$

No...with detailed instructions for your $\left\{\begin{array}{l}\text{immediate}\\\text{prompt}\end{array}\right\}$ attention.

3. Please $\left\{\begin{array}{l}\text{deliver}\\\text{ship}\\\text{supply}\end{array}\right\}$ the $\left\{\begin{array}{l}\text{following}\\\text{below-mentioned}\end{array}\right\}$ $\left\{\begin{array}{l}\text{goods}\\\text{articles}\\\text{items}\end{array}\right\}$

$\rightarrow$ listed in your $\left\{\begin{array}{l}\text{catalog.}\\\text{price list.}\\\text{quotation.}\end{array}\right\}$

4. Please $\left\{\begin{array}{l}\text{book}\\\text{fill}\\\text{execute}\end{array}\right\}$ the following order according to your

$\rightarrow \left\{\begin{array}{l}\text{catalog}\\\text{quotation}\\\text{price list}\\\text{estimate}\end{array}\right\}$ of...(date) and samples submitted.

* book-order:接受訂單; fill...order:配發訂貨; execute...order:執行訂貨。

5. Thank you very much for your letter of May 5 with catalog and price list. We have chosen three qualities and take pleasure in enclosing our order No.1234.

6. Your samples of umbrellas received favorable reaction from our customers, and we are pleased to enclose our purchase order for 1,000 dozen.

7. Confirming our exchange of cables, we are enclosing the following order: Order No. 1234 for 1,000 doz. Nylon umbrellas US$12/doz. CIF

Colon.

8. Please send us by $\left\{\begin{array}{l}\text{post}\\\text{air}\\\text{the first available steamer}\end{array}\right\}\rightarrow$

$\rightarrow$ this order as $\left\{\begin{array}{l}\text{per particulars given below:}\\\text{specified below:}\end{array}\right.$

＊ first available steamer:第一艘開駛船隻; 可利用的第一艘船; 近便的船。

二、條件:（除這裡所列舉者外，讀者尚可參閱第 11 章交易條件的有關部分）

1. Please choose nearest substitute for any article out of stock.

＊ nearest substitute:最近似的代替品。

2. This order must be $\left\{\begin{array}{l}\text{filled}\\\text{executed}\\\text{delivered}\\\text{shipped}\end{array}\right\}\left\{\begin{array}{l}\text{immediately}\\\text{without delay}\\\text{within...days}\end{array}\right\}$ after $\rightarrow$

$\rightarrow$ receipt of L/C, otherwise it will be cancelled.

3. The quality of the order must be the same as that of our sample.

4. The material should be of the finest quality and exactly the same as the sample submitted last week.

5. If you do not have them in stock please send us substitutes of the nearest quality.

6. Prompt shipment is $\left\{\begin{array}{l}\text{very important.}\\\text{essential.}\end{array}\right.$

7. If the goods are not delivered before June 1, we will have

to cancel the order.

8. When the goods are ready for shipment please let us know. We will then send you shipping instructions and other relative information.

9. Upon receipt of your reply we will open an L/C by cable.

10. For payment we have arranged with the XX Bank for a confirmed irrevocable letter of credit in your favour for the amount of US$...

11. The L/C for this order has been opened through Bank of Taiwan and we hope you will receive it before long.

三、結尾句

1. Please confirm $\begin{cases} \text{receipt} \\ \text{acknowledgement} \\ \text{acceptance} \end{cases}$ of this order.

2. Your $\begin{cases} \text{prompt} \\ \text{early} \\ \text{careful} \end{cases}$ attention to this order and

→ $\begin{cases} \text{confirmation} \\ \text{acknowledgement} \end{cases}$ will be appreciated.

3. If this order is satisfactorily $\begin{cases} \text{filled} \\ \text{executed} \end{cases}$, we $\begin{cases} \text{hope to} \\ \text{will} \\ \text{may be able to} \end{cases}$ →

→ place $\begin{cases} \text{further} \\ \text{large} \\ \text{substantial} \\ \text{repeat} \end{cases}$ orders with you.

習　題

一、將下列中文譯成英文。

1. 請依貴公司三月三日價目表接受下列訂貨。

2. 請以航空將本訂單所載下列貨品寄下。

3. 茲檢附本公司第101號訂單連同詳細說明(with detailed instructions)以便貴公司立即處理(prompt attention)。

4. 因為這是試驗性訂單(trial order)，本公司毋庸強調(need hardly impress on you)貴公司謹慎選擇貨物(careful selection of the goods)的重要性。

5. 如承儘速配發下列訂貨(execute...order)，則不勝欣愉。

二、請自(1)至(5)各項中選擇適當的項目，以其編號填入下列各句的括弧中。

項目：(1) Shipping Mark (2) Payment (3) Insurance (4) Shipment (5) Price

(　) By a prime banker's irrevocable L/C payable at sight

(　) During October, 1978

(　) US$ 15 per dozen CIF New York

(　) To cover ICC(B) plus war for 110% of CIF value

(　) 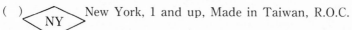 New York, 1 and up, Made in Taiwan, R.O.C.

第十八章　答覆訂貨

(Reply to Order)

第一節　答覆訂貨信的寫法

出口商收到訂單後，應儘速答覆，以免進口商對於進貨或轉售計劃發生困擾。如因存貨關係，無法接受訂貨，也應立即函（電）覆致歉，或推介代替品。對於訂購函電的答覆，可大分為三種：其一為可照訂單全部接受訂購。其二為部分訂貨不能接受或暫時不能供貨。其三為謝絕訂貨。

一、可照訂單全部接受訂購時

答覆時應將交易條件內容詳述。其內容大致如下：

1. 對訂購表示謝意。

2. 表示接受訂購

3. 引述訂單號碼、日期

4. 重述訂單內容（包括商品名稱、品質規格、數量、價格、包裝、刷嘜、保險、裝運、付款等條件）

5. 其他特別事項(Special Instructions)

6. 結尾：結尾句須把握下列幾個原則：

　　(1)希望繼續惠顧

　　(2)保證履行諾言

　　(3)趁機介紹新產品

原則上，進口商以電訊訂貨(telecommunication order)，則出口商應以電訊答覆，隨後以 Seles Note, Sales Confirmation 或書信確認。

訂貨是交易的正式開始，俗語說得好，「好的開始，等於成功了一半」(Well begun is half done.)，能否使進口商滿意，由而繼續惠顧，端視出口商如何處理訂單而定。所以出口商應站在進口商的立場，處處為進口商設想，為進口商做週到的服務。

No.73　接受訂購罐頭洋菇

Dear Sirs,

Subj. Canned Mushroom

We thank you for your letter of January 30, 1990 along with your Purchase Order No.123 for four different qualities of the subject item.

In acceptance of your order, we are returning herewith the duplicate copy of your Purchase Order duly signed.

The first two items are now ready, and we are now arranging shipping space for the first available vessel sailing to Melbourne. Therefore, please have the L/C opened at your earliest convenience in order to enable us to effect prompt shipment. As regards the other two qualities, we will ship them within one month on receipt of your L/C.

As requested, we have today airfreighted one sample can each of these qualities for your evaluation. We are sure that the quality would meet your approval.

We take this opportunity to thank you for your initial order and look forward to receiving more orders from you in the near future.

                                    Faithfully yours,

【註】

1. arrange shipping space:安排艙位, 也可以"book space"代替。

2. "receive more orders":也可以"favor us continuous orders"代替。

No.74 接受訂購電晶體收音機

Dear Sirs,

　　Please accept our sincere appreciation for your letter of August 5, 1990 placing an order for 10,000 sets of transistor radios.

　　We are returning herewith a copy of your order duly signed by us and assure you that the quality of the goods to be delivered will be conforming to the sample and shipment will be made before November 10.

　　To achieve this end, your instructions have been immediatetly passed on to our makers for execution.

　　We look forward to receiving your L/C before the end of September as stipulated in your order.

                                    Yours faithfully,

【註】

1. to achieve this end:爲達成此目的。

2. pass on to＝pass along to: (將東西) 轉給; (將指示) 轉達或交付。

　　例: He passed the jewel on to his wife.

No.75 對初次訂貨的答謝信

Dear Mr. Ling,

It was just fine of you to send us that nice order. Thanks a lot.

For the confidence you have placed in our products, we are very grateful. In return, we shall leave no stone unturned to justify a continuance of that confidence.

You'll always find our organization happy and ready to give you the fullest measure of assistance. Don't feel you'll be putting us to any trouble, because besides selling quality goods, it is our job to create satisfied customers.

We will not forget you after this, your first order. No sir! We could not have carried on for fifty-two years -successfully weathering every business upheaval-were that our policy. This order will be the beginning of a long and pleasant business relationship. Anyway, that's what we shall try to make it.

So kindly think of us the next time you need our goods, and if we can help you in any way, please write us.

Our one desire is to serve you faithfully.

【註】

1. just fine:眞好。

2. thanks a lot:多謝。

3. the confidence you have placed in our products:你對我們產品的信心。

4. in return:以爲報答。

5. leave no stone unturned:盡力。

6. to justify a continuance of that confidence:使（你們）繼續信任（我們的產品）。

7. give the fullest measure of assistance:予以全力協助。

8. don't…trouble:不要覺得會給我們任何麻煩。

9. create satisfied customers:產生滿意的主顧。

10. we will not...this:我們不會在你們第一次訂貨之後就忘了你們。

11. No sir! 不會的!

12. carried on:經營。

13. weathering every business upheaval:渡過各種商業上的動亂。

14. try to make it:努力做到。

二、部分訂貨不能接受，或暫時無法供貨

　　部分訂貨不能接受的原因很多，如庫存賣完，或該貨已停止生產等等。在此種情形，出口商採下列步驟:

　　1.徵求允許先行裝運可供應的部分，並說明另一部分延後交運的理由以及預計何時可出貨。2.徵求允許以同等品代替不能供應的部分。3.如訂購貨物已停止生產，說明原因，並推介代替品。因此撰寫這類信時，其內容大致如下:

　　1.表示謝意。

　　2.重述訂單內容、訂單號碼、日期。

　　3.說明那些貨品何時可以交運。

　　4.對於不能供應部分表示歉意，並說明理由。

　　5.說明不能供應部分須延至何時才能供貨，或推介代替品。

No.76 部分不能供應，另推介代替品

Dear Sirs,

Thank you for your order of 18th September requesting a Rotary Printing Press Model PM-600, PM-800 and PM-1600. We are pleased to book all except the first one which is now under mechanical redesigning.

As you requested us to ship them by the end of November, we have just cabled you two alternatives concerning Model PM-600 as shown in the enclosed cable:

1. As an excellent substitute for PM-600 we recommend 630. This is our latest model and much superior in printing speed, 90 revolutions per minute. Considering your inconvenience, we will make a special price discount to stg. £-1500. If this is acceptable please cable us and we will dispatch all items in November.

2. We may ship PM-800 and PM-1600 during November and PM-600 during January next year upon finishing the redesigning, providing you prefer to import PM-600.

We are very sorry for the inconvenience. However, please note our wish to offer an altenative, especially our special discount price.

We hope this will meet your immediate acceptance so that we can execute the order in a most satisfactory manner.

Yours faithfully,

【註】

1. mechanical redesigning:機械重設計。

2. alternative:代替辦法；代替品(alternative product)。

No.77　無庫存不能立即供應

Dear Sirs,

We thank you for your order for 100 dozen yards of curtain fabric to patterns submitted, but as our stocks have been cleared and the cloth has to be manufactured, we are not in a position to effect delivery in less than a fortnight. It must also be understood that this period will be exceeded if the makers have none of the patterns selected on hand; there is, however, little danger of this.

If your demand is so pressing as to necessitate immediate delivery, we are prepared to supply a material in some respects inferior but satisfactory for most ordinary purposes, as the samples enclosed will show. We suggest, however, that it would pay you to suspend the execution of your order unless such a course is for special reasons inadvisable.

Perhaps you would be good enough to confirm your order subject to these conditions.

Yours faithfully,

【註】

1. curtain fabric:窗簾布。

2. patterns submitted:附來的式樣。curtain fabric to patterns submitted:照所附樣式的窗簾布。

3. little danger of this:無此疑慮，無此危險。

4. it would pay you to...:這樣對你比較划得來。

5. suspend the execution of:擱置配貨。

6. unless such a course is for special reasons inadvisable:除非因特別理由這樣的做法不適當。course:做法，方法，行為。

三、謝絕訂貨(Declining Orders)

　　就出口商而言，當然希望能夠滿足所有來自進口商的訂貨。但有時也會遇到不能不婉謝訂貨的情形。例如無存貨、不再產製、訂貨已過多或已逾接受訂購期限等等，均使出口商不得不拒絕訂貨。在撰寫這種覆函時，必須先表示謝意，然後，說明不能接受訂貨的理由，以獲得進口商的諒解。假如別處能供貨，可告訴進口商向何處訂購，如此不但可獲得進口商的感激，同時還可以給進口商留下好印象，以備將來繼續惠顧。

　　此外，如價錢已變動，比原來發出的 Price List, Proforma Invoice 或 Catalog 所載價格要高的時候，不宜逕予拒絕，而應說明漲價的理由，並希其按調整後的價格訂購。

No.78　已不再生產，推介新產品

Dear Sirs,

　　Thank you for your letter of May 12 enclosing your order for 1,000 sets of "OK"Brand transistor radios.

　　We are sorry we can no longer supply this model, because it is rather old model and the demand for it has fallen to such an extent that we have ceased to manufacture them.

Instead, we should like to recommend our new "YES"Brand transistor radios, which, we think, is more attractive in design and practical for use. The large number of repeat orders we regularly receive from our overseas customers is clear evidence of the wide spread popularity of this brand. At the low price of only US$2. 30 each with a leather case and a strap CIF Singapore is much cheaper than "OK" Brand. The sample with description is being forward to you by air parcel today.

We are looking forward to receiving your comments or order soon.

Yours faithfully,

【註】

1. practical for use:實用。

2. clear evidence:明白的證據。

3. leather case:皮套。

No.79　已停止生產, 推薦新產品

Dear Sirs:　　　　　Subj. Zepher Silk

We thank you for your order of November 14 for the subject silk but regret that we cannot supply this material, as its manufacture has no longer been continued. It was only when the entire absence of inquiries led us to believe that it had been dropped quite out favor of that we decided to take this step.

We are pleased to hear that it has met with your approval and venture to think that you will find our new "GOSSAMER" brand even more satisfactory.

The new cloth is considerably fine, with a dull lustre that is most attractive; the popularity it enjoys among the leading manufacturers is proof enough of its value.

You will find enclosed a price list and a full range of patterns. It seems to us that a trial order would make you share our confidence.

We look forward to receiving your favorable response.

Yours faithfully,

【註】

1. had been dropped quite out of favor＝become unpopular:已不受歡迎。

2. venture to think:膽敢以為...。

   cf. venture to suggest that...:冒昧地提議…。

   venture a guess:大膽地做猜測。

3. lustre＝luster:光澤。

4. dull:暗晦的。

5. a full range of patterns:各種式樣。

6. share our confidence:相信我們（所說的）。

第二節　有關答覆訂貨的有用例句

一、處理訂貨用語

1. To $\begin{Bmatrix} \text{execute} \\ \text{fill} \\ \text{fulfil} \end{Bmatrix}$ an order　　　　執行（處理）訂貨

2. To carry out an order　　　　執行（處理）訂貨

3. To begin the execution of an order　　開始處理訂貨

4. To hurry on the execution of an order　趕快處理訂貨

5. To $\begin{Bmatrix} \text{expedite} \\ \text{hasten} \end{Bmatrix}$ the execution of an

　order　　　　　　　　　　加緊處理訂貨

6. To work overtime with an order　　加班處理訂貨

7. To get an order ready for shipment　備妥所訂貨品待運

8. To ship an order　　　　　　運出所訂貨品

9. To $\begin{Bmatrix} \text{stop} \\ \text{suspend} \\ \text{hold up} \end{Bmatrix}$ an order　　停止訂貨　擱置訂貨　擱置訂貨

10. To postpone an order　　　　延期訂貨

11. To $\begin{Bmatrix} \text{cancel} \\ \text{annul} \\ \text{revoke} \\ \text{countermand} \end{Bmatrix}$ an order　　取消訂貨

12. To withdraw an order　　　　撤回訂貨

13. To hold an order in abeyance 暫時擱置訂單 "hold" 可以 "leave" 代替

二、答覆訂貨的開頭句

1. $\begin{Bmatrix} \text{Thank you} \\ \text{Many thanks} \end{Bmatrix}$ for your $\begin{Bmatrix} \text{order} \\ \text{purchase order} \\ \text{indent} \end{Bmatrix}$ No...of(date)....→

　→ which we confirm herewith as follows:

2. It is a pleasure to receive a $\begin{Bmatrix} \text{trial} \\ \text{sizable} \\ \text{generous} \end{Bmatrix}$ order for...

3. Your order No...of...(date)...,for which we thank you, has been boo-ked as instructed and →

→ we $\begin{Bmatrix} \text{enclose} \\ \text{are enclosing} \\ \text{are pleased to enclose} \end{Bmatrix}$ our confirmation in duplicate.→

→ Please sign and return us the original.

4. $\begin{Bmatrix} \text{We acknowledge} \\ \text{Thank you for} \end{Bmatrix}$ your letter of...(date)...enclosing your →

→ order No...for...(goods), which we have passed on to our →

shipping department for earliest possible shipment.

三、答覆訂貨的文中用語

1. We have instructed our $\begin{Bmatrix} \text{manufacturer} \\ \text{maker} \end{Bmatrix}$ to start manufacture →

→ at once, and you may $\begin{Bmatrix} \text{be} \\ \text{rest} \end{Bmatrix}$ assured that the goods will be →

→ $\begin{Bmatrix} \text{delivered} \\ \text{shipped} \end{Bmatrix}$ $\begin{Bmatrix} \text{within} \\ \text{before} \end{Bmatrix}$ the specified shipment date.

2. You may $\begin{Bmatrix} \text{be sure} \\ \text{be assured} \\ \text{rest assured} \end{Bmatrix}$ that your instructions will be →

→ carefully $\begin{Bmatrix} \text{followed.} \\ \text{adhered to.} \\ \text{observed.} \end{Bmatrix}$

（我們會小心地 $\begin{Bmatrix} 遵照 \\ 忠於 \\ 遵守 \end{Bmatrix}$ 你的指示，請放心。）

3. This order will $\begin{Bmatrix} \text{have} \\ \text{receive} \end{Bmatrix}$ our best attention and we assure you →

→ that we shall do our $\begin{Bmatrix} \text{best} \\ \text{utmost} \end{Bmatrix}$ to $\begin{Bmatrix} \text{fill} \\ \text{execute} \end{Bmatrix}$ it within the →

→ $\begin{Bmatrix} \text{specified} \\ \text{prescribed} \end{Bmatrix}$ $\begin{Bmatrix} \text{time.} \\ \text{date.} \end{Bmatrix}$

4. We will ship the goods as early as possible, and will advise you immediately the shipment is made. Enclosed is our Sales Confirmation No…in duplicate, the original of which please sign and return.

＊這裡的"immediately"是 conjunction,其意為"as soon as"。

四、無法接受訂貨

1. We $\begin{Bmatrix} \text{are sorry} \\ \text{regret} \end{Bmatrix}$ that we are quite unable to $\begin{Bmatrix} \text{fill} \\ \text{execute} \\ \text{carry out} \end{Bmatrix}$ →

→ your order at the present moment, due to →

→ $\begin{Bmatrix} \text{heavy orders.} \\ \text{rush of orders.} \\ \text{heavy demand.} \end{Bmatrix}$

2. We have too many orders on hand and could not effect →

→ shipment before $\begin{Bmatrix} \text{September.} \\ \text{the date specified.} \end{Bmatrix}$

3. We $\begin{Bmatrix} \text{regret} \\ \text{are sorry} \end{Bmatrix}$ that item No…is $\begin{Bmatrix} \text{sold out(售完)} \\ \text{out of stock(缺貨)} \\ \text{out of production} \\ \text{out of make(停製)} \end{Bmatrix}$ →

→ at present.

4. Although we have tried every source of supply we know, we have been unable to get the size you asked for.

5. Although we are unable to fill your order now, we shall be pleased to contact you again, as soon as circumstances allow.

(…情況一允許，我們將再與你聯絡。)

6. We have been compelled to raise our prices by 10%→

→ $\begin{cases} \text{under the pressure} \\ \text{of} \\ \text{owing to} \end{cases}$ $\begin{cases} \text{increased labor costs.} \\ \text{the rise in raw material prices.} \\ \text{havier import duties on raw materials.} \end{cases}$

五、建議訂購替代品

1. The exact $\begin{cases} \text{design} \\ \text{size} \\ \text{model} \end{cases}$ you want is $\begin{cases} \text{out of stock} \\ \text{sold out} \\ \text{no longer available} \\ \text{discontinued to manufacture} \end{cases}$ →

→ $\begin{cases} \text{now} \\ \text{at present} \\ \text{at the moment} \end{cases}$; however, we are sending you by separate →

→ mail a sample similar to it, which we think will be a very good substitute for it.

2. Please $\begin{cases} \text{select} \\ \text{choose} \end{cases}$ $\begin{cases} \text{a suitable substitute} \\ \text{another quality} \end{cases}$ from the enclosed →

→ $\begin{cases} \text{patterns.} \\ \text{samples.} \\ \text{catalogs.} \end{cases}$

3. The quality of the substitute is even better than the one you want, while the price is same.

4. Quality of item 1 is equally $\begin{cases} \text{attractive.} \\ \text{hardwearing.} \text{（耐穿的）} \\ \text{serviceable.} \text{（可用的）} \\ \text{water-repellent.} \text{（不透水的）} \end{cases}$

六、條件

1. Upon receipt of your $\begin{cases} \text{L/C} \\ \text{check} \\ \text{remittance} \end{cases}$, we will $\begin{cases} \text{ship} \\ \text{deliver} \\ \text{despatch} \end{cases} \rightarrow$

   → the goods immediately.

2. Your order No...will be ready for shipment by end of this month. Please let us have your instructions regarding shipping marks together with an L/C to cover this order.

3. In order ot process this order, however, we have to ask you to open L/C immediately.

七、結尾句

1. We hope that this order will be just the $\begin{cases} \text{beginning} \\ \text{first} \end{cases}$ of →

   → many and that we may have many years of pleasant business →

   → relations together.

   (希望本批訂單只是一個開頭，並企望有長年、愉快的業務關係)

2. You may $\begin{cases} \text{rest assured} \\ \text{be sure} \end{cases}$ that we will do our best to →

   → $\begin{cases} \text{fill this order} \\ \text{satisfy you} \end{cases}$ and we hope that this will be the →

   → $\begin{cases} \text{beginning} \\ \text{first} \end{cases}$ of a long and $\begin{cases} \text{happy} \\ \text{pleasant} \end{cases} \begin{cases} \text{cooperation.} \\ \text{association.} \end{cases}$

3. We will be pleased to receive your $\begin{cases} \text{repeat} \\ \text{continued} \\ \text{further} \end{cases}$ orders,which →

→ will always have our $\begin{cases} \text{best} \\ \text{prompt} \\ \text{careful} \end{cases}$ attention.

4. $\begin{cases} \text{Please be assured} \\ \text{You may be assured} \end{cases}$ that we shall $\begin{cases} \text{do our best} \\ \text{spare no effort} \\ \text{make every effort} \\ \text{do everytihing possible} \\ \text{do what we can} \end{cases}$ →

→ to satisfy your $\begin{cases} \text{wishes.} \\ \text{requirements.} \\ \text{customers.} \end{cases}$

* spare no effort:不遺餘力

5. We sincerely regret that we are unable to serve you →

→ $\begin{cases} \text{at this time.} \\ \text{in this instance.} \end{cases}$

* in this instance:在此刻；在這種情況下

6. We trust that you will understand that it is not lack of co-operation and good-will but sheer necessity which makes it impossible for us to meet your wishes in this case.

(我們相信你當會了解此次我們無法達到你的願望，並不是由於欠缺合作與善意，而是完全由於不得已的苦衷。)

7. We believe this treatment of your order will be entirely satisfactory, and hope to be favored with further reguirements.

(我們相信這樣處理你的訂單將使你完全滿意，並希望繼續惠顧。)

習　題

一、將下列中文譯成英文

1. 本公司將全力立即處理這批訂貨(have our best and prompt attention)。

2. 我們會謹慎地遵守貴公司的指示，敬請放心(you may be sure)。

3. 目前第 5 號貨色已經售完，實感抱歉。

4. 因爲第 10 號貨品目前缺貨，本公司以很相似的貨品代替(substitute the very similar article)，按同樣的價格計算，尙希貴公司能表同意。

5. 本公司甚至虧本(even at a sacrifice)接受貴公司限定的價格(your limit)，希望貴公司在不久的將來(in the near future)訂更多貨以補償(compensate)本公司。

二、將本書 No.78 英文信譯成中文。

三、試用下列片語造句

1. to achieve this end

2. pass on to

3. in return

4. venture to think

第十九章　買賣契約書

（Sales Contract）

第一節　貿易契約書的簽立方法

任何一筆國際買賣，經過上述有關各章所述詢價、報價、往返還價，最後經接受；或訂貨確認，買賣契約即告成立。

就契約本身而言，報價與接受，或訂貨與確認，其所用的函件、電報，均爲契約成立的證據（Evidence）。至於買賣契約成立後，買方所發出的訂單，或賣方所發出的售貨確認書，甚至買賣雙方另訂立的買賣契約書（Sales contract）只不過是契約的證據而已。無 Order, Sales Confirmation 或 Sales Contract 固不妨礙契約的成立，但有了這些，則顯得更爲週到。在實務上，買賣雙方對於交易的意思表示，常以 Cable 或 Telex 爲之，而 Cable 或 Telex 中有關交易條件，通常僅限於交易商品、品質、數量、價格及交貨期等而已。對於付款、包裝、刷嘜、保險、檢驗、索賠、仲裁、不可抗力以及其他關係雙方權利義務的各種條件，往往略而不提。因此爲免日後發生糾紛，或解決糾紛時有所依據，進出口雙方均宜在成立契約後，另以書面互爲確認，或另簽訂契約書，以證實或補充契約內容。

實務上，買賣契約書的簽立方法，可分爲二：

一、以書面確認方式代替買賣契約書

交易成立之後，由當事人的一方，將交易內容製成確認書寄交對方。這種確認書由賣方發出的稱爲"Sales　Confirmation"（售貨確認書）或

"Sales Note"（售貨單）。由買方發出的，稱為"Order"，"Order Sheet"、"Purchase Order"或"Purchase Confirmation"。

二、正式簽立買賣契約書

交易成立後，也可由當事人的一方將交易內容製成正式契約書，然後由雙方共同簽署。這種契約書由出口商草擬時，往往稱為"Sales Contract"或"Export Contract"，如由進口商草擬時往往稱為"Purchase Contract"或"Import Contract"。

較單純或金額較小的交易多採用確認方式簽約，但交易性質較複雜或金額較大的則往往採取契約書方式以昭慎重。

關於確認書，在前幾章已陸續敍及，本章擬舉一買賣契約書為例，說明其內容及有關用語。

第二節　貿易契約書實例

No.80　買賣契約書

CONTRACT

This contract is made this 15th day of July, 19⋯ by ABC Corporation (hereinafter referred to as "SELLERS"), a Chinese corporation having their principal office at 19 Wu Chang St., Sec. 1, Taipei, Taiwan, Republic of China, who agree to sell, and XYZ Corporation (hereinafter referred to as "BUYERS"), a New York corporation having their principal office at 30 Wall St., New York, N.Y., USA, who agree to buy the following goods on the terms and conditions as below:

1. COMMODITY: Ladies double folding umbrellas.

2. QUALITY: 2 section shaft, iron and chrome plated shaft, unichrome plated ribs, siliconed coated waterproof plain nylon cover with same nylon cloth sack.

   size:18 ½″×10 ribs

   as per sample submitted to BUYERS on June 30, 19⋯

3. QUANTITY:10,000(Ten thousand) dozen only.

4. UNIT PRICE: US$14 per dozen CIF New York

Total amount: US$140,000(Say US Dollars one hundred forty thousand only) CIF New York.

5. PACKING: One dozen to a box, 10 boxes to a carton.

6. SHIPPING MARK:

NEW YORK

NO. 1 & up

7. SHIPMENT: To be shipped on or before December 31, 19⋯ subject to acceptable L/C reaches SELLERS before the end of October, 1976, and partial shipments allowed, transhipment allowed.

8. PAYMENT: By a prime banker's irrevocable

sight L/C in SELLERS' favor, for 100% value of goods.

9. INSURANCE: SELLERS shall arrange marine insurance covering ICC(B) plus TPND and war risk for 110% of the invoice value and provide for claim, if any, payable in New York in US currency.

10.INSPECTION: Goods is to be inspected by an independent inspector and whose certificate inspection of quality and quantity is to be final.

11.FLUCTUATIONS OF FREIGHT, INSURANCE PREMIUM, CURRENCY, ECT.:

(1) It is agreed that the prices mentioned herein are all based upon the present IMF parity rate of NT$26 to one US dollar. In case, there is any change in such rate at the time of negotiating drafts, the prices shall be adjusted and settled according to the corresponding change so as not to decrease SELLERS' proceeds in NT Dollars.

(2) The prices mentioned herein are all based upon the current rate of freight and/ or war and marine insurance premium. Any increase in freight and/or insurance premium rate at the time of shipment shall be for BUYERS' risks and account.

(3) SELLERS reserve the right to adjust the prices mentioned herein, if prior to delivery there is any substantial increase in the cost of raw material or component parts.

12.TAXES AND DUTIES, ETC.:

Any duties, taxes or levies imposed upon the goods, or any packages, material or activities involved in the performance of the contract shall be for account of origin, and for account of BUYERS if imposed by the country of destination.

13.CLAIMS:

In the event of any claim arising in respect of any shipment, notice of intention to claim should be given in writing to SELLERS promptly after arrival of the goods at the port of discharge and opportunity must be given to SELLERS for investigation. Failing to give such prior written notification and opportunity of investigation within twenty-one (21) days after the arrival of the carrying vessel at the port of discharge, no claim shall be entertained. In any event, SELLERS shall not be responsible for damages that may result from the use of goods or for consequential or special damages, or for any amount in excess of the invoice value of the defective goods.

14.FORCE MAJEURE:

Non-delivery of all or any part of the merchandise caused by war, blockade, revolution, insurrection, civil commotions, riots, mobilization, strikes, lockouts, act of God, severe weather, plague or other epidemic, destruction of goods by fire of flood, obstruction of loading by storm or typhoon at the port of delivery, or any other cause beyond SELLERS' control before shipment shall operate as a cancellation of the sale to the extent of such non-delivery. However, in case the merchandise has been prepared and ready for shipment before shipment deadline but the shipment could not be effected due to any of the abovementioned causes, BUYERS shall extend the shipping deadline by means of amending relevant L/C or otherwise, upon the request of SELLERS.

15. ARBITRATION:

Any disputes, controversies or differences which may arise between the parties, out of, or in relation to or in connection with this contract may be referred to arbitration. Such arbitration shall take place in Taipei, Taiwan, Republic of China, and shall be held and shall proceed in accordance with the Chinese Government arbitration regulations.

16. PROPER LAW:

The formation, validity, construction and the performance or this contract are governed by the laws of Republic of China.

IN WITNESS WHEREOF, the parties have executed this contract in duplicate by their duly authorized representative as on the date first above written.

BUYERS   SELLERS
XYZ CORPORATION   ABC CORPORATION

```
Manager                                    Manager
```

【註】

1. This contract is made...as below: 這一段文字即為 preamble clause."is made" 為締結之意，也可以"is enter into"代替。"this 15 th day of July, 19…"指締約日期，注意前面並無"on"一詞，假如要用"on"則應改為"on　July 15,19…"。"by　ABC　Corporation...,　and　XYZ　Corporation"也 可 以 "between ABC Corporation..., and XYZ Corporation"代替。

   hereinafter:在下文中。cf. hereinafter called "Buyer"「以下稱買方」。principal office:主要辦公室，即指總公司。

2. 本契約條款 1 至 9 即為本契約本文(body)的基本條款(basic terms and conditions)，其內容與確認書的內容並無兩樣。

3. inspection:檢驗。

   independent inspector:獨立公證行。

4. IMF Parity＝International Monetary Fund parity:國際貨幣基金平價。

5. current rate:目前滙率。

6. for buyers' risks and account:風險及費用由買方負擔。

7. reserve the right:保留（調整的）權利。

8. component parts:組件、配件。

9. levies(pl.), levy(sing.):課徵、賦課。

10. impose upon:課徵。

    cf. impose upon 又解作「占…的便宜」;「欺騙」。

    Don't let the children impose on you.別讓小孩占你的便宜（別縱容小孩）。

11. in respect of:關於。

12. given in writing:以書面通知。

13. at the port of discharge:在卸貨港。

14. no claim shall be entertained:索賠不予受理。

15. consequential loss:間接損失。 "consequential damages":為間接損害。

16: defective goods:瑕疵貨物。

17. force majeure:不可抗力。

契約書

本契約由 ABC 公司——總公司設於中華民國臺灣省臺北市武昌街一段十九號 (以下簡稱賣方) 與 XYZ 公司——總公司設於美國紐約州紐約市華爾街三十號 (以下簡稱買方) 於一九…年七月十五日訂定, 雙方同意按下述條件買賣下面貨物:

1. 貨物: 女用雙折洋傘。

2. 品質: 雙節鐵質鍍鉻傘柄, 單面鍍鉻傘骨架, 塗矽銅防水素色尼龍傘布, 同質尼龍布護套。

　　尺寸: 18 ½ 英寸, 10 支骨架。

　　依一九…年六月三十日提供予買方的樣品爲準。

3. 數量: 一萬打。

4. 單價及總金額: 打 US$14 CIF 紐約, 總金額 US$140,000 CIF 紐約。

5. 包裝: 一打裝一紙盒, 十紙盒裝一紙箱。

6. 裝船嘜頭:

NEW YORK

NO. 1 & Up

7. 裝運: 一九…年十二月卅一日前裝運, 但以可接受的信用狀於一九…年十月底前開到賣方爲條件, 容許分批裝運及轉運。

8. 付款: 憑一流銀行的不可撤銷見票即付 (或卽期) 的卽期信用狀付款, 信用狀以賣方爲受益人, 並照貨物金額百分之百開發。

9. 保險: 賣方應洽保水險, 投保 B 款險並加保遺失竊盜險及兵險, 保險金額按發票金額的百分之一一○投保, 並須規定如有索賠應在紐約以美金支付。

10. 檢驗: 貨物須經一家獨立公證行檢驗, 其出具品質及數量檢驗證明書應爲最後認定標準。

11.運費、保險費、幣值等的變動：

(1)茲同意本契約內所列價格全是以目前國際貨幣基金平價滙率臺幣廿六元兌換美金一元為準。倘若這項滙率在押滙時有任何變動，則價格應根據這項變動比照調整及清償，俾賣方的臺幣收入不因而減少。

(2)契約中所列價格全是以目前運費率及（或）兵險和水險保險費率為準。裝運時運費率及（或）保險費率如有增加，應歸由買方負擔。

(3)交貨前如原料及組成配件的成本增加甚鉅，賣方保留調整契約中所列價格的權利。

12.稅捐等：

對於貨物或包件、原料，或履行契約有關活動所課征的稅捐或規費，如由產地國課征，歸由賣方負擔；如由目的國課征，則歸由買方負擔。

13.索賠：

對所裝貨物如有索賠情事發生，則請求索賠的通知必須於貨物抵達卸貨港後卽刻以書面提示賣方，並且必須給賣方有調查的機會。倘若運送船隻到達卸貨港後廿一天內沒有提示這項預先的書面通知以及提供調查機會，則索賠應不予受理。在任何情況下，賣方對於使用貨物引起的損害，或對於間接或特別的損害，或對於超出瑕疵貨物發票金額的款項均不負責。

14.不可抗力：

因戰爭、封鎖、革命、暴動、民變、民眾騷擾、動員、罷工、工廠封鎖、天災、惡劣氣候、疫病或其他傳染病、貨物因火災或水災而受毀壞，在交貨港因暴風雨或颱風而阻礙裝船，或在裝船前任何其他賣方所無法控制的事故發生，而致貨物的全部或一部分未能交貨，這未交貨部分的契約應予取消。然而，在裝運期限截止前，如貨物業經備妥待運，但因前述事故之一發生而致未能裝運，則買方於接到賣方請求時，應以修改信用狀方式或其他方式延長裝運期限。

15.仲裁：

有關本契約買賣雙方間所引起的任何糾紛、爭議或歧見，可付諸仲裁。這項仲裁應於中華民國臺灣省臺北舉行，並應遵照中華政府仲裁法規處理及進行。

16.適用法：

本契約的成立、效力、解釋，以及履行均受中華民國法律管轄。

本契約書兩份業經雙方法定代理人訂定，於前文日期簽署。

買方	賣方
XYZ 公司	ABC 公司
經理	經理

習　　題

一、試述買賣契約的簽立方法。

二、將下列中文譯成英文。

1.本契約自中華民國政府批准之日起生效，契約期間爲一年。

2.由於戰爭，或宣戰，或爆發戰爭，或其他不正規情形而課征加收的任何運費由買方償付。

3.本契約的成立、效力、解釋以及履行均受中華民國法律管轄。

三、試用下列片語造句。

1. impose upon

2. it is agreed

3. for account of

4. in respect of

5. in the event of

第二十章　信用狀交易

(Letter of Credit Transaction)

第一節　信用狀的意義及其關係人

一、信用狀的意義

信用狀是銀行循進口商的要求及指示，向出口商發出的文書，銀行在此文書中，與出口商約定：只要出口商提出合乎該文書中所規定的單據或滙票，即將妥予兌付。信用狀，英文稱為"Letter of Credit"，或簡稱為"Credit"或"L/C"。

二、信用狀交易的關係人(parties concerned under L/C transaction)

信用狀交易的關係人，至少有三方，即開狀申請人，開狀銀行及受益人。除此以外，往往尚有其他關係人介入。

1. Applicant for Credit(開狀申請人)，即向銀行申請開發 L/C 的人，通常即為 Buyer 或 Importer.開狀申請人在 L/C 中尚有下列幾種稱呼：accountee(被記帳人)，grantee(授與人)，accreditor(授信人)，account party(被記帳的一方)，consignee(收貨人)，opener (開狀人)。

2. Opening Bank (開狀銀行) 又稱 Issuing Bank, Establishing Bank, Issuer, Grantor.

3. Beneficiary (受益人) 通常即為賣方或出口商，又稱 Shipper (裝貨人)，Accreditee (受信人)，Addresee (擡頭人)，Accredited

Party（受信的的一方），Drawer, User.

4. Advising Bank（轉知銀行）即轉知 L/C 的銀行，又稱為 Notifying Bank, Transmitting Bank.

5. Negotiating Bank（押滙銀行）即自出口商購入信用狀項下滙票或單據的銀行，又稱 Discount Bank.

6. Confirming Bank（保兌銀行）即應開狀銀行的要求，就所開 L/C 承擔兌付的銀行，保兌銀行一經保兌 L/C，即須承擔與開狀銀行相同的責任。

7. Paying Bank（付款銀行）又稱為 Drawee Bank，即信用狀中規定滙票的被發票人(Drawee)，付款銀行可能為開狀銀行本身，也可能為其他銀行。

8. Reimbursing Bank（歸償銀行）又稱為 Clearing Bank,即應開狀銀行的囑託，對押滙銀行償還其押滙票款的銀行。

第二節　信用狀的種類

一、依可否片面撤銷分：

1. Revocable L/C(可撤銷信用狀)：信用狀開出後，在受益人未押滙前，可隨時片面地予以修改或撤銷，而不需有關當事人同意的信用狀。

2. Irrevocable L/C（不可撤銷信用狀）：受益人收到信用狀後，非經受益人、開狀銀行及保兌銀行（如經保兌）同意，不得隨意修改或撤銷的信用狀。

二、依有無保兌分：

1. Confirmed L/C（保兌信用狀）：經開狀銀行以外的另一銀行保兌的信用狀。

2. Unconfirmed L/C（未保兌信用狀）：未經開狀銀行以外的銀行保兌的信用狀。

三、依滙票期限分：

1. Sight　L/C（即期信用狀）：規定受益人須憑即期滙票(Sight Draft)取款的信用狀。

2. Usance　L/C（遠期信用狀）：規定受益人須憑遠期滙票(Usance Draft)取款的信用狀。

四、依可否押滙（Negotiation）分：

1. Negotiation L/C（可押滙信用狀）：允許受益人向付款銀行以外的銀行請求押滙的信用狀，又可分爲：

   ⑴ General L/C（一般信用狀）：不限定押滙銀行。

   ⑵ Restricted L/C 或 Special L/C（限押信用狀）：限定押滙銀行的信用狀。

2. Straight L/C（直接信用狀）:規定受益人只能向 L/C 所指定銀行提示滙票或單證請求付款的信用狀。

五、依是否須附跟單分：

1. Clean L/C（光禿信用狀）：押滙或請求付款時，不需附上單證的信用狀。

2. Documentary L/C（跟單信用狀）：押滙或請求付款時，須附上一定單證的信用狀。

六、依可否轉讓分：

1. Transferable L/C（可轉讓信用狀）：受益人可將 L/C 轉給他人使用的信用狀。

2. Non-transferable L/C（不可轉讓信用狀）：不可將 L/C 轉給他人使用的信用狀。

七、依用途分：

　　1. Traveler's L/C（旅行信用狀）：專供作旅行用的信用狀。

　　2. Commercial L/C（商業信用狀）：專供貿易用的信用狀。

八、依主從分：

　　1. Master L/C（主信用狀）：由進口商開給出口商的信用狀。

　　2. Back-to-Back L/C（背對背信用狀）：又稱 Local L/C, 出口商憑
　　　 Master L/C 另開給工廠或供應商的信用狀。

九、Stand by L/C（保證信用狀）：專供保證用的信用狀。

十、Red Clause L/C（紅條款信用狀）：受益人在未提示 L/C 所規定單證
　　之前，即可預支款項的信用狀。

第三節　有關信用狀交易的信函

No.81　催促進口商速開信用狀

Gentlemen,

　　We wrote you a letter dated June 25, 19⋯ confirming the receipt of your order for 1,000,000 sq. ft. of Lauan Plywood. In that letter, we enclosed a copy of your Purchase Order duly signed by us.

　　The Purchase Order stipulates shipment to be effected during August and L/C should reach us by the end of July. However, as of this date, we don't appear to have received your L/C. In order to book the shipping space at an earlier date, you are requested to have the L/C opened immediately. For this matter, we have despatched a telex to you today as follows:

"ELCEE FOR LAUAN PLYWOOD UNRECEIVED
YET PLS OPEN IMMDLY"

We appreciate your immediate attention to this matter.

Yours very truly,

【註】

1. as of this date:到今天爲止。

2. We don't appear to have received: 我們似乎還沒有收到; 是一種客氣的措詞。比"We don't have received"要客氣。

3. ELCEE: 卽 L/C 的譯音。

4. unreceived:卽 not received.

5. PLS: 卽"Please"的電報交換文體(telexese)。

6. IMMDLY:卽"immediately"的 Telexese。

No.82　貨已備妥, 催促進口商速開信用狀

Dear Sirs,

Your Order For Canned Asparagus

We refer to our letter to you dated June 1 confirming the captioned Order and trust that you would have already received it. In that letter, we have assured you that we can effect shipment of the total quantity of 900 cartons under the subject order during August provided your L/C reaches us 30 days prior to shipment.

Now the goods are ready for shipment and your L/C has not yet been received. In this regard, we despatched a telex to you on August 4, reading:

> RYL MAY 25 OUR SALES NO. 90/345 CANNED
> ASPARAGUS READY SHIPMENT PLS OPEN LC
> BY CABLE IMMDLY

In order to enable us to ship the goods in time, we shall appreciate your complying with our request to have the L/C opened by cable immediately. For your information, market tone indicates prices are expected to advance before long.

Yours sincerely,

【註】

1. refer to: 關於。

2. provided＝if,以…爲前提; 假使。

3. in this regard: 關於此。

4. RYL＝refer to your letter。

5. mrarket tone＝市況; 市場景況。又寫成"the tone of a market"。

6. before long＝soon: 不久。

No.83　進口商通知出口商已開出信用狀

XYZ Motor Company　　　　　　　　March 28, 19…

Pony Pick UP

Gentlemen,

Your letter of March 10 confirming the order we placed with you for the subject Pick Up has been received. We are pleased to inform you that application for import licence has been approved.

On March 25, an L/C was opened in your favor for an amount of US$55,840.00 to cover the CIF value of this order by Central Trust of China through Korean Exchange Bank, Seoul. For your information, the numbers of the import licence and the L/C are 66 DHI/-003690 and 7 DHI/00123/01 respectively. It will be appreciated if you will arrange to ship the total quantity not later than May 30, 19….

Enclosed for your reference is one copy of the relative L/C we have opened for this order.

> Yours sincerely,

【註】

1. application for import licence: 輸入許可證申請書。在我國，官方的稱法是 application for import permit。輸入許可證在貿易界多稱做 IL (即 Import Licence)或 CBC (即 Central Bank of China 的簡稱)，輸出許可證也常稱爲 CBC。

2. Enclosed for your reference is one copy…: 茲附上 (L/C) 抄本一份供參考。

No.84 國外開來的信用狀

DRESDNER BANK

P.O. BOX 123

Hamburg, West Germany

Taiwan Trading Co., Ltd.	August 6,19…
P.O. Box 123	VIA CABLE THROUGH
Taipei, Taiwan	BANK OF TAIWAN
R.O.C.	TAIPEI, TAIWAN, R.O.C.

Dear Sirs,

IRREVOCABLE LETTER OF CREDIT NO.123

We hereby establish our IRREVOCABLE LETTER OF CREDIT in your favor for account of XYZ AG., Hamburg, up to an aggregate amount of DM 28, 950(DEUTSCHE MARKS TWENTY EIGHT THOUSAND NINE HUNDRED AND FIFTY ONLY) available by your draft(s) at sight drawn on us for 100% of the invoice value accompanied by the following documents:

1. Signed Commercial Invoice in triplicate.
2. Packing list in triplicate.
3. Marine Insurance Policy or Certificate in duplicate endorsed in blank for 110% of the invoice value, covering ICC(B) plus TPND and War.
4. Certificate of Origin in triplicate.
5. Full set of Clean On-Board Ocean Bills of Lading made out to order of Dresdner Bank, Hamburg, Notify Accountee marked "Freight Prepaid".

Evidencing shipment of:

NINE HUNDRED CARTONS OF CANNED ASPARAGUS AI GRADE CIFC 2 Hamburg

Shipment from Taiwan to Hamburg not later than August 31, 19⋯.

Partial shipments are not permitted. Transhipment is prohibited. All drafts so drawn must be marked "Drawn under Dresdner Bank, Hamburg L/C No.123

dated August 6, 19···."

The amount of any draft drawn under this credit must concurrently with negotiation, be endorsed on the reverse side hereof.

Negotiating bank is to forward all documents in one cover direct to us by airmail.

This credit expires on September 15, 19··· for negotiation in Taiwan.

This is a confirmation of the credit opened by cable under today's date through Bank of Taiwan, Taipei.

This credit is subject to Uniform Customs and Practice for Documentary Credits (1983 *Revision*)

International Chamber of Commerce　　Yours faithfully,

*Publication No.*400　　　　　　　　For Dresdner Bank

【註】

1. Dresdner Bank:開狀銀行名稱。

2. Taiwan Trading Co., Ltd: 受益人。信用狀形式上是一封信，受信人一般而言是受益人也卽出口商，但須注意有時以通知銀行爲受信人。

3. August 6, 19···: 開狀日期。

4. via cable through Bank of Taiwan: 說明曾以電報經由 Bank of Taiwan 通知信用狀，所以本 L/C 係電報證實書(cable confirmation)。

5. irrevocable letter of credit: 不可撤銷信用狀（信的標題）。

6. we hereby establish...: 說明開發信用狀的事實，"establish"可以"issue"或"open"代替。

7. in your favor 也可以"in favor of yourselves"代替，「以你爲受益人」之意，如 L/C 的受信人爲通知銀行時以"in favor of…"表示，"in favor of"後面即爲受益人名稱。

8. for account of:「記入（進口商）的帳」之意。"for account of"後面必爲 L/C 申請人，通常爲進口商。

9. up to an aggregate amount of...:總金額以…爲度，爲本信用狀可用金額。

10. available by your draft(s)...on us:得簽發以本行爲付款人的卽期滙票。

11. for 100% of the invoice value:滙票金額按發票金額全額開發，"value"解做「金額」(amount)。

12. accompanied by the following documents:附上以下單證，這裡的"documents"係指"shipping documents"（貨運單證，裝運單證）而言。

13. signed commercial invoice in triplicate:簽了字的商業發票三份。

14. packing list in triplicate:裝箱單三份。注意下列各詞:

duplicate（二份）	sextuplicate（六份）
triplicate（三份）	septuplicate（七份）
quadruplicate（四份）	octuplicate（八份）
quintuplicate（五份）	decuplicate（十份）

15. marine insurance policy...:空白背書海上保險單或保險證明書二份，按發票金額的 110%投保水漬險及竊盜遺失險、兵險。

16. certificate of origin:產地證明書。

17. full set of clean on-board...:全套無瑕疵裝船提單，以 Dresdner Bank 爲收貨人，以 L/C 申請人爲到貨通知人(notify party)，註明「運費付訖」。

18. evidencing shipment of...:證明裝運…（指以上各種單須載明裝運…）

19. CIFC2 Hamburg:貿易條件(trade terms)的一種。

20. shipment from...:由（裝貨港）至（卸貨港）。

21. not later than August 31, 19…:最後裝船日期爲 19…年 8 月 31 日。

22. partial shipments are not permitted:不准許分批裝運。

23. transhipment is prohibited: 不准轉運。

24. all drafts so drawn must be marked "drawn under...": 簽發的所有滙票均須註明「憑漢堡 Dresdner Bank,19…年 8 月 31 日開發的 123 號 L/C 簽發」字樣。

25. the amount of any draft drawn...hereof: 憑本信用簽發的任何滙票金額, 於押滙同時, 必須到本信用狀背面記載。

26. negotiating bank is to forward...by airmail: 押滙銀行應將所有單證一次以航郵寄到本行 (開狀銀行)。

27. this credit expires...in Taiwan: 本 L/C 押滙期限將於臺灣 19…年 9 月 15 日屆滿, 卽本 L/C 的有效期限。

28. this is a confirmation...Taipei: 本信用狀係本日以電報經由臺北臺灣銀行開出的信用狀的證實書。

29. this credit is subject to...No.400: 本信用狀適用國際商會第 400 號公告所載 1983 年修訂的信用狀統一慣例規定。

30. 關於信用狀的問題, 讀者如有興趣作進一步的瞭解, 可參閱本書作者的另著:《信用狀與貿易糾紛》。

   至於國際商會制定的「信用狀統一慣例」, 金融人員研究訓練中心曾予以譯註, 從事貿易的人士宜購讀, 以求進出口押滙業務順利進行。

No.85 我國開往國外的信用狀

The First National Bank of Chicago
INTERNATIONAL BANKING DEPARTMENT EXPORT LETTERS OF CREDIT UNIT
Two First National Plaza/Chicago, Illinois 60670
TELEPHONE (312) 732-5845/48

DATE NOV. 13, 19-

DOCUMENTARY LETTER OF CREDIT	ISSUING BANK'S NUMBER	OUR REFERENCE NUMBER
ISSUING BANK	7246 APPLICANT	CE 20674

ISSUING BANK
BUYERS BANK LTD.
PARIS, FRANCE

APPLICANT
FOREIGN BUYERS LTD.
PARIS FRANCE

BENEFICIARY
EXPORTER FOR PROFIT, INC.
111 N. NORTH
CHICAGO, ILLINOIS

AMOUNT
US$4,875.00 FOUR THOUSAND EIGHT
HUNDRED SEVENTY FIVE U.S. DOLLARS-----

EXPIRY DATE
FEBRUARY 9, 19-

WE HAVE RECEIVED A ☐ CABLE ☒ LETTER FROM THE ISSUING BANK INFORMING US THAT THEY HAVE ISSUED THEIR ☒ IRRE-VOCABLE ☐ REVOCABLE CREDIT IN YOUR FAVOR FOR THE ABOVE MENTIONED APPLICANT THIS CREDIT IS AVAILABLE FOR PAYMENT/ACCEPTANCE, UNLESS OTHERWISE STIPULATED, AGAINST YOUR DRAFT(S) DRAWN AT SIGHT ON THE FIRST NATIONAL BANK OF CHICAGO, CHICAGO, ILLINOIS. DRAFT(S) MUST BE MARKED AS DRAWN UNDER THIS CREDIT, ACCOMPANIED BY THE FOLLOWING DOCUMENTS:

1. DRAFT AT SIGHT IN DUPLICATE DRAWN ON THE FIRST NATIONAL BANK OF CHICAGO, CHICAGO, ILLINOIS.
2. COMMERCIAL INVOICE IN TRIPLICATE.
3. WEIGHT LIST IN TRIPLICATE
4. INSURANCE POLICY OR CERTIFICATE IN DUPLICATE COVERING MARINE AND WAR RISKS.
5. FULL SET OF CLEAN ON BOARD OCEAN BILLS OF LADING ISSUED TO ORDER OF SHIPPER, BLANK INDORSED, MARKED NOTIFY FOREIGN BUYERS LTD., PARIS, FRANCE AND FREIGHT PREPAID.

COVERING SHIPMENT OF: 250 ELECTRIC HAND DRILLS C.I.F. LE HAVRE.

*　　　　　　*　　　　　　*

SHIPMENT FROM NEW YORK

TO LE HAVRE

PARTIAL SHIPMENTS
TRANSSHIPMENTS
ARE NOT
PERMITTED
PERMITTED

INSURED BY ☐ BUYER ☒ SELLER

SPECIAL CONDITION:

SPECIMEN

THE ISSUING BANK ENGAGES WITH YOU THAT ALL DRAFT(S) DRAWN UNDER AND IN COMPLIANCE WITH THE TERMS OF THIS CREDIT WILL BE DULY HONORED ON DUE PRESENTATION AGAINST DELIVERY OF DOCU-MENTS AT OUR OFFICE AS SPECIFIED TOGETHER WITH THIS LETTER OF CREDIT ON OR BEFORE THE EXPIRA-TION DATE.

WHEN CREDITS HAVE BEEN ADVISED TO US BY CABLE PARTICULARS THEREOF ARE SUBJECT TO THE MAIL CONFIRMATION.

INDICATIONS OF ADVISING BANK

☐ THIS LETTER OF CREDIT IS ADVISED TO YOU WITHOUT ENGAGEMENT ON OUR PART.

☒ WE CONFIRM THIS CREDIT AND HEREBY UNDERTAKE THAT ALL DRAFT(S) DRAWN AND PRESENTED AS SPE-CIFIED ABOVE WILL BE DULY HONORED BY US.

THE FIRST NATIONAL BANK OF CHICAGO

AUTHORIZED SIGNATURE

FORM 2640-1-AF (REV. 1-75)　　　　ORIGINAL

No.86　裝船期限內無直航船，要求展延船期

Dear Sirs,

L/C No.123

We have received your letter of August 7, 19⋯ and thank you for the establishment of the subject L/C in our favor. Your prompt compliance with our request is much appreciated.

Upon receipt of the L/C, we have contacted the shipping companies who have direct vessels sailing regularly between Taiwan and Hamburg and regret to inform you that the first available boat to Hamburg is scheduled to leave here on or about September 10. Since the shipment deadline as specified in the subject L/C is August 31 and since transhipment is not allowed, we are unable to ship the goods by a direct vessel before the deadline. Ths is why we despatched the following telex to you today:

"NO DIRECT VESSEL SAILING HAMBURG BEFORE AUGUST 31 PLS EXTEND SHIPT N EXPIRY LC 123 TO SEPT 20 N 31 REPCTVLY"

We are anxiously awaiting the amendment to the subject L/C.

Faithfully yours,

Taiwan Trading Co., Ltd.

【註】

1. direct vessel:直航船。

2. shipment deadline＝latest shipment date:裝船最後日期。

3. 電文中的 telexese 說明:

PLS: please

SHIPT: shipment

N: and

REPCTVLY: respectively

No.87　要求修改准許轉運、分批裝運以及展延有效期限

Dear Sirs,

L/C 90/6053

We have received with thanks your contract No.123 on 1,000 doz. of Umbrellas amounting in total of US$12,000.00 on CIF basis.

The covering L/C No.90/6053 has arrived in the amount of US$12,000.00. However, we are cabling you today asking for the following amendments:

1. Delete "Transhipment is prohibited"

Transhipment will have to be made at Hongkong.

2. Amend to "Partial shipments allowed"

We always like to have this proviso so as to eliminate the possibility of having one or two dozen shortshipped. Which nullify the entire L/C.

3. "Extend expiry and shipment dates 15 days"

The subject L/C requires shipment to be effected by Feb. 5. If we are lucky, we hope to have this shipment go forward on a vessel leaving during the latter part to January; but if not, it will go on a

> vessel that will sail sometime between the 6 th and 9 th of February.
>
> Your prompt attention to the foregoing would be much appreciated.
>
> > Yours faithfully,

【註】

1. amounting in total of 也可以 "amounting to" 或乾脆以 "for" 代替。

2. amend＝amend the L/C

3. proviso: 但書。

4. eliminate the possibility of...: 消除⋯的可能。

5. shortship: 短運。

6. nullifty: 使⋯無效, cf. null and void: 無效。

7. go forward on...: 裝⋯船。

8. at Hongkong＝at the port of Hongkong,照理一都市面前的介系詞應用 "in", 但實際上這裡是指港口, 而港口前面則應用 "at"。例如 "at New York" 是指「在紐約港」之意, 本來 "at" 為 "point of place" 的介系詞, 所以在海上保險, 表示 "claim" 的地點時, 也以類如 "payable at London" 的方式表示。

No.88　要求修改保險條款

Dear Sirs,

　　Thank you very much for your L/C No.123 of the Bank of Montreal covering your No.245. We find it in order except the insurance clause, about which we have sent you a telex as per enclosed copy and are pleased to confirm as follows:

　　In our effort to meet the stipulation that the goods

are to be covered "Against all risks from any cause whatsoever irrespective of percentage of damage" we negotiated with the underwriters but learned that they would not insure the goods of this kind against such extensive risks even at a higher rate, which the bankers insist on the insurance policy strictly conforming to the L/C terms.

As a last resort we have requested you to amend the L/C by replacing the stipulation with "ICC(B) including war risks and TPND". In the light of business practice also, this is the best possible coverage given to these goods, Moreover, all our shipments are so securely packed by our experienced hands that no inconvenience has ever been caused to any of our customers.

Your goods are ready for shipment and can be shipped within this month if the amendment to the L/C is telexed at once. If not, we would have to ask you to extend the validity and shipment time for one month.

We trust that you will see the reason of our request and amend (and, extend, if necessary) the L/C at once.

Yours faithfully,

【註】

1. against all risks…of damage: 投保不論任何原因所致的一切危險，且免計損害所致的百分比。關於"Irrespective of percentage"參閱貨物水險章。

2. underwriters:保險商。cf: insurer:保險人。

3. extensive:廣泛的。　　　　4. insist on:堅持。

5. in the light of: 徵諸; 按照; 根據。

No.89　因颱風無法如期交貨, 要求展期

Dear Sirs,

L/C No.123, Your Order No.321

We are sorry to inform you that it has become impossible for us to complete shipment during September of the captioned order.

In fact, a terrible typhoon struck this part of the country on the 6 th this month and our factory suffered serious damages, making it impossible to ship your order within the validity of the subject L/C which expires on September 15.

Under the circumstances, we hope you will agree to extend the credit till October 26 as we asked you by telex.

Though the delay is beyond our control, we are no less than sorry for it, and will do everything in our power to expedite manufacture. The expected date of shipment will be around October 15.

We trust that you will understand the situation and hope that you will comply with our request.

Yours truly,

【註】

1. complete shipment:完成裝運

2. struck: pp. of strike:侵襲。

3. terrible typhoon:強烈颱風。

4. suffered serious damages:遭受嚴重損害。

5. no less sorry for it：仍感到遺憾。

6. do everything in our power:盡我們所能。

7. expedite:加速；加緊。

No.90　通知出口商已修改信用狀

Dear Sirs,

Re：L/C No.7 DH 1/00123/01

Your letter of April 10, 19… requesting us to extend the shipment date and L/C expiry date to June 30 and July 15, 19… has been received.

To comply with your request, we submitted our Application for Amendment of Credit to the opening bank, i.e. Central Trust of China, on April 14, and trust you will receive from the advising bank the amendment before long.

As we have informed you in our previous letters, we are in urgent need of these pickups. Therefore, please ship them as soon as possible. Enclosed please find a copy of amendment to the credit.

Yours faithfully,

【註】

1. amendment to the credit: 信用狀修改書。注意用"amendment to..."而不用 "amendment of..."。

第四節　有關信用狀交易的有用例句

1. To $\begin{Bmatrix} \text{issue} \\ \text{open} \\ \text{establish} \\ \text{arrange} \end{Bmatrix}$ a credit $\begin{Bmatrix} \text{through} \\ \text{with} \end{Bmatrix}$ a bank.

2. To amend a credit　修改信用狀

3. To extend a credit　展延信用狀期限

4. To $\begin{Bmatrix} \text{increase} \\ \text{decrease} \end{Bmatrix}$ a credit $\begin{matrix} 增加 \\ 減少 \end{matrix}$ 信用狀金額

5. $\begin{Bmatrix} \text{As of this date} \\ \text{Up to now} \end{Bmatrix}$, we have not yet received your L/C. In order → to book the shipping space at an earlier date, you are → requested to have the L/C opened immediately.

6. Your order was confirmed by telex of August 5, subject to arrival of L/C within 20 days from date. The 20 days period having expired on August 25 without receipt of the L/C nor hearing any further advice from you. We telexed you today asking when the required L/C had been opened.

7. With reference to your Order No. 123, the manufacture of the goods has been completed and we are now arranging to ship from Keelung on the first available boat. In the meantime, we await your credit to cover.

8. We wish to inform you that we have $\left\{\begin{array}{l}\text{opened}\\\text{issued}\\\text{established}\\\text{arranged}\end{array}\right\}$ a credit →

→ $\left\{\begin{array}{l}\text{with}\\\text{through}\end{array}\right\}$ ABC Bank, in $\left\{\begin{array}{l}\text{your favor}\\\text{favor of yourselves}\end{array}\right\}$ for US$···→

→ $\left\{\begin{array}{l}\text{covering}\\\text{against}\end{array}\right\}$ our order No.123, $\left\{\begin{array}{l}\text{available}\\\text{in force}\end{array}\right\}$ until Dec.26, 19···.

9. We take pleasure in informing you that we have $\left\{\begin{array}{l}\text{submitted}\\\text{sent}\end{array}\right\}$ →

→ application to our bankers for $\left\{\begin{array}{l}\text{establishing}\\\text{issuance}\\\text{opening}\end{array}\right\}$ of an L/C in →

your favor to cover the total value of this order.

10. We find that the L/C amount is inconsistent with that listed in our proforma invoice. The correct amount should be US$... Possibly, it was a typographical error or an oversight on the part of your bank. Therefore, please request your bank to amend the credit by cable to increase US$...

11. The expiry date of the credit being November 30,19···, we request you to arrange with your bank to extend it till December 20,19···

12. Please amend (adjust) L/C No.123 as follows:

a. Amount to be increased $\left\{\begin{array}{l}\text{up to}\\\text{by}\end{array}\right\}$ US$··· ($\begin{array}{l}\text{增加到}\\\text{增加}\end{array}$)

b. The words "···" are to be $\left\{\begin{array}{l}\text{deleted.}\\\text{replaced by "···"}\end{array}\right.$

c. Validity to be extended to April 30.

13. The L/C has been received this date, but without necessary amend-

ment. Again, we must ask you to refer to our letter of July 3 in which our request was made to you regarding the following clause to be amended:

14. To our regret, however, this credit was found not properly amended on the following points despite our request:

15. As advised you by telex, the sailing of s.s.“⋯” was cancelled because of serious damage she sustained while at the berth of Keelung. This has made us impossible to ship the goods within the life of L/C which expires on June 2. Under the circumstances, we earnestly hope that you will extend the credit to the end of this month.

16. Since this happening has been caused by force majeure over which we have had no control, we hope you will agree with us that we cannot be held responsible for this delay in shipment. We trust we may depend upon your generosity for the extension to shipping date and also of expiry date of the L/C as we requested by today's cable.

17. We will increase the L/C amount by 200 tons at the rate of US$ 150 per metric ton.

18. We have $\left\{ \begin{array}{l} \text{instructed} \\ \text{requested} \end{array} \right\}$ our bankers to amend the clause to → read "partial shipments are permitted."

19. Instead of our arranging for the increase of credit amount at costly cable charges, you may invoice to us through your bankers the said small dffference of amount under your L/C. Meanwhile we are instructing the issuing bank to accept such L/C at the end.

20. We presume the advice of L/C has gone astray in transit from your bankers. If it has not been sent, repeat your instructions to your bankers.

習　題

一、試將下列相關的中英文連起來：

a. grantee   1. 開狀銀行

b. grantor   2. 開狀人

c. accreditee   3. 歸償銀行

d. paying bank   4. 受益人

e. reimbursing bank   5. 付款銀行

二、將下列中文譯成英文：

1. 儘管(in spite of) 本公司一再請求(repeated requests)貴公司備付訂單第 123 號的信用狀迄仍未開達，敬請注意爲荷(invite your attention)。

2. 遵照指示(upon your instructions)：本公司已於今日通知本公司的銀行——臺灣銀行——用電報開出上開訂單 1,000 打洋傘 US$15,000 的信用狀。

3. 貨品現已準備裝運 (ready for shipment)，僅祇等候信用狀開到。

4. 本公司現正等候貴公司信用狀的開達，一旦開到即向船公司預定艙位。

三、臺北貿易公司收到下面信用狀乙張：

　　　　HAMBURG COMMERCIAL BANK

　　　　　　P.O. BOX 123

　　　　Hamburg, West Germany

　　　　　　　　　　　　May 16, 1990

　　　　　　　　　　　　VIA BANK OF TAIWAN

TAIWAN, TRADING CO., LTD.

323, Wu-chang st. Taipei

Taiwan

　　　　IRREVOCABLE LETTER OF CREDIT NO.123

Dear Sirs,

We hereby establish our IRREVOCABLE letter of credit in your favor for account of UNIVERSAL IMPORTERS INC., HAMBURG, up to an aggregate amount of US$7,600.00(US DOLLARS SEVEN THOU-SAND SIX HUNDRED ONLY) available by your drafts at sight drawn without recourse on the accountee for full invoice value accompanied by the following documents:

1. Signed Invoice in Quadruplicate
2. Packing List in Quadruplicate
3. Certificate of Origin in Duplicate
4. Full set of clean on-board bills of lading made out to order of Hamburg Commercial Bank, Hamburg notify accountee marked "Freight Paid"

Evidencing shipment of one hundred and twenty metric tons of Canned Asparagus, AI Grade C & F Hamburg

Shipment from Taiwan to Hamburg not later than July 28, 1990. Partial shipments are not allowed. Transhipment is prohibited. All banking charges outside West Germany are for account of Beneficiary. Drafts drawn under this Credit are to be negotiated on or before August 15, 1990.

We hereby agree with the drawers, endorsers, and bonafide holders of drafts drawn under and in compliance with the terms of this Credit that such drafts shall be duly honored on due presentation to the drawee. This Credit is transferable.

請問:

1. 誰是 Opening Bank?
2. 誰是 Advising Bank?
3. 誰是 Accreditor?
4. 誰是 Drawee?

5. 誰是 Accreditee?

6. 本信用狀可否轉讓?

7. Sight L/C 抑 Usance L/C?

8. 押滙時需提示那些單證?

9. 貿易條件是那一種?

10. L/C 最後有效日是那一天?

第二十一章　貨物水險

(Marine Cargo Insurance)

　　國際買賣是隔地買賣，貨物從出口地運到國外進口商手中這一段運輸期間，難免不遭遇到天災人禍等等意外危險，以致受到損害。因此進出口商為防萬一貨品遭受損害時，可獲得補償，乃有將運輸中的貨物加以保險的必要。

第一節　損害及費用的類型

一、Total loss（全損）

　　即所承保的貨物全部滅失(Loss)的情形，可分為：

　　1. Actual total loss(實際全損)：即投保的貨物已經全部毀滅或受損程度到達已失去原有形態或其所有權被剝奪而不能恢復者。

　　2. Constructive total loss (推定全損)：即投保的貨物全損已無可避免或雖未及全損，但欲由絕對全損保全，其費用將超過其保全後的價值者。

二、Partial loss（分損）

　　即所承保貨物部分損失的情形，可分為：

　　1. General Average（共同海損）：即基於船舶及裝載貨物所有人的共同利益，為避免共同危險，而船長故意及合理對於船舶或貨物加以適當緊急處分而生的犧牲(General average sacrifice)及費用(General average expenditure)。此項犧牲及費用係保全所有利益而生，故應由全體利害關係人分擔，然後由其向保險人要求賠償。

2. Particular Average（單獨海損）：卽指承保貨物因不可預料的危險所造成的滅失或損害，這種損害並非由共同航海的財產共同負擔，而是由遭受損害的各財產所有人單獨負擔者，簡言之，分損無共同海損性質者，卽爲單獨海損。

三、Charges（費用）

1. Sue & labor charges（損害防止費用）：貨物在遇險時，如被保險人或其代理人（如船長）或讓受人爲之努力營救，以減輕損失程度，則保險人對這種費用支出應予賠償，這種費用卽爲 Sue & labor charges.

2. Salvage charges（施救費用）：無契約關係或正式職責關係的第三者對於遭受危險的船貨，自願所使施救工作後，應由被保險人支付的報酬，保險人對此項費用負有賠償之責。

第二節　保險險類

海上貨物保險，根據保險人就海上損害發生的程度，所應承擔的責任加以區分，可分爲下列三種基本險及附加險：

一、基本險

1. 協會貨物保險 A 款險(Institute Cargo Clauses (A))簡稱 ICC (A)。

2. 協會貨物保險 B 款險(Institute Cargo Clauses(B)) 簡稱 ICC (B)。

3. 協會貨物保險 C 款險(Institute Cargo Clauses(C)),簡稱 ICC (C).

二、附加險

常見的附加險有: 1. War（兵險）,2. Strike, riot and civil commotions（罷工、暴動、民衆騷擾）,3. Theft, pilferage and non-delivery（偷竊、拔貨、遺失）,4. Fresh water and rain damage（淡水、雨水損）,5. Breakage（破損）,6. Leakage（漏損）,7. Hook hole（鉤損）,8. Oil damage（油汚）,9. Contamination with other cargoes（汚染）,10. Sweat and heat（汗濕、發熱）,11.Washing overboard（浪沖）,12.Mildew and mould（霉濕及發黴）,13.Rat and vermin damage（鼠蟲害）等。

第三節　有關保險的信函

No.91　向保險公司查詢費率

China Insurance Company Limited

Dear Sirs,

Please quote us your lowest rate for Marine Insurance, ICC(B) plus TPND and war risk, on a shipment of 900 cartons of Canned Asparagus, valued at DM 31,845, by the s/s "Euryphates" from Keelung to Hamburg.

The ship is scheduled to leave Keelung on or about August 20 for Hamburg, and we hope to have your reply at your earliest convenience.

Yours faithfully,

【註】

1. value at:價值爲…（元）。

2. is scheduled:預定。

3. 在我國, 進出口商向保險公司查詢費率時多用電話辦理, 很少用信函查詢。有時, 也可透過保險經紀人(Insurance broker)查詢費率。

No.92　保險公司覆函

Taiwan Trading Co., Ltd.

Dear Sirs,

　　We acknowledge with thanks the receipt of your letter of August 10 inquiring about marine insurance rate to cover shipment of Canned Asparagus from Keelung to Hamburg.

　　In compliance with your request, we hereby quote you our lowest rate at 1% to cover ICC(B) plus TPND and War risks. Please note that this is the lowest rate we are able to offer and there will be no rebate whatsoever.

　　As the shipment is drawing nearer and nearer, we suggest that you contact us at the earliest in order that we may issue the policy in time.

　　We hope that you will pass us your business.

Yours faithfully,

【註】

1. rebate:回扣。按保險界慣例, 保險公司往往按其所開費率, 予投保人若干回扣。

2. pass us your business:將生意交給我們, 意指由我們承保。

No.93　向保險公司投保

China Insurance Company Limited

Dear Sirs,

　　We thank you for your letter of August 15 quoting

us marine insurance to cover ICC(B) plus TPND and War risks for shipment of 900 cartons Canned Asparagus from Keelung to Hamburg.

In view of the fact that the rate you quoted is quite competitive, we have decided to entrust your company with the insurance of this shipment. Enclosed please find a copy of our Application for Marine Insurance duly filled and signed by us. It will be appreciated if you will issue the Marine Insurance Policy as soon as you receive this letter and its enclosure.

As for premium, we will arrange with the negotiating bank to pay you at the time of negotiation of draft.

Yours faithfully,

【註】

1. entrust your company wth the insurance of this shipment:將本批貨交給貴公司承保。

2. application for marine insurance:水險投保書。

3. as for premium: 至於保險費，用"as to premium"較普遍。

4. 在臺灣，以外幣保險的場合，出口商所應付的保險費多由押滙銀行代扣，轉存各保險公司在臺灣銀行的帳戶。保險公司到可憑此辦理結滙。以便理賠時，有外幣可資賠付，所以，本信最後一段說：「關於保險費，將與押滙銀行按排，於押滙時支付給你。」

貿易商有大批貨物分次裝運時，如於每次裝運一批貨物投保一次不僅麻煩，而且可能於裝運時疏忽忘記保險，於是，有預約保險(Open policy)的產生。換言之，貨主可預先購買一定金額的保險，這種保險只訂有概括的條件，至於船名、保險金金額、啓航日期、貨物數量、包裝嘜頭則於每次裝出貨物時由投保人向保險公司申報(Declare)。投保人每次申報保險金額，從預約保險單(Open policy)內扣減，直至 Open policy 所列保險金額用完爲止。這種 Open policy 又稱爲流動保單(Floating policy)。

No.94　向保險公司查詢預約保險條件

Dear Sirs,

　　Please quote your rate for an ICC(A) open policy for US$100,000 to cover shipments of general merchandise by APL's vessel from Taiwan ports to Atlantic ports in Canada and the United States.

　　As shipments are due to begin on June 30, please let us have your quotation by return.

Yours faithfully,

【註】

1. APL'S vessel: American President Line's Vessel: 美國總統輪船公司的船隻。

2. Atlantic Ports:大西洋岸港埠。

3. due to+verb:預定…; 即將…。

No.95　保險公司的覆函（報價）

Dear Sir,

　　We are replying your enquiry of June 10. Our rate for a US$100,000 ICC(A) open policy on general merchandise by APL's vessel from Taiwan ports to Atlantic ports in Canada and the United States is 2% of declared value.

> This is an exceptionally low rate and we trust you will give us the opportunity to handle your insurance business.
>
> > Yours faithfully,

【註】

1. declared value:申報金額。

　　按 FOB 或 C&F 交易，而進口商已購買 Open policy 時，往往要求出口商於起運時替進口商向其保險公司通知(Declare)，其例函如下：

No.96　起保通知書(Insurance declaration)

> To: xxx Marine Insurance Company
> Dear Sirs,
>
> > Re:Your Open Policy No.123
>
> 　　We, on behalf of...(name of assured or importer)...of (place) hereby declare below the particulars of a shipment to be made by us which is covered under the above Policy:
>
> Conveyance:S/S "President"
>
> Voyage: form Keelung to New York
>
> Sailing date: on or about August 5, 19...
>
> Interest: Cold Rolled Steel Sheet
>
> Marking:
>
> Value:US$12,000.00
>
> Insurance coverage: ICC (A) plus war
>
> Packing: in bundle

> Please acknowledge receipt by signing and returning to us the duplicate copy at your earliest convenience.
>
> Yours faithfully,

【註】

1. conveyance:運輸工具。

2. voyage:航程。

3. sailing date:啓航日。

4. interest:保險標的。

第四節　有關保險的有用例句

1. To $\left\{\begin{array}{l}\text{insure}\\\text{cover}\end{array}\right\}$ a thing for US\$…against $\left\{\begin{array}{l}\text{ICC(A)}\\\text{ICC(B)}\\\text{ICC(C)}\end{array}\right\}\rightarrow$

   $\rightarrow\left\{\begin{array}{l}\text{with}\\\text{in}\end{array}\right\}$ China Insurance Company Limited.

   (就某東西向中國產物保險公司投保…元的 A 款險〔等〕)

2. To $\left\{\begin{array}{l}\text{effect}\\\text{cover}\\\text{provide}\end{array}\right\}$ insurance on a thing for US\$… against $\left\{\begin{array}{l}\text{ICC(A)}\\\text{ICC(B)}\\\text{ICC(C)}\end{array}\right\}\rightarrow$

   $\left\{\begin{array}{l}\text{in}\\\text{with}\end{array}\right\}$ Cathay Insurance Co., Ltd.→ for US\$…

3. To insure a thing at a low $\left\{\begin{array}{l}\text{premium}\\\text{rate}\end{array}\right\}\left\{\begin{array}{l}\text{in}\\\text{with}\end{array}\right\}$ the →

   $\rightarrow\left\{\begin{array}{l}\text{insurance company.}\\\text{underwriters.}\end{array}\right.$

4. To $\begin{cases} \text{make out} \\ \text{draw up} \\ \text{issue} \end{cases}$ a policy of US$… $\begin{cases} 掣發 \\ 簽發 \end{cases}$ …元的保單)

5. Please $\begin{cases} \text{quote} \\ \text{let us know} \end{cases}$ your lowest $\begin{cases} \text{ICC(A)} \\ \text{ICC(B)} \\ \text{ICC(C)} \end{cases}$ rate for US$…→

→ on TV SETS from Keelung to New York.

6. Please quote us your best rate for a coverage of ICC(A) including War for the CIF value of US$…plus 10% on electrical apparatus from Keelung to Melbourne.

7. Please effect insurance for US$…on 100 cases of plastic shoes against ICC(B) per s.s.… due to leave Keelung for New York on or about October 15.

8. Attached are details of the shipment to be insured against ICC(B) plus RFWD for 110% of CIF value. The goods leave Keelung per M. S.…for Singapore on or about May 20.

9. We shall shortly be making regular shipments of fancy leather goods to South America by approved ships and shall be glad if you will issue an ICC(A) policy for, Say, US$100,000 to cover these shipments from our warehouse at the above address to port of destination.

10. We ask you to issue a covering note for the insurance.

   * covering note:暫保單。美國稱爲"insurance binder"。

11. We may mention that premiums are very much higher this year, as underwriters are asking increased rates all round.

   * all round:全面地

12. We shall be pleased to know whether you can undertake insurance of wines against ICC(B), including breakage and pilferage.

習　題

一、將下列保險用語譯成中文，並說明其意義。

  1. perils of the seas   2. jettison   3. total loss

  4. partial loss   5. particular average

二、將下列各縮寫字所代表的英文及中文寫出。

  1. ICC(A)   2. ICC(B)   3. SR&CC

  4. RFWD   5. COOC

三、將下列中文譯成英文。

  1. 本公司擬為本批貨載(Shipment)按發票金額加一成投保 A 款險及兵險，請將保
     險單及帳單(Debit note)一併寄下為荷。

  2. 遵囑(As requested)已向中國保險公司就由木星輪(S.S. Jupiter)自基隆運至
     漢堡的 200 箱罐頭蘆筍投保 US$11,000 的 A 款險及兵險。

第二十二章　海上貨物運輸

(Ocean Transportation)

第一節　定期船運輸

在海上運輸，如買賣貨物爲一般雜貨(General Cargo)或零星貨物，通常多委託定期船(Liner, Liner Vessel)以搭載方式裝運。如係大宗散裝貨(Bulk Cargo)，則選擇不定期輪(Tramper, Tramp Vessel)以傭船方式裝運。

在定期船運輸，出口商備妥貨物後，即可根據船公司(Shipping Company)或船務代理行(Shipping Agent)所印發的船期表(Shipping Schedule)或報紙船期欄廣告，選擇適當的船隻，向船公司或其代理行洽訂艙位(Booking Space)。洽訂艙位或以 Booking Note 爲之，或以信函方式爲之。

在我國，洽訂艙位通常多採用 Booking Note 方式，即由出口商（或委託報關行）向船公司索取空白的 Booking Note、Shipping Order 及 Mate's Receipt 成套格式，依式套打塡就後送請船公司就 Shipping Order(S/O)予以簽署，日後憑此 S/O 將貨物交給船長，貨物裝上船後，大副(Mate)在 Mate's Receipt(M/R)上簽字，並退還出口商，以便由其向船公司換領 B/L。

以信函方式向船公司或其代理行洽訂艙位的方式在臺灣並不流行，但在英美等國則常採用此方式，茲舉一例於下：

No.97 向船公司查詢運價及洽訂艙位

Dear Sirs,

We shall be pleased if you will quote us the lowest rate of freight for 100 bales of Cotton Yarn weighing 400 lbs. each, to be shipped to Hamburg direct before September 25 from Keelung.

As the L/C stipulates that transhipment is not allowed, therefore, you must quote us for a vessel sailing from Keelung to Hamburg direct before the deadline as mentioned above.

Your early quotation will be appreciated.

Yours faithfully,

No.98 船公司對於查詢運價及洽船的覆函

Dear Sirs,

In response to your inquiry of...(date)... concerning the ocean freight to Hamburg for 100 bales of cotton yarn, we are pleased to inform you that the rate of freight is US$70 per cubic meter or 1000 kilos, which is the agreed minimum rate of freight of the European Route Conference.

As to the steamers which will sail from Keelung to Hamburg, there are two regular liners, viz., s.s. "A" due to leave Keelung on September 15, to be followed by

s. s. "B" on the 24th of the month.

We hope to have the pleasure of dealing with your shipment.

Yours truly,

【註】

1. US$70 per cubic meter or 1000 kilos: 意指一立方公尺或 1000 公斤運費美金七十元，至於按一立方公尺 (CBM) 或按 1000 公斤計收，將視那一種對船方有利而定。Kilos 為 Kilograms 之意。

2. European Route Conference: 歐洲航線運費同盟。

No.99　申請加入運費同盟

Far Eastern Freight Conference

Taipei, Taiwan

Dear Sirs,

In order to avail ourselves of the good and swift services of vessels of the conference lines, we are writing you to apply for membership to the Conference and shall appreciate your passing our application on to your Hongkong Office for consideration.

For your reference, we are makers of umbrellas and, during the past three years, have exported large quantity of our products to North and South America, South and West Africa, as well as to Europe. The total tonnage of umbrellas exported amounts approximately 3,000~35,000 measurement tons per annum, which is substantial enough to warrant us to join the Conference as a member.

Your prompt screening of our application and approval will be appreciated.

Yours faithfully,

【註】

1. freight conference＝shiping conference: 運費同盟是在一特定航線上有定期輪行駛的船公司，為限制或消除彼此間的競爭，而以協定方式結合而成的一種卡特爾(Cartel)組織。如果貨主加入運費同盟，則可享受運價上的優待。

2. Conference Lines: 參加運價同盟的船公司。

3. tonnage: 噸數；噸位。

4. warrant: 證明……為正當，即"give a right to"或"give a good reason for"之意。

參加運費同盟的會員貨主，原則上必須將其貨物交給同盟輪運輸，否則將受到處分。然而，有時因特殊原因無法交給同盟輪運輸，在此場合，應向運費同盟備案，取得諒解。

No.100　通知運費同盟無法將貨交同盟輪裝運

Far Eastern Freight Conference

Taipei, Taiwan

Dear Sirs,

We are very much pleased to be admitted as a member of the Conference and assure you that we will, from now on, abide by all its regulations and rules.

Now, we are facing a quandary: our customers in Lagos insisted that the umbrellas we are to deliver to them be shipped by vessels of ABC Line, which is a non-conference line because they have contractual obligatons with that particular line to have all the goods shipped to them per their vessels. Meanwhile, the L/C our customers have opened to us also calls for the goods to be shipped by vessels of the ABC Line.

Since this case is beyond our control, we can do nothing but to comply with our customers' request. As this is an exceptional case, we report it to you just for your information and reference.

Yours faithfully,

【註】

1. abide by: 遵守。
2. quandary: 窘境; 困惑。cf. in a quandary: 進退兩難 (維谷)。
3. call for: 規定; 要求。

第二節　傭船運輸

　　大宗散裝物資的運輸多利用不定期輪。不定期輪的洽船與定期輪不同。一般而言, 貨主僱傭不定期輪時, 須先向船公司或透過傭船經紀人(Chartering Broker)詢價, 然後由船公司報價, 經過討價還價之後, 傭船契約才告成立。契約一旦成立後, 在雙方或經其代理人在未簽定正式傭船契約書(Charter Party, 簡稱 C／P)之前, 常先簽立成交書或訂載書(Fixture Note), 並由船方、貨方、經紀人共同簽名後分別保存。隨後即根據這成交書作成正式的傭船契約。但如船方與貨方均在同一地方時, 往往免簽成交書，而逕行協商簽立正式的傭船契約書。

No.101 貨主向傭船經紀人詢價

Dear Sirs,

We have about 6,600 metric tons of Ammonia Sulfate packed in bags for shipment from Keelung to Bangkok and wish to charter a ship of about 3,300 metric tons. Would you please arrange a vessel and quote us the best charter rate for the shipment.

We should add that the vessel must be lying on the berth at Keelung on or before August 10, 19… ready to ship the cargo.

Please let us know whether you can arrange this for us and, if so on what terms.

Yours faithfully,

【註】

1. ammonia sulfate: 硫酸亞。

2. charter rate: 傭船價。

3. lying on the berth: 停泊在碼頭船席。lying 爲 lie 的現在分詞。

No.102 傭船經紀人的報價

OFFER SHEET

Taiwan Fertilizer Co., Ltd.   June 15, 19…

Dear Sirs,

In reply to your enquiry of June 10, we are pleased to offer you subject to the following terms and conditions:

1. Vessel: to be nominated later.

2. Cargo & Quantity: 6,600 metric tons （±10%） of Ammonium Sulfate packed in bags.

3. Loading Port(s)： one safe berth of one port Keelung, Taiwan.

4. Discharging Port(s)： one safe berth of one port, Bangkok, Thailand.

5. Loading Date:

   a) Laydays： not to commence before July 10, 19….

   b) Cancelling date: July 31, 19….

6. Freight & Payment： At the rate of @US$7.50 per metric ton based on FIO terms, @US$9.00 per metric ton based on FO terms, and @US$10.50 per metric ton based on Berth Terms.

   Freight payable at Taipei upon completion of loading.

7. Lighterage & Stevedorage： If any, for charterer's account and risk at both ends.

8. Loading & Discharging rate： Loading rate: 200 metric tons per gang per WWDSHEXUU.

   Discharging rate： CQD

9. Demurrage & Despatch money： Demurrage at US$2,000 per day or pro rata for any part of a day and half of despatch.

10. Commission & Brokerage : Owner's agents at both ends.

11. Agents:

12. Other terms & Conditions : a) A/P, if any, for charterer's account.

b) Other terms and conditions as per "GENCON" C/P revised in 1922.

This offer is subject to your reply reaching us before noon, June 20, 19….

Yours truly,

Confirmed and accepted Tien Kuang Shipping Co., Ltd.

By: By：

【註】

1. to be nominated later：船名另定。

2. cargo & quantity：貨物名稱及其數量。

3. safe berth：指可以安全停靠的碼頭、船席。

4. loading date：裝貨日期。

5. laydays：到裝期限。

6. not to commence before…：某月某日以後開始…。

7. cancelling date：指傭船人有權解約的日期。

8. freight & payment：運費率及其付款方法。

9. FIO：為"free in and out"的縮寫，係裝卸條件的一種。關於裝卸條件有下列幾種：

liner terms：又稱為"berth term"定期輪或埠頭條件。即裝卸費用均由船方負擔。

FI(free in)：裝船費用船方免責（即由貨主負擔）。

FO(free out)：卸貨費用船方免責（卽由貨主負擔）。

FIO(free in and out)：裝卸費用船方免責（均由貨主負擔）。

FIOS(free in and out and stowed)：裝船卸貨以及艙內堆積費用船方免責（均由貨主負擔）。

FIOT(free in and out and trimmed)：裝船、卸貨以及平艙費用船方免責（均由貨主負擔）。

10. lighterage & stevedorage：駁船費及船上裝卸費。

11. at both ends：在兩端，卽指在裝貨港及卸貨港。

12. loading & discharging rate：裝貨率及卸貨率（速度）。

13. WWDSHEXUU: Weather Working Days Sundays & Holidays Excepted unless used：爲裝卸期限(laytime or layday)計算方式的一種，意指「星期例假除外，除非照常裝卸的天氣適宜工作日」，此外尚有：

WWD(Weather Working Day)：天氣適宜工作日。

WWDSHEX(Weather Working Days Sundays and Holidays Excepted)：星期例假除外之天氣適宜工作日。

CQD(Customary Quick Despatch)：照港口習慣速度。

14. demurrage & despatch money：延滯費及快速獎金。

15. agents：指船務代理。

16. commission & brokerage：手續費及經紀人佣金。

17. A/P 爲"age premium"或"additional premium"的縮寫，卽逾齡船保險費或額外保險費之意。

18. GENCON C/P：一般傭船契約。

No.103 訂載書

Taiwan Fertilizer Co., Ltd.　　　　　　June 20, 19…

Taipei

Dear Sirs,

FIXTURE NOTE

This confirms engagement of freight spaces with you for a trip charter of M.V."LICHON" or substitute as per following terms and conditions:

1. Cargo & Quantity: Ammonium Sulfate in bags for total 6,600 metric tons net (plus or minus 10%).

2. Type & Name of Vessel and Shipment:

   Tramp M.V. "LICHON" with Singapore flag building 1980 or substitute with two shipments to be carried into execution during August through October of 19 with each 3,300 metric tons 10% more or less at carrier's option.

3. Loading Port: one safe berth of one safe port, Keelung.

4. Discharging Port: one safe berth of one safe port, Bangkok.

5. Loading Rate: 200 metric tons per gang per WWDSHEXUU.

6. Discharging Rate:CQD.

7. Demurrage/Despatch at loading port: Demurrage rate at the rate of US$2,000 per day or pro rata of a

day. Despatch at the rate of US$1,000.00.

8. Laytime: Laytime to commence at 8 a. m. next regular working days after vessel's arrival at loading berth or anchorage within the commercial limits of the port and tender of Notice of Readiness during regular working hours.

9. Lighterage & Stevedorage: If any, at loading port to be for charterer's account, and at discharging port to be for consignee's account.

10. Freight Rate: US$7.50 per metric ton NET FIOS. Taiwan Transportation Tax, if any, for shipper's account. A/P, if any, for account of cargo.

11. Payment of Freight: Freight to be fully prepaid in NT Dollar or in U.S. Dollar at Taipei upon completion of loading on the Bill of Lading quantity, discountless and non-returnable whether ship and/or cargo lost or not. Exchange rate shall be official rate at the time of freight payment.

12. Brokerage: No brokerage.

13. Other Terms: As per Gencon Charter Party with modifications as revised 1922. New Jason Clause, Both to Blame Collision Clause, Institute War Risks Clause 1 &2, and P&I Clause to be deemed fully incorporated in this Fixture Note.

Please sign the duplicate herewith enclosed and return to us at your earliest convenience.

```
Accepted by:

                                Yours very truly,

                        Tien Kuang Shipping Co., Ltd.
```

【註】

1. M. V. "Lichon" or substitute：裝載船名為"Lichon"輪或其代替船。

2. with Singapore flag：掛新嘉坡旗的，即船籍為新嘉坡。

3. 10% more or less at carrier's option:承運船有多運或少運 10%的選擇權。

4. one safe berth of one safe port:一個可以安全停靠碼頭船席的安全港口。

5. laytime:裝卸時間。

6. regular working days：正常工作日。

7. notice of readiness：完成準備通知書。

8. for account of cargo：歸貨主負擔。

9. discountless：無折扣。

10. non-returnable：不退還運費。

11. official rate：官定滙率；官價。

12. New Jason Clause：新傑遜條款。

13. Both to Blame Collision Clause：雙方過失碰撞條款。

14. deemed：視為。

15. incorporated：併入。

第三節　裝運通知

　　出口商於辦理裝運手續時，應將辦理裝運的事宜向進口商發出通知，這種通知稱為裝運通知(Shipping Advice, Notice of Shipment,英國則稱為Advice of Shipment)。裝運通知的內容包括：Order No.(或L/C No.)貨名、數量、船名、裝船日期(或預定開航日期, estimated time of departure, ETD)、裝貨港、預定抵埠日期(estimated time of arrival, ETA)、以及嘜頭等。這種通知在 FOB, C&F 交易時尤爲重要，通常多在裝船前以 Cable、Telex 或 FAX 向進口商發出通知，以便進口商可及時購買保險。

No.104　裝運通知確認書並附上抄本貨運單證

Dear Sirs,

Re: <u>Your order for canned asparagus</u>

We are pleased to inform you that we have shipped today, as per copy of our cable enclosed, by the M. S. "Suez Maru" 900 cartons canned asparagus ordered by you on May 25. The boat is scheduled to arrive at Hamburg on September 30.

In order to cover ourselves for this shipment, we have drawn on Dresdner Bank, Hamburg, by sight draft for DM28, 950.00 under the L/C No.123 and negotiated it with City Bank of Taipei, Taiwan.

To facilitate your taking delivery, we are enclosing the following non-negotiable shipping documents, each in duplicate:

Commercial Invoice No. 123

Bill of Lading No.231

Insurance Policy No. 567

Packing List No.123

Certificate of Origin NO. 378

We trust that you will find all these documents in

order and the goods will arrive at Hamburg in perfect condition.

End:A/S   Yours faithfully,

【註】

1. The M.S."Suez Maru":"M.S."爲"motor ship"的縮寫。船名前可以有"the"，也可以沒有。但在一般文字裡通常有"the"。用了"the"往往可以避免誤解。如"Queen Mary has just started"裡的"Queen Mary"也許指人，也許指船，但"The Queen Mary has just started"裡的"The Queen Mary"一定指船。準此，M.S. "Suez Maru"或S. S."Taiwan"前面可以有"the"，也可以沒有"the"。

2. in order to cover ourselves for this shipment:爲了收回這批貨款。

3. to facilitate your taking delivery:爲便利您的提貨。

4. non-negotiable shipping documents＝copy shipping documents：抄本貨運單證，因不能做爲押滙之用，所以稱爲"non-negotiable"。

茲舉一 Cable shipping advice 於下：

No.105

YOUR ORDER 123 CANNED ASPARAGUS 900 CARTONS SHIPPED PER MS SUEZ MARU ETD KEELUNG JULY TEN ETA HAMBURG SEPTEMBER THIRTY COPY DOCUMENTS AIRMAILING

第四節　有關貨物運輸的有用例句

1. We are pleased to $\left\{\begin{array}{l}\text{inform}\\\text{advise}\end{array}\right\}$ you that we have shipped → your order

No...for...(goods)... $\begin{Bmatrix} by \\ per \\ on\ board \end{Bmatrix}$ the s.s"..."which sailed →

→ here on...(date).

2. We have today $\begin{Bmatrix} forwarded \\ shipped \end{Bmatrix}$ per s.s."London"which $\begin{Bmatrix} sailed \\ left \\ cleared \end{Bmatrix}$ here

→ today.

　＊clear:出港；啓碇；開出，即"leave port"之意。

3. We $\begin{Bmatrix} confirm \\ have\ the\ pleasure\ of\ confirming \\ are\ pleased\ to\ confirm \end{Bmatrix}$ our $\begin{Bmatrix} telex \\ cable \end{Bmatrix}$ just despat-

ched → informing you that we have shipped...

4. Your order No...has been forwarded, as per copy of our → $\begin{Bmatrix} cable \\ telex \end{Bmatrix}$

enclosed, by the s.s."Taiwan"which left this port → today and which

is to arrive at your port on...(date)

5. We are $\begin{Bmatrix} sending\ you\ herewith \\ enclosing \end{Bmatrix}$ invoice and $\begin{Bmatrix} air\ waybill \\ B/L \end{Bmatrix}$ in →

duplicate with insurance policy.

6. We have $\begin{Bmatrix} drawn \\ valued \end{Bmatrix}$ on $\begin{Bmatrix} Bank\ of\ America \\ you \end{Bmatrix} \begin{Bmatrix} at\ sight \\ at\ 30\ d/s \end{Bmatrix}$ for \$···→

→ under L/C No.123 issued by Bank of America, and negotiated it

thru $\begin{Bmatrix} our\ bankers. \\ Bank\ of\ Taiwan. \end{Bmatrix}$

7. In order to cover $\begin{cases} \text{this shipment} \\ \text{ourselves for this shipment} \end{cases}$, we have →

$\begin{cases} \text{valued} \\ \text{drawn} \end{cases}$ on...Bank at sight for US$···under their L/C No.123 →

→ through our bankers, Bank of Taiwan.

8. We $\begin{cases} \text{trust} \\ \text{hope} \end{cases}$ the $\begin{cases} \text{goods} \\ \text{shipment} \end{cases}$ will reach you $\begin{cases} \text{in perfect condition} \\ \text{safe} \\ \text{safely} \\ \text{in good condition} \end{cases}$ →

→ and open up to your $\begin{cases} \text{complete} \\ \text{entire} \end{cases}$ satisfaction.

→ $\begin{cases} \text{expect} \\ \text{await} \end{cases}$ your further $\begin{cases} \text{order.} \\ \text{patronage.} \end{cases}$

9. We thank you for this order and trust the goods will reach you promptly and in good order.

10. We trust the high quality of our...(name of article)...will prove an inducement of further business.

11. We trust this purchase will bring you a good profit and result in your further orders.

12. If you have further need of this item, we shall be very much pleased to accept a reorder.

13. We are looking to $\begin{cases} \text{the continuation of our pleasant business} \\ \text{relations.} \\ \text{a continued and increasing business with} \\ \text{you.} \end{cases}$

14. We shall be pleased to receive your further orders, which will always receive our utmost careful attention.

習　題

一、試寫出下列各縮寫字所代表的英文及並譯成中文。

　1. S/O　2. M/R　3. C/P　4. FO　5. CQO

二、將下列中文譯成英文。

　1. 謹通知貴公司本年一月十日第123號訂單所列1,000打棉內衣(cotton under-wear)已裝上太平洋之熊輪(S.S. Pacific Bear)於二月十日自基隆啓航(sailed from Keelung)開往洛山磯(Los Angeles)。

　2. 隨函附上下列貨運單證副本各兩份，請查收。

　3. 對於本批貨載，本公司已憑洛山磯美國商業銀行(Bank of America, Los Angeles)所開第123號信用狀透過臺北臺灣銀行國外部讓售滙票。

　4. 本公司相信本批貨物將以良好情況(in good condition)到達並令貴公司完全滿意(given you complete satisfaction)，多謝惠顧並盼繼續訂購(further order)。

　5. 請於該滙票提示時惠予承兌，至爲感激。

第二十三章　貨運單證

(Shipping Documents)

第一節　國際貿易貨運單證的種類

國際貿易貨運單證(Shipping Documents)繁多，但可分為基本單證與附屬單證兩大類。

1. Fundamental Shipping Documents
(1)買賣方面：Commercial Invoice　　　(商業發票)
(2)運輸方面
- Ocean Bill of Lading　　　(海洋提單)
- Air Waybill　　　　　　(航空提單)
- Receipt for Parcel Post　(郵政包裹收據)

(3)保險方面
- Insurance Policy　　　(保險單)
- Certificate of Insurance　(保險證)

2. Subsidiary Shipping Documents
- Packing List　　　　　　　　　　(裝箱單)
- Weight／Measurement list　　　　(重量尺碼單)
- Consular Invoice　　　　　　　　(領事發票)
- Customs Invoice　　　　　　　　(海關發票)
- Certificate of Origin　　　　　　(產地證明書)
- Inspection Certificate　　　　　　(檢驗證明書)
- Phytosanitary Certificate　　　　(檢疫證明書)
- Fumigation Certificate　　　　　(薰蒸證明書)
- Health Certificate　　　　　　　(衛生證明書)

第二節 各種貨運單證實例

一、商業發票(Commercial Invoice)

　　商業發票（簡稱發票）爲貨物運出時，賣方向買方所開出說明所出售貨物名稱、價格及有關費用的文件，國際貿易上使用的商業發票雖無一定的格式，但就其內容而言，通常都由首文、本文及結尾文構成。

No.106　商業發票

```
                           A B C CO. LTD.
                         P. O. box 123, Taipei, Taiwan
                           INVOICE

No. ___(1)123___                         Date: Oct. 30,(2) 1984

INVOICE of  10 sets(3) Spare Parts fo K Car's Bogie Truck

For account and risk of Messrs. X Y Z  (4) Co. Ltd, Sydney


Shipped by A.B.C.Co.,(5) Ltd.              Per S.S. Hai (6)Tai
sailing on or about Oct. 30,(7) 1984  From Keelung,(8) Taiwan to  (9) Sydney
L/C No. 567  (10)              Contract No.  (11) 456
```

Marks & Nos.	Description of Goods	Quantity	Unit Price	Amount
(12) 456 SYDNEY PKG 1--10 MADE IN TAIWAN R. O. C.	(13) Spare parts fo K car's bogie truck Item No. 17 roller bearing 　　　　　　BT23 Item No. 18 ball bearing 　　　　　　BT30	(14) 4 sets 6 sets	(15) FOB Keelung US$ 1,000.- 120.-	(16) 4,000.- 720.- US$4,720.-

```
Say USDOLLARS(17) Four Thousand Seven Hundred and Twenty Only.
Drawn under:L/C(18) No.567 issued by ANZ Bank, Sydney, dated Oct.
15, '84.   (19)
Insurance :buyer's care      (20) A B C CO.,LTD

                (21)               Export Manager
```

關於商業發票的用語

1. signed manually signed visaed legalized sworn certified	commercial invoice in	duplicate triplicate quadruplicate quintuplicate sextuplicate septuplicate octuplicate decuplicate

2. signed commercial invoice indicating IL No…

3. signed commercial invoice duly countersigned by Mr. A.

二、海洋提單(Ocean Bill of Lading, Marine Bill of Lading)

　　海洋提單爲船公司或其代理人所簽發證明收到貨物，並約定將貨物自一地運至另一地，交給提單持有人的一種物權證書(Document of Title)。海洋提單本質上爲載運貨物的化身，係各種貨運單證中最重要者。

　　1. 提單例示及說明

　　說明：(編號如例示)

⑴ Shipper：即託運人，通常即爲出口商。

⑵ Consignee：受貨人，通常在此欄填上"to order"(待指定)；"to order of shipper"(待託運人指定)；但也有以"to order of xxx Bank"表示者。

⑶ Notify Party：被通知人，通常爲進口商或其代理人。

⑷ B/L No.：提單編號；由船公司編列。

⑸ Shipped on：表明貨物已裝上船。所以本 B/L 爲 Shipped on Board B/L。

⑹ Vessel：填入船名。

(7) Port of Loading: 裝貨港。

(8) Port of Discharge: 卸貨港。

(9) Place of Delivery by on-carrier: 轉運港。

(10) Marks & Nos.: 嘜頭及件號。

(11) Number and Kind of Package; Description of Goods: 件數及包裝種類; 貨物名稱。

(12) Gross Weight: 毛重，卽貨物總重量。

(13) Measurement: 尺碼，卽貨物體積。

(14) Total Number of Packages: 總件數。

(15) Specification of Freight and Charges: 運費計算說明。

(16) Freight Payable at: 塡上運費支付地點。

(17) Place and Date of Issue: 簽發（提單）地點及日期。

(18) Number of Original Bs／L: 正本提單份數。

(19) Signature: 簽發，提單人簽字。

(20) Acceptance Clause: 表明託運人同意接受本 B/L 各項規定的條款，稱爲同意條款。

No.107　海洋提單

INTERASIA LINES, LTD.

B/L NO　PNTA-901

Interasia Lines
BILL OF LADING

Shipper Marubeni Corporation, London Branch, New London Bridge House, London Bridge Street, London SE1 9SW, United Kingdom.	RECEIVED by the Carrier from the Shipper in apparent good order and condition unless otherwise indicated herein, the Goods, or the container(s) or package(s) said to contain the cargo herein mentioned, to be carried subject to all the terms and conditions provided for on the face and back of this Bill of Lading by the vessel named herein or any substitute at the Carrier's option and/or other means of transport, from the place of receipt or the port of loading to the port of discharge or the place of delivery shown herein and there to be delivered unto order or assigns. In accepting this Bill of Lading, the Merchant (as defined by Article 1 on the back hereof) agrees to be bound by all the stipulations, exceptions, terms and conditions on the face and back hereof whether written, typed, stamped or printed, as fully as if signed by the Merchant, any local custom or privilege to the contrary notwithstanding, and agrees that all agreements or freight engagements for and in connection with the carriage of the Goods are superseded by this Bill of Lading. In witness whereof, the undersigned, on behalf of INTERASIA LINES, LTD. the Master and the owner of the Vessel, has signed the number of Bill(s) of Lading stated above, all of this tenor and date, one of which being accomplished, the others to stand void.
Consignee Order of Central Trust of China, Banking & Trust Department, 49, Wu Chang Street, Sec. 1, Taipei, Taiwan 100.	

Notify party Taiwan Machinery Mfg. Corpn/ Ministry of Economic Affairs 15, Foo-Chow Street, Taipei, Taiwan.	**Ocean vessel** (Liner Vessel) Asian Princess	**Voy. No.** 84	**Pre-Carriage by**
	Port of loading Penang		**Port of discharge** Kaohsiung
	Place of receipt Penang　CFS		**Place of delivery** Kaohsiung　CY
Final destination (for the Merchant Reference)			

ORIGINAL
FIRST

Marks and Numbers	No. of Containers or Pkgs. Kind of Packages	Contents/Container No. & Seal No.	Gross weight & Measurement
C　　　T GF4-671400 78-GF4-1739 MOEA (249) P　　　D KAOHSIUNG BUNDLE NOS:　1-35　(S.T.C.) 　　36-115　(Escoy) 　　116-150　(Escoy) 　　151-180　(Escoy) 　　181-250　(S.T.C.) 　　251-325　(S.T.C.) Corrective Approved 7.3.8.	9,130 326-340　(S.T.C.) 341-415　(S.T.C.)	Ingots Straits Refined Tin (In Bundles of 22 Ingots each bundle) Letter of Credit No. 8DH1/02141/01 CTC Invitation No.　GF4-671400 Contract No.　78-GF4-1739 Import Licence No.　67RA1-002393 (Kaohsiung) 　　　　　　　　　　67RA1-002391 (Kaohsiung) "Freight Prepaid"	415,000 Kgs.

TOTAL NO. OF CONTAINER OR PACKAGES　Nine Thousand, One Hundred & Thirty Ingots Only.
(Four Hundred & Fifteen Bundles only).

FREIGHT & CHARGES	Revenue tons	Rate	Per	Prepaid	Collect

Ex rate	Prepaid at	Payable at	Dated Kuala Lumpur　28.10.79
			INTERASIA LINES, LTD. For and on behalf of INTERASIA LINES INTEGRATED FORWARDING & SHIPPING SDN. BHD?
Total prepaid	**No. of original Bs/L signed** Three (3)	**Place of Bs/L issue** Kuala Lumpur.	As Agents.

DULY ON BOARD
DATE: 28 OCT 19—
SIGNATURE

2.提單種類

(1) 按貨物是否已裝船分 $\left\{\begin{array}{l}① \text{Shipped B/L}（裝運提單）\\② \text{Received B/L}（備運提單）\end{array}\right.$

(2) 按是否轉運分 $\left\{\begin{array}{l}① \text{Direct B／L}（直達提單）\\② \text{Through B/L}（聯運提單）\end{array}\right.$

(3) 按受貨人表示方法分 $\left\{\begin{array}{l}② \text{Straight B/L}（直接提單）\\① \text{Order B/L}（指示提單）\end{array}\right.$

(4) 按是否有瑕疵批註分 $\left\{\begin{array}{l}① \text{Clean B／L}（清潔提單）\\② \text{Unclean B/L}（不潔提單）\end{array}\right.$

(5) 按是否詳載運輸條款分 $\left\{\begin{array}{l}① \text{Short form B/L}（簡式提單）\\② \text{Long form B/L}（長式提單）\end{array}\right.$

(6) 其他提單 $\left\{\begin{array}{l}① \text{Container B/L}（貨櫃提單）\\② \text{Combined Transport B／L}（聯合貨運提單）\end{array}\right.$

3.提單上的各種批註

Top(Cap, Head)off	蓋子鬆落
Nails off	鐵釘鬆落
Seal broken	封印破裂
Seals off	封印鬆落
Bundle off	捆紮鬆弛
Hoop off	箍帶鬆落
Rattling	裡面商品動搖發出聲音
Leaking	漏出
Bented	彎曲
Hoop rusty	鐵及帶生銹
Bag torn	袋子破裂
Bale torn	袋子破裂

Renailed	補釘
Staves off	標蓋鬆落
Split	裂開
Mark mixed	嘜頭混淆
Mark indistinct	嘜頭不清楚
Dented	凹下
Cover chafed	擦傷
Old (Second-hand) case	舊箱
Used drum	舊鐵桶
Loose packing	包裝鬆弛
Crushed	壓壞
Label missing	標籤脫落
Two cartons in wet condition	兩箱潮濕
Two cases broken	兩箱破裂
Contents crushed	壓碎

N/R for
$\begin{cases} \text{breakage} \\ \text{mortality} \\ \text{wither} \\ \text{melting} \\ \text{perishing} \\ \text{damage} \end{cases}$
$\begin{cases} \text{破損} \\ \text{死亡} \\ \text{枯萎} \\ \text{溶解} \\ \text{腐敗} \\ \text{損害} \end{cases}$ 不負責

〔註〕N／R 爲 Not responsible 的縮寫。

三、保險單(Insurance Policy)

保險單是證明保險契約成立的正式憑證，載明雙方當事人約定的權利及義務，如與保險公司訂有流動保單(Floating Policy)，則對各批貨物的保險, 保險公司另簽發保險證明書(Certificate of Insurance)。保險單依其分類標準的不同, 有如下的分類:

  1.依船名是否已確定分

  (1) Named policy (船名確定保單), 又稱爲"Definite policy"。

  (2) Unamed policy (船名未確定保單), 又可分爲:

    ① Floating policy (流動保單): 一種僅列明概括的條件, 而船名和其他條件留待事後通知 (Declare)的保險單。

    ② Open policy (預約保單): 以預約方式一次承保未來多批貨物的保險單。

  2.依保險金額是否確定分

    ① Valued policy (定值保單): 即保險單上載明保險標的物價值 (即保險價額 Insured value)者。

    ② Unvalued policy (不定值保單): 即保險單上僅訂有保險金額 (Insured amount)的最高限度, 而將保險價額留待日後保險事故發生時再予補充確定者。

四、裝箱單（Packing List）

No.108　保險單

華僑產物保險股份有限公司
MALAYAN OVERSEAS INSURANCE CORPORATION

NO. 56, Tun Hwa North Road, Taipei, Taiwan, Republic of China
Telex: 11122 MILCINC, 21672 MILCINC, Cable: MILCINC TAIPEI
Telephone (02)7752088 Facsimile 886-02-7416591(Marine)

POLICY NO.

Claim, if any, payable at Taipei
in N.T. Dollars

MARINE CARGO POLICY

ASSURED

特約條款
IMPORTANT CLAUSE

Invoice No

Amount insured

Per Approved Vessel of which Name and Sailing Date are TO BE DECLARED subject to
Institute Classification Clause as per back hereof

From   To

SUBJECT-MATTER INSURED

Conditions

Marks and Numbers as per Invoice No. specified above.   Valued at the same as Amount insured.

Place and Date signed in   Number of Policies issued

☞ The Assured is requested to read this policy and if it is incorrect return it immediately for alteration.

IMPORTANT

INSTITUTE REPLACEMENT CLAUSE (applying to machinery)

LABEL CLAUSE (applying to labelled goods)

CO-INSURANCE CLAUSE (applicable in case of Co-Insurance)

For and on behalf of MALAYAN OVERSEAS INSURANCE CORPORATION

Charles Wang

President

Not valid unless Countersigned by ....
05-056-2
78. 4. 10. 000份

FORM NO. MPUE 00

又稱爲包裝單、花色碼單, 英文又稱爲 Packing Specification, Specification of Packages 或 Specification of Contents, 係說明一批包裝貨物, 逐件內容的文件, 爲商業發票的補充文件。

包裝貨物如每件內容不同時, 必須有裝箱單, 否則受貨人除非拆包, 將無法知悉各件包裝的內容, 海關、公證行抽驗時也將因而遭遇困難。

No.109　裝箱單

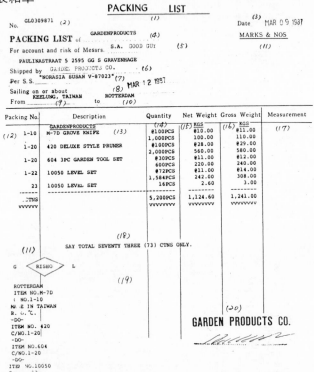

五、重量尺碼證明書 (單) (Weight/Measurement Certificate/List)

係記載裝運貨物每一件重量及體積的文件, 這種文件或由賣方出具或由公共丈量人(Public weigher)出具。

No.110　重量尺碼證明書（單）

```
Shipper                              NIPPON KAIJI KENTEI KYOKAI
KAISEI SANGYO CO., LTD.              (JAPAN MARINE SURVEYORS & SWORN MEASURERS' ASSOCIATION)
                                     FOUNDED IN 1913 & LICENSED BY THE JAPANESE GOVERNMENT
OKAMOTO FREIGHTERS LTD.
Certificate No.
1300-10420-0005386 (03)              CERTIFICATE AND LIST
Sheet                                        OF
        1
Certificate issued                   MEASUREMENT AND/OR WEIGHT
YOKOHAMA  NOV 4, 19—
                                         HEAD OFFICE:
Ref. No. (For our reference)         KAIJI BLDG., No. 5-7, 1-CHOME, HATCHOBORI
M2090-1538  004  (0339)  0239                CHUO-KU, TOKYO-104
                                                 JAPAN
Ocean Vessel       Port of Loading
VIRGINIA              YOKOHAMA
Port of Discharge  Date & Place of Measuring and/or Weighting
Kaohsiung            NOV. 3, 19—, YOKOHAMA
```

Marks & Nos.	No. of P'kgs.	Kind of P'kgs,	Description	G. W.	Meast.
T F M C					
KAOHSIUNG					
KSCL-F116					
MADE IN TAIWAN					
C/NO. 1-2					
			"GOLD" VENEER ROTARY KNIFE	KG	CU.METER
	2 CASES			680	0.265
	******				
DETAILS:	CASE	M CM M CM M CM		KG	CU.METER
		L W H			
	1	2 89 0 27 0 18		360	0.140
	2	2 58 0 27 0 18		320	0.125

```
We hereby certify that the above measurements
and/or weights of the goods were taken by our
measurers solely for reasonable ocean freight
in accordance with the provisions of recognized
rules concerned.                                 N.K.K.K.
```

六、領事發票(Consular Invoice)

又稱爲領事簽證貨單，是由出口商向駐在輸出國的輸入國領事申請簽發的一種單證，其作用爲：1.供作進口國海關和貿易管理當局統計之用。2.代替產地證明書之用。3.限制或禁止某些非必需品或未經批准的貨品隨便進口。4.藉簽證課征規費，做爲領事辦公費。

No.III 巴拿馬領事發票（正面）

REPUBLICA DE PANAMA
MINISTERIO DE HACIENDA Y TESORO
DIRECCION GENERAL DE CONSULAR Y DE NAVES

Nº 00338   E

VALOR B/.3.00 EL JUEGO
LEY 52, ENERO 4, DE 1963
F.C.77 (90.000)

FACTURA CONSULAR
CONSULAR INVOICE

DISTRIBUCION

ORIGINAL : DEBERA SER PRESENTADO A LAS AUTORIDADES ADUANERAS
DUPLICADO : DIRECCION GENERAL DE CONSULAR Y DE NAVES
TRIPLICADO : CONTRALORIA GENERAL
CUADRUPLICADO ARCHIVO CONSULADO

POR EFECTOS EMBARCADOS EN:

PAIS DE ORIGEN: Taiwan, R.O.C.
COUNTRY OF ORIGIN

PUERTO DE EMBARQUE: KEELUNG
PORT OF SHIPMENT

PUERTO DE LLEGADA: CRISTOBAL
PORT OF ARRIVAL

VAPOR: NDLLOYD ELBE
NAME OF VESSEL

FECHA DE ZARPE: DATE OF SHIPMENT
Jan. 5. 19-

VENDEDOR (ES): SELLERS OR SHIPPERS MAIN ...
ENTERPRISE CO.,LTD.

CONSIGNADO A: RODIVAN S.A.
CONSIGNEE

LUGAR DE DESTINO: FINAL DESTINATION

NO. CONOCIMIENTO DE EMBARQUE Y FECHA: BILL OF LADING & DATE
Jan. 5. 19-

NO. DE MANIFIESTO Y FECHA
NO. OF MANIFIEST & DATE

IDENTIFICACION NUMBER	BULTOS NUMBER OF PACKAGES	CLASE DE BULTOS KIND OF PACKAGES	CAPACIDAD EN LITROS CAPACITY IN LITRES	PESO EN KILOS NETO NET WEIGHT	BRUTO GROSS WEIGHT	DESCRIPCION DE LAS MERCADERIAS DESCRIPTION OF THE MERCHANDISE	VALOR PARCIAL US$	VALOR TOTAL US$
C/No.		W/cases	doz.	Kgs.	Kgs.	Cosmetics: "EXOTICA" BRAND	(per doz)	
1-16	16		64	82	96	NP-900/F Make-up Kits	9 00	564 00
17-50	34		136	177	224	NP-555/B Make-up Kits	10 00	1,360 00
51-80	30		120	148	180	MSA-124 Mascara	7 64	916 80
81-92	12		48	73	102	M-108 Mascara	5 20	145 60
93-100	8		40	60	79	NP-252/B Make-up Kits	4 50	180 00
101-108	8		48	84	127	M-173/C Mascara	6 70	321 60
	108	W/cases	456	624	808	TOTAL		3,488 00
	vvvvvv	vvvvvvv	vvv	vvv	vvv			vvvvvvvv

SAY: US DOLLARS THREE THOUSAND FOUR HUNDRED EIGHTY EIGHT ONLY.

SAY: TOTAL ONE HUNDRED AND EIGHT WOODEN CASES ONLY.

SAY: TOTAL FOUR HUNDRED FIFTY SIX DOZEN ONLY.

TOTAL	108 W/cases	VALOR TOTAL DE LA MERCANCIA	US$3,488.00

MARCAS MARKS			USO DE ADUANAS FOR CUSTOM ONLY
RODIVAN CRISTOBAL PANAMA C-2307 C. 1-108 IN TAIWAN MADE IN CHINA	FLETE INTERNO INLAND FREIGHT	TOTAL F.O.B. US$3,488.00	
	MUELLAJE Y MANEJO HANDLING CHARGES	FLETE FREIGHT 536.00	NO. DE LIQUIDACION
	DEMORAS ZARPE DELAY IN SHIPMENT		
	OTROS GASTOS OTHER CHARGES	SEGURO INSURANCE Nil	OFICIAL DE ADUANAS
	DESCUENTO DISCOUNT	TOTAL CIF: (C&F) US$4,024.00	AUDITOR:
	COMISION COMMISSION		

NOTA: LAS FACTURAS CONSULARES Y DEMAS DOCUMENTOS QUE SE PRESENTEN PARA LA CERTIFICACION CONSULAR 8 DIAS HABILES DESPUES DE LA FECHA DE EXPEDICION DEL CONOCIMIENTO DE EMBARQUE, PAGARAN UN RECARGO DEL 1% ANTE EL CONSUL RESPECTIVO. (ART 447 DEL CODIGO FISCAL)

IMP. CERVANTES 52431

七、海關發票(Customs Invoice)

　　目前貨物銷往美國、加拿大、紐西蘭、澳洲、南非聯邦、西非各國時，出口商須提出海關發票，其作用：1.供作進口國海關統計之用。2.供作產地證明之用。3.供作進口國查核有無傾銷之用。

No.112　美國特別海關發票

DEPARTMENT OF THE TREASURY UNITED STATES CUSTOMS SERVICE 19 U.S.C. 1481, 1482, 1484	SPECIAL CUSTOMS INVOICE (Use separate invoice for purchased and non-purchased goods.)	Form Approved. O.M.B. No. 48-RO342

1. SELLER
Burda Enterprises Inc.
1 Shingshen S. Road, Sec. 1
Taipei, Taiwan,
Rep. of China

2. DOCUMENT NR.*

3. INVOICE NR. AND DATE*
No.123, May 10, 1986

4. REFERENCES*

5. CONSIGNEE
The L.P. Henryson Company, Inc.
18W, 33rd St., New York, N.Y. 10001

6. BUYER *(if other than consignee)*
same as consignee

7. ORIGIN OF GOODS
Taiwan, Rep. of China

8. NOTIFY PARTY*
Same as consignee

9. TERMS OF SALE, PAYMENT, AND DISCOUNT
CIF, By Sight L/C

10. ADDITIONAL TRANSPORTATION INFORMATION*

11. CURRENCY USED　US Dollar

12. EXCH. RATE *(if fixed or agreed)*　US$1=NT$38

13. DATE ORDER ACCEPTED　April 10, 86

14. MARKS AND NUMBERS ON SHIPPING PACKAGES	15. NUMBER OF PACKAGES	16. FULL DESCRIPTION OF GOODS	17. QUANTITY	18. HOME MARKET	19. INVOICE	20. INVOICE TOTALS
L.P. NEW YORK C/1 – 100 MADE IN TAIWAN REP. OF CHINA	100	STRETCH NYLON YARN DYED 100/24 /2 100 ctns. each contains 60 lbs.	6,000 lbs	US$1.50 per lb.	US$1.50 per lb.	C.I.F. New York US$9,000.–

21. ☐ If the production of these goods involved furnishing goods or services to the seller (e.g., assists such as dies, molds, tools, engineering work) and the value is not included in the invoice price, check box (21) and explain below.

27. DECLARATION OF SELLER/SHIPPER (OR AGENT)

I declare:

(A) ☐ If there are any rebates, drawbacks or bounties allowed upon the exportation of goods, I have checked box (A) and itemized separately below.

(B) ☐ If the goods were not sold or agreed to be sold, I have checked box (B) and have indicated in column 19 the price I would be willing to receive.

I further declare that there is no other invoice differing from this one (unless otherwise described below) and that all statements contained in this invoice and declaration are true and correct.

(C) SIGNATURE OF SELLER/SHIPPER (OR AGENT):
Burda Enterprises Inc.

28. THIS SPACE FOR CONTINUING ANSWERS

22. PACKING COSTS	US$100.
23. OCEAN OR INTERNATIONAL FREIGHT	US$400.–
24. DOMESTIC FREIGHT CHARGES	US$10.–
25. INSURANCE COSTS	US$90
26. OTHER COSTS (Specify Below)	

THIS FORM OF INVOICE REQUIRED GENERALLY IF RATE OF DUTY BASED UPON OR REGULATED BY VALUE OF GOODS AND PURCHASE PRICE OR VALUE OF SHIPMENT EXCEEDS $500. OTHERWISE USE COMMERCIAL INVOICE.

*Not necessary for U.S. Customs purposes.

Customs Form 5515 (12-20-76)

No.113 檢驗報告

TMS

TEL: 53827 / 52003
CABLE ADD. "ASIALINE"

This report is issued in good faith and to the best of our knowledge and ability but without responsibility on our part.

寶島海事檢定有限公司
Taiwan Marine Survey, Ltd.
2FL. NO.10 LIN HAI 2ND ROAD.
KAOHSIUNG. TAIWAN.

REPORT OF SURVEY

NO.... TMS-76-008
PAGE............1...
DATE 23rd Dec. 1976

CERTIFICATE OF INSPECTION

CARGO: 162 Bales - 275,000 yards of Grey Polyester 65% Cotton 35% Cloth 45/45 88/64 38 Inches for Dyeing.

MARKS:
FRONT
CH65-03535
ANTWERP
B/NO.1-162
NW KGS
GW KGS

REAR
POLESTER/COTTON
BLENDED SHIRTING
KEEP DRY
DONT HOOKS

VESSEL: Per s.s. "Orient Chief" from Keelung to Czechoslovakia Antwerp, Sailing on about 23rd December 1976.

SHIPPERS: Messrs. Spinning Co., Ltd.

BUYERS: Messrs. Del Gottardo.

THIS IS TO CERTIFY THAT we, the undersigned General Marine Surveyor, did, at the request of Messrs. Spinning Co., Ltd. attended on 22nd December 1976, at the Central Freight Terminal Co., Ltd. Keelung, for the purpose of inspecting the above mentioned cargo, and report as follows:-

PACKING: The grey cloth press-packed in gunny bales, reinforced with wooden strips and secured by iron bands, marks as above.
162 bales in total quantity presented for inspection.

CONDITION: All bales appeared in good order and condition at time of our inspection.

INSPECTION: All bales of outside, as far as could be ascertained externally, were found to the bales of outside without marking the country of origin and the name of manufactory or the name of shippers.

A number of bales taken at random and opened in our presence and the contents were inspected and found to the grey cloths contained therein was yardlapped into bolts and the bolts of both ends found to without any marking stencilled and stamped on the pieces.

Among the inspection, the bales outside and the bolts opened were verified to be in conformity with the specification of requirement and the goods are 100% neutral both inside and outside.

TAIWAN MARINE SURVEY, LTD.

Director

八、檢驗證明書(Inspection Certificate)

　　爲防止出口商裝出的貨物品質不合契約規定或短裝，進口商常要求出口商須提出檢驗證明書。此外，爲符合輸入國海關規定，進口商也要求出口商提供檢驗證明書。

　　檢驗證明書除爲保障出口品質，維護國家信譽或依輸入國海關的規定須由輸出國政府機構（例如我國檢驗局）簽發外，大多由1.製造廠或同業公會。2.公證行。或進口商指定的代理人。

九、產地證明書(Certificate of Origin)

　　產地證明書的作用有四：

　　1.供作享受優惠稅率的憑證。2.防止貨物來自敵對國家。3.防止傾銷。4.供作海關統計。

　　產地證明書通常由商會簽發。

　No.114　產地證明書

臺灣省商業會

Taiwan Chamber of Commerce

CABLE ADDRESS

P. O. BOX 1762, TAIPEI, TAIWAN "TCOC" TAIPEI

THE REPUBLIC OF CHINA TEL. 313144, 314152, 319668

臺北市公園路三號三泰大樓七樓

CERTIFICATE OF ORIGIN　　　　　Sept. 9, 19…

To whom it may concern:

　　We hereby certify the under-mentioned commodity is of Taiwan Origin.

Commodity: Tee Shirts for Men's Size Assortment from 32-42 White

Quantity: 1,500 doz.(38 Cartons)

Shipment: Shipped per s.s. "PONAPE MARU"/"TA-HITI MARU" from Keelung Taiwan to Papeete with transhipment at Yokohama

Credit: L/C No. 942/74 issued by Banque De L'indo-chine, Papeete, Tahiti

Shipper: TAIWAN Textile Manufacturing Co., Ltd., Taipei, Taiwan

Shipping Marks:

PAPEETE

VIA

YOKOHAMA

C/No. 1-38

MADE IN TAIWAN

REPUBLIC OF CHINA

習　　題

一、國際貿易貨運單證有那些？試列舉重要者並說明其用途。

二、將下例單證譯成英文。

　　1.重量尺碼單　　　　2.裝箱單　　　　　　3.特別海關發票

　　4.產地證明書　　　　5.領事發票

三、將下列英文譯成中文。

1. straight B/L

2. unclean B/L

3. combined transport B/L

4. N/R

5. open policy

6. insured value

7. public weigher

第二十四章　滙　　票

(Bill of Exchange)

第一節　滙票的意義

滙票(Bill of Exchange, Draft, 簡稱爲 Bill, Exchange, 縮寫爲 B/E)，依英國票據法，其意義爲：

"A bill of exchange is an unconditional order in writing addressed by one person to another, signed by the person giving it, requiring the person to whom it is addressed to pay on demand or at fixed or determinable future time a sum certain in money to the order of a specified person, or to bearer."

（所謂滙票，乃經一人簽署，給他人的無條件書面命令，要求受命人於見票時或在未來一定日期或在將來可確定的日期，支付一定的金額給一特定人或其指定人或執票人。）

我國票據法則謂：「稱滙票者，謂發票人簽發一定之金額，委託付款人於指定之到期日，無條件支付與受款人或執票人之票據。」

第二節　滙票格式

國際貿易使用的滙票一般都用英文，其格式如下：

No.115　滙票

```
Draft No.(1)              BILL OF EXCHANGE
                                (2)
For      (4)                        Taipei,     (3)
At      (5)      sight of this FIRST of Exchange(Second
                                (6)
the same tenor and date being unpaid)Pay to the order
of HUA NAN COMMERCIAL BANK, LTD.
the sum of   (7)
_____ value received
                                            (8)
Drawn under   (9)
Irrevocable L/C No._____ dated _____
To _____(10)_____
                                  (11)
```

【說明】：

(1) 滙票號碼。

(2) Bill of Exchange:表明其爲滙票。

(3) 表明發票地點及日期。

(4) Exchange for:用阿拉伯字表示滙票金額。

(5) 填上滙票期限，例如，卽期時，"at sight"，90 天到期則"at 90 days' sight"。

(6) 其大意爲：「憑本滙票第一聯(以旨趣及發票日期相同之第二聯滙票尙未付訖爲限) 見票 （或見票後……天) 支付華南商業銀行或其指定人……。」

⑺ 填上金額 (以文字表示)。

⑻ valued received:價款收訖，意指發票人對付款人承認收到本滙票所載金額。

⑼ "Drawn under"後面填上開狀銀行名稱。"Irrevocable L/C NO."後面填上信用狀號碼。"dated"後面填上開狀日期。本條款稱爲"Drawn Clause"(發票條款)，在憑 L／C 開發滙票時，都須填上，如非憑 L／C 者，免填。

⑽ "To"後面填上付款人名稱。

⑾ 填上發票人名稱及發票人簽字。

第三節　滙票的種類

滙票從不同的角度來看，有種種的分類。

一、依滙票的發票人及付款人身分分

　　1.銀行滙票(Banker's draft or bill)：銀行向銀行發出的滙票。通常用於順滙。

　　2.商業滙票(Commercial draft, Trade bill)：商人向商人或銀行發出的滙票。

二、依滙票是否附有貨運單證分

　　1.跟單滙票(Documentary draft or bill)：附有貨運單證的滙票。這種滙票可向銀行申請押滙，所以又稱押滙滙票。

　　2.光票(Clean draft or bill)：未附有貨運單證的滙票。

三、依滙票期限分

　　1.卽期滙票(Sight draft or bill, Demand draft or bill)：見票(sight)卽付或要求(on demand)卽付的滙票。

　　2.定期滙票(Time draft or bill)或遠期滙票(Usance draft or bill)：卽將來某一時日付款的滙票，可分爲

　　⑴發票後定期付款滙票：卽發票日後一定日期付款的滙票，例如

　　　"ninety days after date"，即以發票日後九十天付款。

　　(2)見票後定期付款滙票: 即見票日後一定期間付款的滙票，如
　　　"ninety days after sight"，即以見票日後九十天付款。

　　(3)定日付款滙票: 即以某特定日付款的滙票。

四、依交付單證方式分

　1.付款交單滙票(Documents against payment draft or bill,
　　D/P bill): 付款人付清票款後才交付貨運單證的滙票, 又稱爲付款
　　押滙滙票(Documentary Payment Bill)。

　2.承兌交單滙票(Documents against acceptance draft or bill,
　　D/A bill): 即滙票經付款人承兌(accept)後即交付貨運單證的滙
　　票, 又稱承兌押滙滙票(Documentary Acceptance Bill)。

第四節　滙票關係人及票據行爲

一、滙票關係人

　　Drawer　發票人

　　Drawee
　　Payer ｝被發票人(付款人)

　　Payee　受款人

　　Bearer　　執票人

　　Endorser　背書人

　　Endorsee　被背書人

　　Holder　　持票人

　　Acceptor　承兌人

　　Surety　保證人

　　Payment agent　擔當付款人

Referee in case-of-need　預備付款人

Acceptor for honor　　參加承兌人

Payer for honor　參加付款人

The party for whose honor acceptance or payment is made
被參加人

Prior party

Preceding endorser ｝前手

Subsequent party　後手

【註】

1. Holder:為持有票據，有權要求支付票據金額的任何人。

2. Acceptance:遠期滙票的付款人（被發票人）同意發票人的支付命令而在滙票正面簽名的行為，稱為承兌(Accept)，被發票人在滙票上簽名承兌時，此被發票人即為承兌人(Acceptor)。

3. Payment agent:即代替付款人擔當支付票據的人。

4. Reference in case-of-need:指除付款人外，發票人或背書人另行指定的付款地的人，使其在付款人拒絕承兌或付款時，承擔承兌或付款責任。

5. Acceptor for honor:當遠期滙票被"Drawee"拒絕承兌時，由另一個人代付款人承兌的行為，叫做參加承兌(Acceptance for honor)，這另一個人稱為"Acceptor for honor"，原來的"Drawee"稱為被參加承兌人。

6. Payer for honor:包括預備付款人和其他任何人，在滙票遭拒付時，代發票人、背書人等對執票人付款的行為叫做"Payment for honor"(參加付款)，原來的"Drawee"就叫做被參加付款人。

7. Surety:票據債務人以外的第三者，在滙票上簽字保證票款的支付的人。

8. Presentment:又稱為"Presentation"，即執票人將滙票送請"Drawee"承兌或付款的行為。

9. Recourse:票據不獲承兌或付款時，執票人對前手要求償還票款及費用的行爲。

10. Protest:執票人可獲承兌或付款時，要求製作拒絕證書機構出具拒絕證書的行爲稱爲"Protest"。所作成的文書，叫做拒絕證書(Protest)。

二、票據行爲

Draw, Drawing	發票
Endorse, Endorsement	背書
Negotiate, Negotiation	轉讓；讓購
Present, Presentment	提示
Accept, Acceptance	承兌
Pay, Payment	付款
Payment, Domiciled	擔當付款
Non-acceptance	拒絕承兌
Non-payment	拒絕付款
Acceptance for Honor	參加承兌
Payment for Honor	參加付款
Recourse	追索
Protest	做成拒絕證書

第五節　有關滙票的有用例句

一、動詞

1. To draw a $\left\{ \begin{array}{l} \text{bill} \\ \text{draft} \end{array} \right\} \rightarrow$ ·簽發滙票

$\rightarrow \left\{ \begin{array}{l} \text{on} \\ \text{upon} \end{array} \right\} \left\{ \begin{array}{l} \text{City Bank} \\ \text{a person} \end{array} \right\} \rightarrow$ ·on 後面爲被發票人

→ $\left\{ \begin{array}{l} \text{in favor} \\ \text{to the order} \end{array} \right\}$ of some person →

・in favor ⎫ of 後面
　to the order ⎭ 爲收款人

→ at $\left\{ \begin{array}{l} \text{sight} \\ \text{90 days' sight} \end{array} \right\}$ →

・at 後面爲滙票期限

→ for US\$12,345.00 →

・for 後面爲金額

→ against $\left\{ \begin{array}{l} \text{documents} \\ \text{shipment} \end{array} \right\}$ →

・against 爲「憑……交換」之意，即 in exchange for 之意

→ covering $\left\{ \begin{array}{l} \text{order No.567} \\ \text{the CIF invoice amount} \end{array} \right\}$ →

・covering 爲抵付之意

→ under L/C No.12 $\left\{ \begin{array}{l} \text{opened} \\ \text{issued} \\ \text{established} \end{array} \right\}$ by City Bank →

・under 爲「依據」,「憑」之意

→ $\left\{ \begin{array}{l} \text{through} \\ \text{with} \end{array} \right\}$ Bank of Taiwan

・through 爲「經由，透過」之意

2. To accept(sign)a bill　　　承兌滙票

3. To $\left\{ \begin{array}{l} \text{honor a bill} \\ \text{protect a bill} \\ \text{take up a bill} \end{array} \right.$　　兌付滙票
　　　　　　　　　　　　兌付滙票
　　　　　　　　　　　　兌付滙票

To $\left\{ \begin{array}{l} \text{give} \\ \text{afford} \end{array} \right\}$ one's protection　兌付滙票

4. To meet a bill upon presentation at maturity　　到期提示付款

5. To $\left\{ \begin{array}{l} \text{provide for a bill} \\ \text{prepare due honor for a draft} \end{array} \right.$　準備兌付滙票
　　　　　　　　　　　　　　　　準備兌付滙票

6. To $\begin{cases} \text{recommend} \\ \text{commend} \end{cases}$ a bill to one's protection   請某人兌付滙票

7. To ask one's protection for a bill   請某人兌付滙票

To solicit the favor of one's acceptance of a draft   請某人承兌滙票

To request a person to protect a bill   請某人兌付滙票

8. To endorse a bill   將滙票背書

9. To $\begin{cases} \text{discount} \\ \text{negotiate} \end{cases}$ a bill   將滙票 $\begin{cases} \text{貼現} \\ \text{讓購} \end{cases}$

10. To $\begin{cases} \text{withdraw} \\ \text{retire} \end{cases}$ a bill   贖票

11. To collect a bill   將滙票託收

12. To encash a bill   將滙票兌現

13. To present a bill for $\begin{cases} \text{acceptance} \\ \text{payment} \end{cases}$ 提示滙票承兌
提示滙票付款

14. To dishonor a bill   拒絕兌付滙票

To dishonor a bill by $\begin{cases} \text{non-acceptance} \\ \text{non-payment} \end{cases}$ 拒絕滙票的 $\begin{cases} \text{承兌} \\ \text{付款} \end{cases}$

15. To return a bill $\begin{cases} \text{accepted} \\ \text{unaccepted} \\ \text{unpaid} \end{cases}$ $\begin{cases} \text{承兌後} \\ \text{拒兌後} \\ \text{拒付後} \end{cases}$ 退還滙票

16. To $\begin{cases} \text{protest} \\ \text{note} \end{cases}$ a bill   作成拒絕證書

To get a bill protested   使…作成拒絕證書

To cause a ptotest to be made   使…作成拒絕證書

二、有關滙票的例句

1. We would $\begin{cases} \text{ask you to give} \\ \text{commend} \end{cases}$ our draft your kind protection.

2. We would $\left\{ \begin{array}{l} \text{solicit the favor of} \\ \text{claim} \end{array} \right\}$ $\left. \begin{array}{l} \text{your kind protection} \\ \text{to our draft} \end{array} \right.$

　　(惠請予以兌付〔承兌，支付〕我們所開的滙票。)

3. We have the pleasure to inform you that we have $\left\{ \begin{array}{l} \text{drawn} \\ \text{valued} \end{array} \right\}$ on

→ you this day for US$…, at 3 m/s, to the order of xxx Bank, which we commend to your kind protection.

　　＊ 3 m/s: 3 months after sight

　　　3 m/d: 3 months after date

4. Against this shipment, we have drawn on you at sight for the invoice amount of US$…, in favor of Bank of Taiwan, which please protect on presentation.

5. We have valued on your goodselves by 60 days' sight draft for US$…, under L/C No. 123 issued by Bank of America in favor of Bank of Taiwan, to whom we handed full set of shipping documents.

6. We have drawn a sight draft on you for the payment of the invoice value US$…under L/C 234 issued by City Bank and negotiated through Central Trust of China, Taipei.

7. According to the terms agreed upon, we have drawn on you at sight against the shipping documents through Bank of Taiwan. We ask you to protect the draft upon presentation.

習　　題

一、解釋下列名詞

　　1. Documentary Draft　　　　　2. Sola Draft

3. Acceptance   4. Acceptance for Honor

二、請根據下列資料，將下面滙票空格填入適當事項：

A. Beneficiary of L/C: ABC Trading Co., Ltd., Taipei.

B. Accountee:X. Y. Z. & Co., London.

C. Opening Bank: Westminster Bank, London.

D. Negotiating Bank: Bank of Taiwan.

E. Invoice Amount: Stg. £ 200,000.

F. Date of Negotiation: June 30, 1990.

G. L/C No. 123, dated May 1, 1990.

H. Some clauses in L/C.

We hereby authorize you to draw on us at 90 days after sight for account of X. Y. Z. Co., London...

This L/C expires on July 30, 1990

Bill of Exchange   (1) 19··········

For (2) _____

At (3) _____ sight of this First of Exchange(Second of the same tenor and date being unpaid) pay to the order of

(4) _____

the sum of (5) _____

Drawn under (6) _____

To:(7) _____   (8)

第二十五章　收款與付款

(Collection and Payment)

第一節　概説

　　一般而言，國際貿易貨款的清償通常多以押滙收款的方式完成。換言之，賣方於貨物運出後，開具滙票連同提單、商業發票、保險單等向外滙銀行押借款項，銀行則將匯票及貨運單證寄往進口地委託代理銀行收取貨款，這種押滙收款的方法，又可分爲「憑 L／C 收款」及「憑 D／P, D／A 滙票收款」兩種。

　　買賣雙方如往來密切，信用可靠，或雙方互有進出口的買賣關係時，也有採用「記帳」(Open Account)方式或分期付款(Instalment)等方式，清理帳款者。

　　憑 L／C 收款時, 出口商於發出裝運通知時, 順便通知將憑 L／C 前往銀行辦理押滙外, 通常多不另向進口商發出收款函。但是如採用 L／C 以外的方式收款, 或結算因貿易而生的附帶費用時, 通常多須發出收款函甚至催款函。而欠人的一方爲清付帳款, 也多須撰寫付款函。

第二節　收款信與付款信的寫法

一、憑 L／C 收款的函件

　　出口商在貨物裝出後, 卽可開具滙票備妥 L／C 規定的單證, 連同 L／

C 向當地外滙銀行申請押滙 (Negotiation)。向外滙銀行申請押滙時, 出口商須提出押滙申請書, 茲舉一例於下:

No.116 向銀行申請押滙

Huan Nan Commercial Bank

Foreign Department

Taipei   September 15,19···

Dear Sirs,

We are presenting to you herewith for negotiation our draft No.Y123 for DM28,950.- drawn under L/C No. 123 issued by Dresdner Bank accompanied by the following documents:

Commercial Invoice in triplicate

Packing List in triplicate

Marine Insurance Policy in duplicate

Certificate of Origin in triplicate

Bill of Lading in duplicate

For the proceeds, please have it settled in accordance with the regulations governing foreign exchange transactions.

In consideration of your negotiating the above documentary draft, we guarantee that you can receive the proceeds within two months, and further undertake to hold you harmless and indemnified against any discrepancy which may cause non-payment and/or non-acceptance of the said draft, and we shall refund you in original currency the

whole and/or part of the draft amount with interest and/or
expenses that may be accrued and/or incurred in connection
with the above on receipt of your notice to that effect, and
also verify that all advices relative to credit instruments
including amendment advice(s), if any, have been submitted to you without failure.

Faithfully yours,

【註】

這是一份出口押滙申請書，每家外滙銀行都備有這種空白申請書供出口商索取。

1. in consideration of:因為。

2. proceeds:票款。

3. undertake to hold you harmles:保證不使你受到損害。

4. indemnify against:使…免受損失、損害等。

5. non-payment:拒付。

6. non-acceptance:拒絕承兌。

7. to that effect:意旨；大意。notice to that effect:這種意旨的通知。

　例: I have received a cable to the effect that:我收到一封電報,大意是說…。

　He wrote to that effect:他寫的大意如此。

　　事實上，出口商初次向外滙銀行申請押滙時尚須簽具質押權利總設定書(General Letter of Hypothecation)。如出口商所提出的貨運單證與 L／C 不符，換言之，單證有瑕疵(discrepancy)，則外滙銀行可能不願接受押滙，但如出口商信用良好，可著其提供 Letter of Indemnity 之後，斟酌情形接受押滙。

二、憑 D／P、D／A 滙票收款的函件

No.117 出口商函告進口商已開出滙票並請其承兌—D／A

Dear Sirs,

Your order No.123

We take pleasure in informing you that, in accordance with your order No.123 of April 20, we have shipped you today by s.s. "President" from Keelung to New York, 500 cases of Fancy Goods, which we trust will reach you in good order and condition.

Herewith we hand you the following non-negotiable shipping documents for the shipment:

2 copies of commercial invoice

1 copy of bill of lading

2 copies of packing list

1 copy of certificate of weight and measurement

1 copy of special customs invoice

To cover this shipment we have drawn on you a 90 d/s draft for US$20,000 through our bankers. The original shipping documents are attached to the draft, and will be handed to you by our bankers through Bank of New York on your acceptance of the draft.

We shall be glad if you will duly protect the draft on presentation,and look forward to being favored with a continuance of your order.

Encl:A/S   Faithfully yours,

【註】

1. fancy goods:精美貨品；新奇貨品。

2. duly protect the draft:妥予兌付滙票。"protect"可以"honor"或"take up"代替。

3. to be favored with a continuance of your order:繼續惠賜訂單。

4. "and look forward...order"可以下列句子代替：

"and assure you that any further orders you may place with us will always be carefully attended to."

No.118　出口商通知進口商已開出滙票並請其付款—D／P

Gentlemen:

We thank you for your letter of December 4, giving us shipping instructions together with marks and numbers, for 10 cases Cotton Goods on your order of November 5. We now have the pleasure of apprising you that the goods are being shipped per N.Y.K. steamer "KONGO MARU" sailing on the 14th December, as per our Invoice enclosed.

As arranged, we are drawing a draft upon your good-selves at sight, D／P, for the amount, viz., US$5,000, and ask you to protect it. The shipping documents will be surrendered by the Oriental Bank against your payment of our draft.

The present market is a trifle easier, but this is owing more to the time of the year, which is between seasons, than to any other cause, and no doubt prices will be firm again in about a month's time.

We trust our shipment will give you every satisfaction and induce you to place with us your renewed orders, to which we shall at all times be most pleased to give our every attention.

Yours very truly,

Inc. 1 Invoice

【註】

1. apprising:報告；通知。cf. apprising one of something:告知某人某事。

2. to surrender＝to give up:交付。

3. a trifle easier＝somewhat easier:有一點疲弱(軟)，"easier"為"easy"的比較級，easy＝weak,意指行情趨下。

4. firm:意指堅穩。

   cf. fluctuating:恍惚，意指漲跌不定。

   strong:堅挺或放長。

   steady:穩定，意指價格一時不致有巨幅漲落。

No.119　滙票到期未獲兌付發出警告

Dear Sirs,

We confirm our letter of 15th July informing you that according to advice from the Bank of America, our draft No.123 for US$10,000, due on May 30, against Invoice No.123 and order 77/123, had not yet been paid.

We had asked to attend to this matter at once, but to our astonishment we were informed by the Bank that, in spite of repeated requests, you still had not paid.

Through such delay considerable expenses are incurred, such as rent, moratory interest, additional insurance, etc., for which you are responsible.

We would remind you that our agreement with you is worded as follows: "We hereby agree to accept your Draft promptly and to meet same when due without regard to objections respecting this parcel, which shall be referred to arbitration. Should we not honor the Draft when presented or due, we hereby authorize you or your representatives to resell the goods by auction, private contract, or in such manner as may appear best to you, after giving us ten days' notice. We also agree to make good, on demand and without dispute, any deficiency arising from such sale."

In view of the above agreement recognized by you, we must ask you to honor our draft for US$10,000 within 10 days after receiving this letter, failing which we shall be compelled to resell the goods and hold you responsible for any loss.

We trust that you will now not neglect this last warning.

Yours faithfully,

【註】

1. attend to:注意。

2. rent:倉租，即因滙票拒付未能提貨而存儲於碼頭倉庫所生的費用。

3. moratory interest:延付所生利息。

4. remind:提醒。

5. is worded as follows：以如下的文字表達, 卽"is expressed as follows"之意。

6. when due:到期時。

7. without regard to objections:無異議。

8. shall be referred to arbitration:交付仲裁, 卽使有異議也須交由仲裁判斷。

9. parcel:一批貨, 卽"a portion of a shipment of goods", 例如："We have been able to obtain the excellent prices of this parcel."

10. in such manner as may appear best to you:依你們認爲最佳的方法。此句子形容"to sell the goods"。

11. to make good:補償, 接"any deficiency"（任何不足之數）。

12. on demand:一經要求。

13. failing which:否則。

14. hold responsible:負責。

15. warning: 警告。

三、 帳目的清理函件

1.帳目的清理(Settlement of Account)：在現代國際貿易, 貨款的收付固然大多以 L／C 或 D／P、D／A 方式了結, 但如雙方交易淵源深厚, 或雙方互有進出口的買賣關係時, 也間有採用 Open Account 的結帳方式。卽賣方將貨物運出後, 將貨款記入買方帳, 於一定時間後, 雙方結算清理。除此之外, 買賣雙方間或與代理商之間也有佣金以及其他種種債權債務暫予記帳, 於一定時間後, 再清理。

在清理時, 貿易習慣上, 一切人欠帳項可開掣借項清單(Debit Note)寄交對方, 一方面通知該項欠項已借入其帳戶, 他方面則尚含有索欠之意。欠人者也宜開掣貸項清單(Credit Note)表示誠意。如買賣雙方互有人欠、欠人帳款, 則可繕製往來帳單(Statement of Account)以供核對及清付帳款之用。

No.120　賣方寄出借項清單

Dear Sirs,

　　We regret to have to inform you that an unfortunate error in our invoice No.832 of August 18 has just come to light. The correct charge for nylon shirts, medium, is US$15 per dozen and not US$14.50 as stated.

　　We are therefore enclosing a debit note for the amount undercharged, namely US$500.-

　　The mistake is due to a typing error and we are sorry it was not noticed before we sent the invoice.

　　　　　　　　　　　　　　　　Yours very truly,

【註】

1. unfortunate error:令人遺憾的錯誤。

2. has just come to light:剛剛發覺。

3. correct charge:正確的價款。

4. medium:中號。

5. undercharge:少計價款。

6. typing error＝typographical error:打錯。

No.121　借項清單

DEBIT NOTE

Messrs. John Huges & Co.　　　　　　　August 22,19…

112 Kingsway

Liverpool

　　We hereby advise you that we have placed the under-mentioned amount to the debit of your account with us.

ORDER NO.	DESCRIPTION	AMOUNT
123	To 1,000 doz. nylon shirts, medium charged on invoice No. 832 at US$14.50 per doz. should be 15.00 per doz. difference	US$500.-
The amount debited to your account		US$500.-
For Taiwan Trading Co., Ltd.		

【註】

在 Debit note 中的 Description 欄以"To"開始，意指"Dr. to"（即 debit to）。

No.122　貸項清單

CREDIT NOTE

Messrs. John Huges & Co.　　　　September 20,19···

112 Kingsway

Liverpoll

We hereby advise you that we have placed the undermentioned amount to the credit of your account with us.

ORDER NO.	DESCRIPTION	AMOUNT
123	By ten doz. returned. Charged to you on Invoice No.832	US$150.-

The amount credited to your account	US$150.-
	For Taiwan Trading Co., Ltd.

【註】

Credit Note 中的 Description 欄則以"By"開始，意指"Cr. to"（即 credit to）。

No.123 賣方寄出往來帳單

Dear Sirs,

We are enclosing a Statement of your account up to and including May 31,19...,showing balance in our favor of US$125.40

We suppose you will prefer to send us a check, as this is too small to draw a bill for.

Yours very truly,

【註】

1. in our favor:有利於我方，即欠我方。

balance in our favor:我方順差。

2.催款函(Collection Letter)：如果發出"Statement of Account"或請求付款函，未獲覆或逾期未付款時，應發出催款函。寫催款函是一種藝術，寫得不夠藝術就傷感情(hurt feeling)。欠人者，不付款大約有三種原因，⑴忘記，⑵裝聾作啞，⑶財務困難。不論是那一種，都可採下列三步驟：

（1）催促付款(Urge Payment)：這種信函為催付函(Reminder)，語氣略帶堅定，但仍應有禮，其內容包括：⑴提及前次發出的 Statement of Account,或 Letter of Asking for Payment。⑵假定發生了特殊事故，才使付款延誤。⑶附上前寄信函副本。

(4)請滙下帳款。

No.124　催促付款信

Dear Sirs,

　　As you are usually prompt in settling your account, we wonder whether there is any special reason why we have not received any information with regard to the statement of account submitted on May 31.

　　We think you may not have received the statement of account and we enclose a copy and hope it may have your early attention.

　　　　　　　　　　　　　　　　　　Yours faithfully,

【註】

1. in settling your account:可以"in paying your bills"代替。

2. 第二段"we think……":說「以爲你沒有收到我們的帳單」是一種有禮貌的表現法。

　　（2）堅持要求付款(Insist on Payment)：如果發出催付函，仍不發生作用，只好再寫第二次措詞較強硬的信，信中應以堅強的語氣要求對方付款，信中內容包括：(1)重述以前所寄的收款函。(2)提出一個付款的最後期限，堅持要求在此期限內付訖。

No.125　堅持要求付款信

Dear Sirs,

　　We are disappointed not to have received any word from you in answer to our letters concerning the bill of US$100. —— which you owe us.

As you know, the terms of our agreement extend credit for one month only. This bill is now two months overdue. Surely you don't want to lose your credit standing with us and with others —— nor do we want to lose you as a customer.

We therefore insist on receiving a check at once. We urge you to keep your account on the same friendly and pleasant basis it has always been in the past.

Of course if there is some reason why you cannot pay this bill, or can pay only part of it now, we would be very happy to talk it over with you...and perhaps we could be helpful.

Yours very truly,

【註】

1. owe us:欠我們。
2. as you know:如你所知。
3. the terms...only:欠帳的期限只有一個月。
4. overdue:逾期
5. lose credit standing:失去信用。
6. lose you as a customer:失去你這麼一個客戶。
7. urge:催促。
8. keep your account...basis:保持你的帳戶建於友善而愉快的基礎。
9. perhaps...helpful:或許我們可以幫忙。

（3）要求立即付款(Demand Payment)：三催四請，對方仍然裝聾作啞，相應不理時，祇好發出最後通牒，但不要向對方恐嚇、威脅、漫罵，以免犯法。這種信的內容包括：(1)表示幾次催款皆未獲覆甚感遺憾。(2)要求立即付款。(3)如果仍不付款，將採法律行動（交給律師處理）。

No.126　要求立即付款信

Dear Sirs,

We are much disappointed in your ignoring our repeated requests for payment of your outstanding account of US$124.50.

It is with the utmost regret that we have now reached the stage when we must demand immediate payment. We have no wish to be unreasonable, but failing payment by September 15, we are afraid you will leave us no choice but to place the matter in the hands of our attorney, but we sincerely hope this will not become necessary.

Yours faithfully,

【註】

1. ignore:忽視。

2. we have no wish...unreasonable:我們無意不講理。

3. leave us no choice＝leave us no alternative:使我們別無其他辦法。

4. but to place...our attorney:除將本案交給我們律師處理外（別無他法）。

　3.催付樣品費、佣金等: 在國際貿易，昂貴的樣品，往往向進口商索取樣品費。然而，進口商往往裝糊塗。再者，出口商應付佣金者，也往往拖延不付。於是，不得不寫信催付，以下二例，可供參考。

No.127 催付樣品費

Dear Sirs,

Subject:Sample Charges

On October 12,19··· we submitted to you a Debit Note No.123 for the sample charges. However, up to this date, we don't appear to have received your remittance.

In the circumstances we are forced to conclude that our Debit Note has not reached you, and we now enclose a duplicate of the Debit Note for reminding you of this outstanding account.

May we draw your immediate attention to this matter? Your prompt remittance to us will be appreciated.

Yours faithfully,

【註】

1. sample charges:樣品費。

2. are forced to conclude:不得不推斷；不得不作···的結論。

No.128 催付佣金

Dear Sirs,

We wish to draw your attention to the fact that our commission of $2,100 on the two sales effected through us and shipment was made during May as per your Invoices 101 and 102, totaling $42,000 in value, has not been remitted into our account with our bank, Bank of Taiwan, Taipei, though it was due in March. We believe that the relative drafts were negotiated at your bank in due course.

In the past you were always punctual in remittance, and therefore the delay in this instance is presumed to have been caused by some clerical error. At any rate, we are inconvenienced by the existence of the outstanding account.

If you have not yet arranged for remittance, please do so at once. If you have already done, you may disregard this letter. Your prompt reply will be appreciated.

Yours faithfully,

【註】

1. is presumed:假定。

2. in this instance:在這種情況下。

3. clerical error:小錯誤，另有「筆誤」之意。cf. clerical mistakes:筆誤。

4. at any rate=in any case:總之；無論如何。

4.請求延付(Customer Requests Time to Pay):如因某種原因無法按期付款，則宜說明理由，取得對方的諒解。

No.129　請求延緩付款

Dear Sirs,

Everyone has his individual problems, and today, on account of changing conditions, each of us is seemingly confronted with more financial problems than ever.

We have been having one bad break after another over the past few months. On account of the shortage of raw material, we have not been working in high gear. As a result, we have no alternative but to ask you forget our account for thirty days. Things are looking up and, barring unforeseeable difficulties, we should be able to pay you before...

The world has lucky and unlucky people. It is too bad that misfortunes came to us in a row, but it is our good fortune to have friends like yourselves, whose unfailing assistance in time of need has helped us to tide over our difficulties.

With very good wishes.

Very sincerely yours,

【註】

1. everyone has his individual problems:每個人都有他自己的問題，即家家有本難念的經。

2. on account of changing conditions:由于情況的改變。

3. confront with:遭遇到。

4. than ever:比以前。

5. one bad break after another:一連串的不幸。cf. one after another.

6. working in high gear:全速生產。

7. forget our account for thirty days:忘記我們的帳款三十天。意指延遲三十天付款。

8. things are looking up:情形好轉中。

9. barring:除非。

10. unforeseeable difficulty:不可預料的困難。

11. the world has lucky and unlucky people:世上有幸運的人和不幸運的人。

12. It is too bad that misfortunes came to us in a row:很不幸我們遭遇到一連串的不幸。"in a row"接連，一連串之意。

13. unfailing assistance:無止境（不斷）的協助 cf. unfailing friendship:永久的友誼。an unfailing friend:一個可靠（忠實）的朋友。

14. in time of need:在困難之時。cf. A freind in need is a friend indeed:患難之交始爲眞朋友。

15. to tide over difficulties:渡過困難。

　　5.請求先付部分款：如因某種原因無法一次付清，而請求先行支付部份款時，也應說明理由，以求對方的同意。

No.130　請求先付部分款

Dear Sirs,

We have your letter of December 20, and note with regret your views on the subject of our slow payment.

To some extent we feel that these are justified, but we would point out that we are not entirely to blame. Our inability to settle our account is due to the difficulty of marketing the goods bought from you, a direct result of the high figures at which you priced them and the poor condition in which arrived. It is not our wish to emphasize the justification for our action, but we must remind you that it does not seem quite fair that the entire onus of responsibility in what is, after all, a joint transaction, should be thrown on us.

In the circumstances we trust you will accept our banker's draft for US$···on account, subject to payment of the balance next month.

　　　　　　　　　　　　　　　　Yours faithfully,

【註】

1. the subject of slow payment:遲付的問題。

2. to some extent:在某種程度內。

3. we are not entirely to blame:不能完全歸咎於我們。

4. marketing the goods:銷售貨物。

5. the high figures at which you priced them:你們所訂價格過高。

6. justification:理由；口實；辯解。

7. it does not seem quite fair:不似很公平。

8. the entire onus of responsibility in what is:關於……的全部責任。

9. after all:究竟。

10. a joint transaction:共同的交易。

11. throw on us:推諉於我們。

6. 對方先償還部分款要求的答覆

No.131　同意先償還部分款的要求

Dear Sirs,

We have your letter of December 27 and thank you for the draft for US$···enclosed. In view of the fact that hitherto your payments have been made promptly, we agree to your proposal that payment of the balance of US$···should be postpones.

It must be understood, however, that our agreement does not amount to an acceptance of your opinion of the goods supplied. While regretting your failure to effect a sale, we take no responsibility for it. Further, we do not regard our prices as excessive, and would point out that adequate allowance was made for damage in transit. It

must be clearly understood also that we dissociate ourselves from the loss on resale, which definitely forms a
new transaction.

A receipt for your payment is enclosed, and we must
insist on receiving the balance strictly in accordance
with your undertaking.

Yours truly,

【註】

1. in view of the fact that:鑒於（你在此以前→ hereto）。

2. amount to:等於。

3. effect a sale:銷售。

4. allowance:補償

5. dissociate...from:不相涉；無關。

6. undertaking:諾言。

No.132　不同意先償還部分款的要求

Dear Sirs,

We thank you for your letter of November 5 enclosing a draft in part-payment of your account, but would
point out that the sum still outstanding in considerable.

As we work, to a very great extent, on a small-profit
basis, extended credit with the consequent loss of interest tends to absorb the small figure which accrues to us.

In the circumstances we think you will agree that
long-term credit is impracticable and that in asking for
an immediate settlement we are not making an excep-

tional request.

                                    Yours faithfully,

【註】

1. in part-payment of your account:支付你的部分帳款。

2. sum still outstanding:尚欠付的帳款。

3. considerable:相當可觀。

4. to a very great extent:極大部分（的交易）。

5. work on a small-profit basis: 以薄利的基礎（交易）。

6. consequent loss of interest:利息的損失。

7. absorb the small figure which accrues to us:奪去我們所獲得的微薄利潤。

8. long-term credit:長期賒欠。

9. impracticable:不切實際。

10. not making an exceptional request:非不情之請。

　　7.函送帳款

No.133　以銀行滙票償付帳款

Dear Sirs,

　　We are pleased in enclosing a bankers' draft No.123 for US$100 dated October 15 drawn by Bank of Taiwan on the City Bank, New York in settlement of your commission for the order No.321.

　　We shall be pleased if you will send us your official receipt.

                                    Yours faithfully,

【註】

1. 爲了萬一遺失時易於掛失，宜將滙（支）票號碼、金額、日期、發票人、付款人等均予以寫明。

2. bankers' draft＝bank draft:銀行滙票。

3. in settlement of:用以清償。

4. official receipt＝formal receipt:正式收據。

8.函送收據

NO.134　寄送收據

Dear Sirs,

Thank you very much for your letter of October 15 enclosing a bankers' draft No.123 for US$100.00 on the City Bank, New York in settlement of our commission for order No.321, and we are pleased to send herewith an official receipt.

We hope you have been satisfied with our service and that we may solicit more orders for you.

Yours faithfully,

OFFICIAL RECEIPT

No.123

US$100.00　　　　　　　　　　　　　　October 20,19···

RECEIVED from Messrs. Taylor & Co., New York, N.Y. the sum of U.S. Dollars One Hundred Only in settlement of commission for order No.321.

Stamp
Taiwan Trading Co., Ltd.

(signature)

第三節　有關收款與付款的有用例句

一、關於 L／C、D／P、D／A

1. We are surprised to receive a notice from our bankers here to the effect that the draft for US$…drawn on ABC Bank under LC No.321 covering your order No.123 has been dishonored (unpaid) for the following reason which we cannot admit. The said draft has already negotiated and we are in a very embarrassing situation.

2. We are surprised to receive a notice from our bankers that draft for US$…drawn on you have been dishonored by you without any reason.

二、關於帳目清理

1. We have pleasure in enclosng
   We are enclosing ⎫
   We enclose ⎭ a statement of account, showing →

   → a balance in our favor of US$…

2. We { are enclosing / enclose } a statement of your account up to and →

   → including June 30, showing a balance in our favor of US$…,→

   → which we hope you will find { correct. / in order. }

三、關於催收帳款

第一次催收

開頭句

1. We notice that your account, which was due for payment on...,

$\left\{\begin{array}{l}\text{is still outstanding.}\\ \text{still remains unpaid.}\end{array}\right.$

2. We shall be glad if you will give attention to our account dated...,

which still remains $\left\{\begin{array}{l}\text{unsettled.}\\ \text{unpaid.}\end{array}\right.$

3. We wish to draw your attention to our invoice No.···,for the sum of

$\$$···which $\left\{\begin{array}{l}\text{we have not yet received.}\\ \text{is still unpaid.}\end{array}\right.$

4. We are writing to remind you that we have not yet received the balance of our September statement, amounting to US$···,payment of which is now more than a month overdue.

5. It is doubtless that the rush of business at the period of the year has caused you to overlook the payment of our account $···, which was due on March 2.

結尾句

1. We hope to receive your check $\left\{\begin{array}{l}\text{by return.}\\ \text{within the next few days.}\end{array}\right.$

2. We look forward to your remittance $\left\{\begin{array}{l}\text{by return.}\\ \text{within the next few days.}\end{array}\right.$

3. As our statement may have gone astray we enclose a copy and shall be glad if you will deal with it promptly.

4. $\left\{\begin{array}{l}\text{Kindly}\\ \text{Please}\end{array}\right\}$ $\left\{\begin{array}{l}\text{inform us}\\ \text{let us know}\end{array}\right\}$ $\left\{\begin{array}{l}\text{immediately}\\ \text{promptly}\end{array}\right\}$ when we may expect the

$\left\{\begin{array}{l}\text{settlement of your outstanding account.}\\ \text{your remittance for the commission.}\end{array}\right\}$

5. We shall thank you to kindly $\left\{\begin{array}{l}\text{send us your check for}\\ \text{remit us}\end{array}\right\}$ →

$$\rightarrow US\$\cdots \begin{Bmatrix} \text{in settlement of} \\ \text{in payment of} \end{Bmatrix} \begin{Bmatrix} \text{the agent commission} \\ \text{our invoice} \end{Bmatrix} \rightarrow$$

→ at your earliest convenience.

* payment of our invoice:支付發票上所載貨款

第二次催收

開頭句

1. We do not appear to have had any reply to our request of...for settlement of the amount due on our invoice No...of...

2. We regret not having received a reply to our letter of...reminding you that your account, already more than a month overdue, had not been settled.

3. We are at a loss to understand why we have had no reply to our letter of asking you to settle the amount outstanding on our November statement.

結尾句

1. We trust you will now attend to this matter without further delay.

2. We must now ask you to settle this account $\begin{Bmatrix} \text{by return.} \\ \text{within the next few days.} \end{Bmatrix}$

3. We regret that we must now press for immediate payment of the amount still owing.

4. As the amount owing is considerably overdue, we shall be grateful to receive your check by return.

第三次催收

開頭句

1. We note with surprise and disappointment that we have had no replies to our two previous applications for payment of your account.

2. As we have had no reply to our previous requests for payment of our invoice dated..., we must now ask you to remit the amount due by the end of this month.

3. Owing to the fact that you have ignored our repeated application for settlement of your outstanding balance US$···we can only assume that you are not prepared to effect a payment.

結尾句

1. Unless we receive your $\left\{\begin{array}{l}\text{check} \\ \text{payment}\end{array}\right\}$ in full settlement by the end of this month,→

$\rightarrow\left\{\begin{array}{l}\text{we shall instruct our solicitors to recover the amount due.} \\ \text{we shall take legal proceedings.} \\ \text{we shall have no choice but to take other steps.} \\ \text{we shall take legal action for the recovery of the amount due.} \\ \text{we shall have no alternative but to recourse to legal proceedings.}\end{array}\right.$

2. This is our third application for payment of the enclosed account, and unless a remittance is received within...days, we shall be compelled to place →

$\rightarrow\left\{\begin{array}{l}\text{the account into other hands for collection.} \\ \text{the matter in the hands of our attorneys for collection.}\end{array}\right.$

3. Previous applications for payment of the enclosed account having been ignored, we must definitely inform you that unless our claims are satisfied by October 30, we shall immediately →

$\rightarrow\left\{\begin{array}{l}\text{take legal action} \\ \text{have no alternative but recourse to legal proceedings}\end{array}\right\}$ for → the recovery of the amount.

四、對於催款的覆函

1. In settlement of your account of commission, we enclose a check No...for US$⋯issued by Bank of Taiwan, on Bank of America, for which please send receipt.

2. We enclose a cheque for US$⋯to settle the accounts to the end of May, and →

    → $\left\{ \begin{array}{l} \text{shall be glad if you will acknowledge its receipt.} \\ \text{your acknowledgement of receipt will be appreciative.} \end{array} \right.$

3. We are enclosing a $\left\{ \begin{array}{l} \text{sola draft} \\ \text{check} \end{array} \right\}$ No.123 for US$540.00 issued by Irving Trust Company, payable to your order, on Bank of Taiwan, which we $\left\{ \begin{array}{l} \text{hope} \\ \text{trust} \end{array} \right\}$ you will find in order.

4. Covering the amount of commission, we have today remitted →

    → US$1,520 by $\left\{ \begin{array}{l} \text{mail payment order} \\ \text{cable} \end{array} \right\}$ on Bankers' Trust →

    → Company in New York in favor of yourselves through our

    → Bankers, Taipei City Bank,

    → enclosed form and return it to us.

5. The financial difficulties from which we are suffering at present are cause of our inability of meeting your draft at maturity. Would you kindly allow us a further extension of the payment, say another one month?

    ＊ at maturity: (支票, 滙票) 到期, 不是"on" maturity。

習　題

一、解釋下列術語

1. General Letter of Hypothecation　　2. Debit Note

3. Credit Note　　　　　　　　　　　4. Open Account

二、將下列中文譯成英文

1. 茲奉上貴公司十月份帳單乙紙，敬請查收滙付爲荷。

2. 茲奉上支票乙紙，計美金 50 元，用以支付貴公司十月份帳單(statement of account)。

3. 茲附奉至 6 月 30 日止(made up to and including the 30th June)帳單乙份。從本帳單可明瞭(observe)貴公司尙欠本公司美金 200 元(a balance in our favor of US$200)。

4. 謹歉告倘貴公司不立即淸付單帳，嗣後無法再接受訂單。

5. 除非十日內獲得臺端答覆，否則將不得不把單帳交給律師代爲收取。

第二十六章 索賠與調處

(Claims and Adjustments)

第一節 索賠的概念及其種類

一、索賠的概念

從事貿易，自尋找交易對象，洽談交易以至交貨、付款完成一筆交易，通常多需要經過一段漫長的時間。而且買賣雙方遠隔兩地，其交易的過程大多有賴於函電的往返接洽。而往返的函電有不明、錯誤、誤解之時，貨物自出口地運到進口地有曠日費時，貨物輾轉搬運，途中難免遭遇到意內或意外之事；因語言殊異，法律風俗習慣及傳統心理的不同，買賣雙方之間，難免發生齟齬。

由於上述的種種情事，而受到委屈或損失的一方，自將向對方提出抱怨(complaint)或要求賠償等等。所謂索賠(claim)廣義地說，就是指這些抱怨及要求賠償等而言。

二、索賠的種類

造成委屈或發生損失的原因，或由於故意，或由疏忽或由於意外事故，或由於不可抗力。所以其責任，並不以進口商爲限，大體說來，貿易上的索賠可大別爲兩類。

    1.貨物損害索賠(claim for loss and damage of goods)又可分爲：

      (1)運輸索賠(transportation claim)

      (2)保險索賠(insurance claim)

2.買賣索賠(trade claim or business claim)

以下分別介紹有關買賣索賠信、運輸索賠信及保險索賠信的寫法。

第二節 買賣索賠信的寫法

買賣索賠是買賣雙方當事人間的索賠，以相對人為索賠對象。在商業往來上，因不滿意而引起爭議乃是家常便飯而不可避免的。在進口商方面，發現貨物品質低劣(inferior quality)、規格不符(different specifications)、數量短少(shortage)、包裝不良(bad packing)、貨物破損(breakage)、延誤裝運(delay shipment)等延不交貨等情事時，自必去函(電)責問或要求賠償損失，出口商對於這些事件，則有予以補正或解釋的必要。

反之，假如進口商簽發訂單之後，遲遲不開發信用狀或開出不正確的信用狀，或任意取消訂單或拒不付款等等則出口商也必將去函(電)責問進口商或要求損害賠償，進口商對於這種索賠，自有處理的必要。

一、撰寫索賠信的要領

索賠人(claimant)的態度應力求平心靜氣，措詞應委婉適當，切忌肆意謾訕，或詞鋒激厲，否則不但轉圜無望，而且可能使雙方決裂，惡感愈深，結果無補實益，徒遭損失。撰寫索賠信時，應注意下列七點：

1. 迅速提出索賠，尤應在規定索賠期限內提出。延誤索賠，將削弱索賠的立場，並且將使對方難於找出發生事故的原因。

2. 索賠理由務求明確。具體指出索賠的原因，如有證明文件尤佳。

3. 引用案號、日期及有關資料，以便其查辦。

4. 提出希望解決的方法。

5. 不要貿然認為對方有錯，也許對方有理由。

6. 措詞誠摯有禮，意誠詞婉，避免粗魯以免僨事。

7.要求早日解決。

二、答覆索賠信的撰寫要領

處理索賠有兩大目標：一是提出使對方滿意的解決辦法；二是排除不愉快，維持良好的關係。當接到令人不快的索賠信時，應以冷靜的態度，就事論事，不感情用事。須知得罪客戶容易，爭取客戶卻不簡單。茲將撰寫答覆索賠信的要領，簡述於下：

1.假如錯在我方時：

⑴儘速答覆：可使對方覺得受到重視，由而減少不悅的情緒。

⑵表示謝意：收到客戶的索賠信，仍宜表示高興，並致謝意。

　①因爲由於客戶的來函，可以知道對方有什麼不滿之處。

　②使我方有解釋的機會，或糾正錯誤，由而保持良好關係。

　③使我方獲得改善的機會。

⑶對於所發生的問題表示遺憾，並表示願意解決。

⑷解釋所以發生問題的原因。

⑸敍述有條有理，避免含糊其詞。

⑹顧客永遠是對的(customer is always right)，因此即使對方出言不遜，也應謙恭地作答。

⑺最後，保證今後不再發生此類情事，並表示仍願繼續合作。

2.假如錯不在我方時：

⑴儘速答覆。

⑵對於提醒我方注意其不滿表示謝意及遺憾。

⑶表示同情對方的處境。

⑷提示有關資料，將事實作充分的解釋，促其了解責任的歸屬。

⑸表示我方雖無責任，但仍很關心，並表示願協助解決。

3.雖然我方無錯，但仍同意酌給撫慰金(consolation money)：有時錯誤雖不在我方，但基於同情，願酌給付撫慰金，以安慰對方，由而維持良好的關係。在此情形，Adjustment Letter 的寫法，應注意下列各點：

(1)感謝對方的提醒我方注意其不滿，及表示對這件事感到遺憾。

(2)表示同情及體諒。

(3)把事實真相有條有理地說明，使對方了解責任不在我方，不必告訴對方本應怎麼做，或本不應怎麼做，或未曾怎麼做。

(4)表示將願意酌給撫慰金。

(5)建議對方將來應如何防範此類事情的再發生。

No.135　責問賣方逾期交貨

Dear Sirs,

We confirm that we have sent you today a telex which reads as follows:

"TELEX REPLY WHEN U SHIPD ODR 123 IF UN-SHIPD YET SEND GOODS BY AIR BEFORE MAY 15 N 70% AIRFREIGHT SHALL BE UR A/C"

Your attention is invited to our order which stipulates shipment must be effected within 60 days after receipt of our L/C, and please also pay your careful attention to our previous communications of April 2 and 12 stressing that punctual shipment is essential. In spite of this ten days have passed from the latest shipment date as stipulated in the L/C and we have not yet received your shipment advice.

If the umbrellas have not yet been shipped, we must request you to ship them to us by air before May 15, instead of by ship and 70 percent of the airfreight shall be for your account.

Thank you for your cooperation and look forward to your immediate reply by telex.

Yours faithfully,

【註】

1. which reads as follows:內容如下。注意這種句子的用法, reads 的"s"不能省, "as follows"不能改爲"as following"。

2. 關於 telex 的簡體字說明:U:you; SHIPD＝Shipped; UNSHIPD＝un-shipped;N＝and;UR＝your;A／C＝account。

3. your attention is invited:敬請注意。

4. in spite of:儘管; 雖然。

5. for your account:算你的帳; 歸你負擔。

No.136　覆責問逾期交貨函

Dear Sirs,

This is to confirm that we have sent you today a telex which reads as follows:

"YX 2 UR ODR 123 FOR 1000 DZ UMBRELLAS WILL BE SHIPD BEFORE MAY 15 PER AIR DETAILS AIRMG"

We are very sorry for the delay which has occurred in the shipment of your order ♯123 occasioned by a serious breakdown in our machinery which brought all our work to a standstill for nearly two weeks. The damage, however, has been repaired, and to recoup for the loss in time we will ship the goods to you by airfreight before May 15 as requested. You may expect to receive the goods by May 20 or so. We are also agree to absorb 70% of the airfreight.

Meanwhile, please extend the shipment date and expiry date of the L／C to May 15 and May 25 respectively. Of course, amendment charges, if any, shall be for our account.

We regret the inconvenience you have sustained, but trust this unavoidable accident will not influence you unfavorably in the matter of future orders.

Yours faithfully,

【註】

1. 關於 telex 的內容:

YX＝your telex;2＝2 日, DZ＝dozen; AIRMG＝airmailing。

2. occasioned by:肇因於…

3. breakdown:損壞。

4. to recoup for the loss in time＝to atone for the delay in delivery:為了彌補時間上的損失。

5. or so＝about. cf. one hundred or so:一百左右。

6. absorb:負擔。

7. unavoidable accident:不可避免的意外事件。

8. bring our work to standstill:使我們的工作停頓。

9. the inconvenience you have sustained:你遭受的不便。

10. influence...order:對日後的訂貨引起不良影響。

No.137　關於短裝的索賠

Gentlemen:

Re:1,000 cartons Canned Mushroom

shipped per s.s. "NISHO MARU"

　　Please be informed that the subject goods shipped by the captioned vessel arrived at New York on April 20.

　　Upon taking delivery of the cargo, we have found that there were only 920 cartons against 1,000 cartons shipped by you. When checking with the shipping company, we were told that only 920 cartons had been loaded on board the carrying vessel. Since the loss is not negligible you are hereby requested to make up the 80 cartons shortshipped when you deliver the last two items to us. In the meantime, it will be appreciated if you will check at your end and to make sure if all these 1,000 cartons had been loaded.

　　We are looking forward to receiving your findings soon.

Yours faithfully,

【註】

1. at New York:照文法來說，大都市的前面應用"in"，這裡用"at"，乃係指港口之故，其意爲"at the port of New York"。

2. take delivery of cargo:提貨。

3. loaded on board the carrying vessel:裝上載貨船。

4. not negligible:非同小可。

5. make up: (將不足的予以) 補償。"make up"的意義很多，例如

  (1) The committe is made up of seven wembers. (組成)

  (2) He made up an excuse. (捏造藉口)

  (3) You have to make up your 2 nd year English. (重修二年級英文)

6. shortshipped:短裝。

No.138　對短裝索賠的覆函

Dear Sirs,

Re: 80 cartons of Canned Mushroom

<u>shortshipped per s.s Nisho Maru</u>

Your letter of April 25,19⋯ concerning the subject shipment has received our immediate attention. We regret very much that this incident has caused you much inconvenience. Much as we are eager to help you straighten out this matter, however, we regret to inform you that we are not the party to blame. According to the B/L issued to us by the shipping company, there is noted clearly that 1,000 cartons have been loaded on the carrying vessel. It is quite a regrettable matter that the New York Office of the shipping company has failed to tell you the fact, thus causing you to file the claim against us.

According to the clauses on the back of the B/L which forms the base of freight contract entered into between the carrier and us, the shipping company is responsible for any cargo shortlanded at the port of discharge. In view of this, you are requested to lodge your claim with the shipping company immediately and ask them to compensate you therefor. We trust that the carrier certainly will make up the losses incurred by you under this shipment.

Please let us know as soon as the case has been satisfactorily settled and if there is anything we can do to help settle this case, please just write us. We will comply with your instructions wholeheartedly.

Yours faithfully,

【註】

1. eager:渴望。接"for"，"after"，"about"或不定詞"to"。

2. straighten out＝settle:解決。

3. we are not the party to blame:我們不是要負責的一方；不能歸咎於我們。

4. file the claim against us＝lodge the claim with us＝lay claim to us＝render the claim against us＝make the claim on us＝set up the claim to us.

5. shortland:短卸。

6. the losses incurred by you:你所遭到的損失。

7. wholeheartedly:全心盡力地。

No.139　向賣方提出貨物損壞的索賠

Dear Sirs,

Subj: Damage on Canned Asparagus

Shipped per M.S. "Suez Maru"

We acknowledge with thanks the receipt of your

telex and your letter dated September 1 and 2 respectively together with the shipping documents for the subject shipment.

Upon the arrival of the goods at Hamburg we have taken delivery of these 900 cartons and have asked Far East Superintendence Co., Ltd. to have them inspected. It is very regrettable that many of the cartons have been found broken, thus causing a large number of cans inside the broken cartons hollowed. We are sure that the asparagus contained therein would have been damaged. We enclose a copy of a survey report, No.999 issued by Far East Superintendence Co., Ltd. from which you will note the details of the damage inflicted on the goods under this shipment.

We shall appreciate it if you will compensate us the CIFC 2 value of all the hollowed cans as shown is the report.

Since this is an urgent matter, your early attention thereto is hereby requested.

Faithfully yours,

【註】

1. hollowed:凹陷。

2. Far East Superintendence Co., Ltd.:遠東公證公司。

3. the damage inflicted on the goods:貨物遭受的損害。

4. under this shipment:本批貨載。

No.140　覆貨物損壞索賠函

Dear Sirs,

Re: Damage on Canned Asparagus
shipped per M.S. "Suez Maru"

Your letter of October 5,19… along with a copy of survey report issued by Far East Superintendence Co., Ltd. has been received and we have given it our immediate attention.

Although we are very regretful to be advised that damage has been inflicted on the shipment we made to you, we, however, are not responsible for the losses and, consequently, are not in the position to compensate you therefore as requested. As arranged with the insurance company, the marine insurance coverage for this shipment includes all risks and the damage as mentioned in the survey report has already been covered by the insurance policy. Therefore, instead of claiming against us, you are requested to submit your claim together with all necessary supporting documents to the insurance company-China Insurance Co., Ltd.,whose address is as follows:

China Insurance Co., Ltd.

Wu Chang St., Sec. 1

Taipei, Taiwan,

R.O.C.

In case you need our further assistance regarding this case, please let us know. We will do everything we can to help you.

Faithfully yours,

【註】

1. consequently＝therefore; as a result。
2. supporting documents:供佐證的單證。
3. in case＝if

No.141 責問賣方運交錯誤貨物

Gentlemen:　　　　Indent No.15

With reference to the 5 cases of assorted goods ex s.s. "Arabia" we are greatly surprised at the unbusinesslike way in which you are handling our order. You have sent us, instead of the "Commonwealth" Rubber Boots with black-fleeced lining which we distinctly ordered, 1,000 doz. pairs "Ideal" Rubber Boots with cloth lining, which are quite unfit for the market they are intended for.

By referring to the above Indent dated June 21, you will find that we have impressed upon you the especial importance of the boots being of warm lining, and that any delay in delivery will have serious cosequence upon us. You made a similar mistake in the last shipment putting us to considerable inconvenience and annoyance. As you are well aware, competition of the home-made goods is so keen that should we miss the best season for the sale of these goods, there is a fear that they will hang upon our hands dead stock until next year. What is worse still, we shall lose our ground by continually disappointing our customers, and indeed we fear that our reputation with our clients for prompt execution of orders in now at stake, and this entirely through your neglect.

If you will value our further orders, you will please oblige us by sending, with all speed, the "Commonwealth" Rubber Boots with warm lining on receipt of this letter, at the same time cabling us the approximate date of their arrival in Kobe. When writing please also give us your instructions as to the disposal of the goods sent in error.

We trust you will pay more attention to our commands in the future, otherwise we shall have to go elsewhere for our future supplies.

Faithfully yours,

【註】

1. to be surprised 不宜以"to be astonished"或"to be astounded"代替。
2. unbusiness-like:無效率的；無條理的。
3. black-fleeced lining:黑色羊毛質襯裏；黑絨襯裏。
4. cloth lining:布質襯裏。
5. unfit for＝unsuitable for
6. impress upon:使銘記；使記住。

7. warm lining:暖質襯裏。

8. home-made goods:手工製品；本國製品。這裡指後者而言。

9. hang upon our hands:留存手頭中。

10. dead stock:dead 爲"unproductive"之意。"dead stock"未售出的存貨。

11. to lose one's ground:失去信用。

12. value:重視。

13. with all speed:全速。

14. commands＝orders，現在很少用"commands"此字。

No.142　對運交錯誤貨物索賠的覆函

Gentlemen:

Your Indent 15

We sincerely regret the mistake we have made in the execution of your order. There is no doubt that, according to your Indent, the "Common-wealth" Rubber Boots with black-fleeced lining were ordered. We have therefore hastened to make the exchange and the right goods have just been sent forward by express to Keelung for shipment per s.s. "Fushimi Maru" which is scheduled to sail on October 3, arriving at Kobe on the 7 th of the same month; at the same time we have cabled you accordingly.

As we do not wish to let you bear the consequences of an error on our part, we are sending you herewith the invoice corrected and would ask you to dispose the goods sent to you by mistake at the best possible prices on a consignment basis. Should you have any stock remaining on your hands at the end of December, you may ship back to us the balance at our expenses.

As you know, our shipping department fall into somewhat disorganized condition in consequence of the sudden and untimely death of our head clerk at the busiest season of the year,but fortunately the vacancy left by him was filled a few days ago to our satisfaction.

Your cooperation is this instance is very much appreciated and we trust the replacement will reach you in due course.

Yours faithfully,

【註】

1. hasten to make exchange:趕快掉換。

2. sent forward:運出。

3. express:捷運。

4. bear the consequences of:承擔…的後果。

5. dispose:處分。

6. on a consignment basis:以寄售方式。

7. at our expenses:費用由我們負擔。

8. shipping department:貨運部門。

9. fall into:陷入…狀態；變成…。

10. in consequence of:由於;因爲…的緣故。

11. somewhat:有一點。

12. disorganized condition:紊亂狀態。

13. untimely death:死得非其時。

14. vacancy：遺缺。

15. filled:補實。

16. replacement:掉換的貨物。

No.143   與樣品不符的索賠

Dear Sirs,

Re: Your contract No.122 for

100 B／S Wool Yarns

Under the captioned contract, we have taken delivery of 100 bales of wool yarns shipped per s.s. "Eugene Lykes".

Upon unpacking the bales, we have found that 20 bales of the lot are much inferior to your sample. This error has been apparently made by the carelessness of your shipping clerk, and can be easily found out by examining the remaining stock on your side. But, for your reference, we have asked FESCO to draw out samples from the 20 bales in question and have airmailed them to you today by parcel post.

For the inferiority in quality, you are requested to make a compensation allowance in price, the amount of which you will please telex to us, after examination of the sample sent you.

Yours faithfully,

【註】

1. wool yarns: （羊）毛紗。

2. out of the lot: lot 爲"shipment"（貨儎）或"consignment"（貨物）之意。

out of＝from

3. inferior to:注意用"to"解做"than"。

4. apparently: no doubt

5. shipping clerk:貨運承辦人。

6. remaining stock:（手邊）剩下的商品。

7. FESCO: Far East Superintendence Co., Ltd.

8. draw out:抽取。

9. inferiority in quality:也可用"inferior quality"代替。

No.144　對貨樣不符索賠的覆函

Dear Sirs,

We confirm we have telexed to you today the following message:

"YOUR CLAIM SAMPLE UNDER CONTRACT NO. 123 RECEIVED UPON EXAM THE GOODS ARE WELL UPTO STANDARD GRADE IN QUALITY EXCEPT A LITTLE MORE DUST THAN ORDINARY CASES WE OFFER ONE PERCENT ALLOWANCE"

We have duly received your samples of the wool yarns shipped by s.s. "Eugene Lykes", which you sent to us protesting that they are different from our samples previously sent.

On careful examination of the samples sent by you, the wool yarns in question have been found well up to the standard grade in quality, except the dust existing in them is a little more than in ordinary cases. We should like, therefore, to make an allowance of 1% for the excess presence of dust. Please note that this is the best allowance we can offer.

If, however, you are not satisfied with this offer, we wish to submit the case to an arbitration and to abide by its decision.

Yours faithfully,

【註】

1. sent to us protesting:提出抗議。"protesting"也可以"complaining"代替。

2. sample previously sent:也以"previous sample"代替。

3. in question:繫爭的;爭議中的。

4. well up to the standard grade:很夠標準級。

5. in ordinary cases:普通的場合。

6. not satisfied=not content with, 也可用"not agreeable to"。

No.145　貨物受損的索賠

Dear Sirs,

Re:Shipment of our order No.123

per M.V. "Oriental Despatcher"

The goods you shipped against our order No.123 per M.V. "Oriental Despatcher" arrived at Keelung on May 15.

Upon examination immediately after taking delivery, we found that many of the goods were severely damaged, though the cases themselves showed no trace of damage.

Considering this damage was due to the rough handling by the shipping company, we claimed on them for recovery of the loss, but investigation made by the surveyor has revealed the fact that the damage is attributable to the improper packing. For further particulars, we refer you to the surveyor's report enclosed.

We are, therefore, compelled to claim on you to compensate us for the loss, US$250, which we have sustained by the damage to the goods. We trust you will be kind enough to accept this claim and deduct the sum claimed from the amount of your next invoice to us.

Yours faithfully,

【註】

1. no trace:無跡象。

2. considering＝thinking:以爲。

3. rough handling:粗魯的處理。

4. to claim $\begin{cases} \text{on} \\ \text{upon} \\ \text{against} \end{cases}$ a person:向某人索賠。

5. has revealed the fact:揭開了某事實，即發現。

6. attributable:歸因於…。

7. improper packing:包裝不當。

8. for further particulars:進一步的詳情。

9. to be compelled to＝to be forced to:不得不。

10. which 指 loss 而言。

11. to accept this claim＝to admit this claim:承認此索賠，即同意賠償。

cf. $\left.\begin{array}{l}\text{advance a claim}\\\text{put forward a claim}\end{array}\right\}$ 提出索賠。

entertain a claim:受理索賠。

to dismiss the claim of...on the ground that...; 基於…原因, 駁回索賠。

relinquish (withdraw) a claim:撤回索賠。

12. sustained by=suffered from:蒙受。

13. next invoice to us:下次開給我們的發票。

No.146 對貨物受損索賠的覆函

Dear Sirs,

We have received your letter of May 17 informing us that the goods shipped to you against your order No.123 arrived damaged on account of the imperfectness of our packing.

This is the first time that we have received such a complaint from our customers, although we have been shipping the goods for five years in the past, packing them in the similar manners as we shipped the goods to you.

Furthermore, we would point out that we hold a copy of clean B/L from shipping company, which relieves us of all responsibilities. We are, therefore, convinced to think that the present damage was due to extraordinary circumstances under which they were transported to you. We are, therefore, not responsible for the damage, but as you must have insured the shipment at your end, we would suggest that take up the matter with the shipping company or lodge your claim with the insurance company if you have insured the goods against All Risks. We shall, of course, place at your disposal any documents necessary to substantiate your claim.

While we are sorry for the inconvenience you have suffered, we believe the above explanation will prove satisfactory to you.

Yours very truly,

【註】

1. arrived damaged=reached you damaged 即"arrived in damaged condition"。

2. imperfectness:不完善。

3. in the similar manners=in the same manners

4. to relieve...responsibilities:免除…的責任。

5. are convinced to think:堅信。

6. due to=casued by

7. extraordinary circumstances:特殊情事。

8. under which:在此狀態下，which 指上述「特殊情事」。

9. take up the matter with:將此事向…提出。

10. to lodge a claim:提出索賠。

11. All Risks:全險。

12. to place at your disposal:聽你使喚；聽你使用；聽你支配。

13. to substantiate:作證，供作證明。

cf. substantiated claim:正當的索賠（要求）。

14. prove:成為…；使（你滿意）。

No.147 將索賠案件提交仲裁

Dear Sirs,

We hasten to inform you that we have instructed our Hongkong Branch to adjust your claim for the defects in our goods shipped in execution of your order No.50 of the 3rd May. We are now surprised to note from your letter of the 7th June that you are not prepared to consider the offer of a 15% allowance made by our Branch to compensate you.

Though we consider our offer adequate, and even generous, we extremely regret that our offer has been refused. As it is not likely to come to amicable settlement between us, we suggest that we have to submit the matter to arbitration, according to the stipulations in Business Agreement.

We would recommend on the ground of economy a joint arbitrator, but should you prefer to have one appointed by each of us, and a third called in with a casting vote in the event of disagreement, we would be prepared to fall in with your wishes.

Yours faithfully,

【註】

1. amicable settlement:友善的解決；和解。

2. arbitration:仲裁。有關仲裁的用語，列舉若干於下：

arbitrator:仲裁人;umpire:判斷人、公斷人。當仲裁人有兩人而其意見不一致時，由他們再選一人，作最後的決定，此人稱為"umpire"。arbitration award:仲裁判斷書。

to settle the matter by arbitration:以仲裁解決事件。to submit (refer) the matter to arbitration:將事件提交仲裁。

3. Business Agreement:「交易條件協定書」請參閱第十一章。

4. joint arbitrator:（由爭執的買賣雙方協議選定的）共同仲裁人。

5. a third＝a third person,第三人，即上述的 umpire.

6. to call in:聘請。

7. casting vote:裁決權。指二位仲裁人意見不一致時，判斷人（umpire)的最後決定權。

8. disagreement:（二位仲裁人的意見的）不一致。

9. to fall in with＝to meet; to agree to:同意。

第三節　運輸索賠信的寫法

一、貨物短損形式

　　貨物於運輸過程中、裝卸作業時均有發生破損及短少的可能。對於這些短損，船方是否應予賠償，胥視損毀或短少發生的原因，是否為船方依法應負責而定。

　　貨物發生短損的情形，大致可歸納下列幾類:

1. Shortland（短卸）2. Shortage（短失）3. Damage:(1) Breakage (2) Sweat（汗濕）(3) Rain and Fresh Water Damage (4) Wet by Sea Water (5) Cover　Torn (6) Scratch (7) Rust (8) Leakage (9) Scorch (10) Stain (11) Bending & Denting (12) Collapse of Cargo

二、索賠的要領

　1.一發現貨損害、滅失應即以書面通知。提貨時:

⑴如貨物的損害、滅失顯著者，受領人應於受領貨物時即刻以書面通知。

⑵如貨物的損害、滅失不顯著者，受領人應於提貨後，三日內，以書面通知。

⑶受領人也得不以書面通知，而在收貨證件上證明損害或滅失。

⑷將通知副本抄送出口商、保險公司等有關方面。

2.索賠時應備齊有關文件：

⑴Claim Letter⑵B／L⑶Invoice⑷Packing List⑸Damage & Shortage Report（短損報告）⑹Debit Note⑺Survey Report

3.不斷催請處理結案。

4.提出訴訟；應於卸貨後一年內爲之。

No.148 初步索賠通知(Preliminary Notice of Claim)

Dear Sirs,

Preliminary notice of claim for

stain damage to 200 bales cotton

ex s.s. "President"

We regret to inform you that stain damage is found in connection with the shipment of the below-mentioned cargoes:

B／L	MARKS	NO. OF PACKAGE	DESCRIPTION OF GOODS
123	CTC LOT 1 1/200	200 bales	AMERICAN RAW COTTON

shipped by Hohenberg Bros. Co., from Galveston on board s. s. "President" consigned to us under B／L No. 123, and arrived at Keelung on August 5,19…。

We hereby declare that we reserve the right to file a claim with you for this damage when the details and amount of the damage are ascertained. We will apply to Lloyd's Agents for survey on August 8,19…。

Please acknowledge the receipt of this letter.

Very truly yours,

【註】

1. stain damage:油損。

2. arrive at Keelung:船到達某港埠時用"at"不用"in"。

3. reserve the right to file a claim $\begin{Bmatrix} \text{with} \\ \text{against} \end{Bmatrix}$ you:保留向你索賠之權。

4. ascertained:確定。

No.149　請求公證行做公證

Dear Sirs,

Application for survey

You are hereby requested to conduct a survey of the following goods consigned to us which have arrived damaged:

Shipper: Hohenberg Bros. Co., Memphis, Tenn., USA

Description of goods: 200 B／S American Raw Cotton

Marks & No.:　CTC

Lot 1

1/200

Date of landing: August 5,19···

Date of delivery to us:August 10,19···

Nature of damage: stain damage by oil

Numbers of packages/units to be examined: 200

bales

Location stored: No.18 warehouse, Keelung Harbour

Your attention to this matter and issue to us your

Survey Report in triplicate will be appreciated.

Yours faithfully,

【註】

1. conduct a survey:查勘; 鑑定。也可以"make a survey"代替。

2. location stored:存儲地點。

3. 貨物如有受損，進口商應請求保險公司及船公司同意的公證行做公證，並取得公
證報告(survey report)以便索賠。

No.150　正式索賠──污損

Dear Sirs,

Re: Claim for stain damage to 200 B／S

cotton ex s. s. "President"

With reference to our letter of August 11 we now

submit our claim for the captioned amounting to

US$340.25 as per Debit Note No. C-123 attached, and

shall be glad to receive settlement at your earliest convenience.

In support of this claim, we also enclose one copy each of the following documents:

1. Shipper's signed invoice.
2. Shipper's packing list/weight list.
3. Damage certificate issued by Keelung Habour Bureau.
4. Survey Report issued by Robert W. Hunt Co.
5. Receipt for inspection fee.
6. B/L.

Yours faithfully,

【註】

1. settlement:此處解釋做「賠款」。

2. damage certificate:短損證明書。

3. receipt for inspection fee:檢驗費收據。

4. 當進口商提貨時，如發現貨物有受損應及時向船公司提出索賠。但由於公證費時，所以，通常先提出初步索賠通知(preliminary notice of claim)，俟取得公證報告後，再提出正式索賠(final claim)。

No.151　正式索賠──破損

Gentlemen:

Subject: claim for loss of soybean oil
per s.s. "Pioneer" arrived
at Keelung July 25,19…

We regret to inform you that upon taking delivery of

the 50 drums of the captioned goods which were discharged at Keelung from s.s. "Pioneer", we have found drum
No.8,11,12,26,44 were partly broken, resulting in a loss of
180 lbs. or 81 kgs. of its contents. We have, therefore, to
claim on you for compensation for this loss.

In support of our claim, we enclose the relevant
documents as follows:

4 copies of debit note.

1 original survey report made by the China Survey
   Co., Lti.

1 copy of invoice for inspection fee.

1 copy of shipper's invoice.

1 B/L No. 15 of s.s. "Pioneer".

1 damage certificate issued by Keelung Harbour
   Bureau.

We shall appreciate it if you will kindly let us have
your check for US$304.20 in settlement of the above
claim at your earliest convenience.

Yours very truly,

【註】

1. in support of our claim:爲了支持我們的索賠。

2. relevant:相關的, 卽 relative。

No.152   船公司的理賠函

Gentlemen:

S.S. HAWAII BEAR VOY. 16—W

B／L No.19 & B／L No.34;

Claim a/c Alleged Shortage of 8 p'cs Hides.

Reference is made to your claim letter of March 3 and our reply of the 6 th, captioned as above.

Please be advised that our cargo tracers on eight pieces Hides have been returned stating that same were not overlanded at any of the vessel's ports of call.

We therefore acknowledge our liability and responsibility for shortage of(8)pieces Hides in amount of $74.36 and are enclosing our check in that amount in full settlement thereof.

Very truly yours,

United States Lines Company

Agents:PACIFIC FAR EAST LINE, INC.

【註】

1. cargo tracers:貨物追查信。

2. overland:誤卸。卽"misland"之意。

3. port of call:停靠港。

No.153　船公司拒絕賠償函

Dear Sirs,

S. S. "TJIPONDOK"/47

BELAWAN DELI/KEELUNG B/L No.2

332.2 M/T PALMOIL IN BULK, 2.7838 m/t shortage

We acknowledge receipt of your letter dated 10 th August, 19···dealing with abovementioned shortage.

We may invite your kind attention to clause 9(a) printed on the reverse side of the covering Bill of Lading which we quote below:

"As the carrier has no reasonable means of checking the weight of bulk cargo, any reference to such weight in this Bill of Lading shall be deemed to be for the convenience of the shipper only, but shall constitute in no way evidence against the carrier."

Furthermore we may refer to the clause printed on the face of the Bill of Lading reading in part:

"Contents and...weight...unknown, any reference in this Bill of Lading to these particulars is for the purpose of calculating freight only."

We regret that in view of the above we are not in a position to assume liability for the shortage.

Please note that in passing the above information on to you, we do so without prejudice to our defences under the terms and conditions of the contract of affreightment.

Yours faithfully,

【註】

1. bulk cargo:散裝貨。

2. contents and weight unknown:內容及重量不詳。

3. deemed:視為。

4. constitute:構成。

5. in no way:絕不。

6. against carrier:對抗運送人。

7. without prejudice=without detriment to existing right or claim:對於現存權利或要求無影響或無損；不侵害權利；不使權利受到損害。

8. defences:抗辯。

9. contract of affreightment:運輸契約。

第四節　保險索賠信的寫法

　　進出口商所以將貨物投保保險，目的在於貨物受到損害時可由保險公司獲得補償，所以懂得如何投保而不懂得如何索賠，仍不實用。

一、處理保險索賠前，應注意：1.取得公證報告(Surveyor's Report)、2.迅速通知保險人或其代理人(Notice of Claim)、3.掌握索賠權時限及時向事故責任人索賠、4.備齊索賠文件。

二、全損索賠應提出的文件

1. Claim Letter 2. Insurance Policy 3. Commercial Invoice

4. Packing List／Weight Certificate

5. Sea Protest Copy（海灘證明書副本）

6. Certificate of Total Loss from the Carrier（運送人全損證明書）

7. Survey Report 8. B／L

三、單獨海損索賠應提出的文件

1. Statement of Claim（索賠計算書）　2. Insurance Policy　3. Claim Letter　4. Survey Report　5. B／L　6. Commercial

Invoice　7. Packing List／Weight Certificate　8. Damage and Shortage Report　9. Others

四、共同海損索賠應提出的文件

申請保險公司繳納保證金或簽發保證函應提出的文件：

1. Copy of Notice from Carrier　2. Insurance Policy　3. Commercial Inovoice　4. B／L　5. Packing List／Weight Certificate　6. Average Bond（共同海損保證書）　7. Others

No.154　向保險公司提出油污索賠通知(Notice of Claim)

Dear Sirs,

Re: Preliminary Notice of damage of

American Raw Cotton per s.s. "President"

Your marine insurance policy No.123

We regret to inform you that the American Raw Cotton consigned to us and covered by the subject policy arrived damaged.

Shipped from: Galveston, U.S.A.

On board: s.s. "President"

Arrived at: Keelung on October 5,19⋯

Nature of loss: stain damage by oil

If you have no objections, we will apply to Lloyd's Agents for survey and documents in respect of the formal claim will be forwarded to you in due course.

A copy of our preliminary notice of claim against the carrier is enclosed and your attention to this matter will be appreciated.

> Yours faithfully,

【註】

1. apply to＝ask:要求。
2. stain damage by oil:油污。
3. preliminary notice of claim:初步索賠通知，與"formal claim"相對而言。
4. Lloyd's Agents: Lloyd's 爲 The "Corporation of Lloyd's" (勞依茲公司) 的簡稱，是英國保險市場中一個特殊組織。Lloyd's Agents 爲勞依茲公司分布全球各處重要商埠的代理處，協助處理報導海上運輸或保險有關的業務（查勘公證等）或消息。

No.155 向保險公司提出油污正式索賠(Formal Claim)

Dear Sirs,

Re:Claim for stain damage to American
Raw Cotton under your policy No.123

With reference to the captioned claim, of which we sent you a preliminary notice of claim on October 6, we hereby file the claim amounting to US$512.10 as per the enclosed Debit Note No.345.

In support of our claim, we enclose also the following documents:

1. one original policy No.123 duly endorsed
2. one original survey report No.456 with a receipt for survey fee
3. one copy of shipper's invoice
4. one copy of B／L No.789 of s.s. "President"
5. one copy of our letter to carriers and their reply

6. one copy of packing/weight list

7. statement of claim

We shall appreciate it if you will investigate the matter promptly and let us have your cheque in settlement of the above claim at your earliest convenience.

Yours faithfully,

【註】

就索賠程序而言，應先向船公司索賠，如船公司不予理賠，則憑其覆函，向保險公司提出索賠。萬一船公司相應不理，不覆函也不理賠，則貨主可免附船公司覆函，而以其致船公司的索賠函副本代替之。

No.156　包裝破損，請保險公司理賠

Dear Sirs,

Re: 100 bags ABS Resin
shipped per s.s. "May Flower"
Your Policy No.123

Enclosed please find one copy of the following documents with regard to the abovementioned shipment:

1. Survey Report issued by INTECO.

2. Invoice issued by American Chemical Co.

3. Weight list issued by the same supplier.

4. Insurance policy

5. B/L issued by APL.

6. Damage report issued by Keelung Harbour bureau.

From the survey report enclosed herein, you will

note that there are ten bags of the 100 bags of this shipment with cover torn, and contents partly exposed and split.

In view of the damages, you are requested to compensate us the CIF Value plus ten percent of these 10 broken bags at your earliest convenience.

Yours faithfully,

【註】

1. cover torn:包裝破裂。

2. exposed:暴露。

3. split:散失。

No.157　保險公司對於包裝破損索賠的覆函

Dear Sirs,

Re: Your claim for loss of ABS Resin

shipped per s.s. "May Flower"

covered by our policy No.123

We have received your letter of May 5,19⋯ submitting the subject claim together with the relative documents and have duly noted the contents thereof.

After our careful deliberation of this case, we regret to reply that you should file the claim first against the carrier. In case shipping company refuses to compensate you for the loss and their refusal found justified, we will have the case our further consideration. Returned herewith are the relative documents submitted by you.

Please be assured that, if the loss is found within the coverage of our policy, we will compensate you for the loss immediately.

Yours faithfully,

【註】

1. have duly noted the contents thereof:妥予注意到其內容。舊式用法。

2. deliberation＝careful thought:熟慮。因此"after careful deliberation"中的 "careful"毋寧是多餘的。cf. after long deliberation。

3. justified:證明爲正當；有道理。

4. compensate you for 可以"make up the loss","make up for the loss"代之。

No.158　經船公司拒賠後向保險公司索賠

Dear Sirs,

Re:<u>Your policy No.123</u>

We acknowledge the receipt of your letter of May 8, 19…concerning our claim under the subject policy and have duly noted its contents.

Following your instructions, we have referred this case to the shipping company and filed our claim against them for the loss we have suffered. Now we have received their reply and enclose one photostatic copy thereof for your reference and perusal.

Considering that the coverage of the captioned policy includes risks against leakage and breakage, you are requested to give our claim further and favorable consideration. In order to facilitate you to settle this claim, we submit again all the relative documents required.

　　We are looking forward to your early settlement of this case.

Faithfully yours,

【註】

1. considering＝in view of the fact:鑑於…。
2. coverage:擔保範圍；承保範圍。
3. risks against leakage and breakage:漏損及破損險。
4. early settlement 也可以"speedy settlement"代替。

第五節　有關買賣索賠的有用例句

一、提出買賣索賠的有用例句

1. The goods $\begin{Bmatrix} \text{we ordered} \\ \text{which you shipped} \end{Bmatrix}$ on May 7 have arrived →

　　→ $\begin{cases} \text{damaged.} \\ \text{in damaged condition.} \end{cases}$

2. We have duly received...(names of goods)...ordered from you on→

　　→ March 4, but $\begin{cases} \text{we find to our regret} \\ \text{regret to say} \end{cases}$ that...

　　＊ find to our regret:歉然發現　　＊ regret to say:遺憾的是

3. We have to inform you that...(goods)...ordered from you on August 7 has not arrived here. Nor have we heard anything from you concerning the shipment.

　　（茲通知貴公司關於本公司 8 月 7 日訂購的……迄未抵達本地，關於本批貨的裝運情形也未收到貴公司任何通知。)

4. With reference to our order ♯123 dated June 9, for...(name of goods), we shall be $\begin{Bmatrix} \text{glad} \\ \text{pleased} \end{Bmatrix}$ to know when we may →

$\begin{Bmatrix} \text{expect shipment} \\ \text{expect delivery} \end{Bmatrix}$,as the goods are $\begin{Bmatrix} \text{most urgently.} \\ \text{in most urgent need.} \end{Bmatrix}$

5. As the goods are urgently $\begin{Bmatrix} \text{needed} \\ \text{required} \end{Bmatrix}$,we must ask you to

→ $\begin{Bmatrix} \text{dispatch} \\ \text{ship} \end{Bmatrix}$ them $\begin{Bmatrix} \text{by air.} \\ \text{without further delay.} \\ \text{immediately.} \\ \text{on or before August 10.} \end{Bmatrix}$

＊without further delay:勿再延誤；勿再稽延

6. When we placed our order with you, we pointed out that prompt →

→ $\begin{Bmatrix} \text{delivery} \\ \text{shipment} \end{Bmatrix}$ was $\begin{Bmatrix} \text{essential} \\ \text{absolutely necessary} \end{Bmatrix}$. However, we have →

not yet received the goods or any advice when we may expect →

→ $\begin{Bmatrix} \text{delivery} \\ \text{shipment} \end{Bmatrix}$. Your delay will threaten the loss of one of our →

→ $\begin{Bmatrix} \text{old} \\ \text{new} \end{Bmatrix}$ → customers.

＊threaten the loss:使失去……

7. $\begin{Bmatrix} \text{Upon} \\ \text{On} \\ \text{When} \end{Bmatrix}$ unpacking the $\begin{Bmatrix} \text{consignment} \\ \text{cases} \\ \text{packages} \end{Bmatrix}$, we found that the goods→

→ did not agree with the original $\begin{Bmatrix} \text{sample.} \\ \text{pattern.} \\ \text{swatch.} \end{Bmatrix}$

　　* unpacking:拆開　　　　* did not agree with:與……不符

8. You used a much inferior quality of the stuff, which makes the products look very clumsy and not at all as fine as the original sample.

　　* stuff:原料　　　* look very clumsy:看起來很差

　　* not at all:毫不

9. Frequent complaints have been received from our customers to the effect that the pens leak and will not write without blotting. Quite frankly, they fall far below our standard.

　　* to the effect that＝purporting that:大意是說

　　* will not write without blotting:一寫就沾污紙張

　　* quite frankly:坦白地說

10. A comparison of the cuttings enclosed will convince you of the reasonableness of our proposition.

　　(請比較隨函附上的剪布，即可使你認爲我們的提議是合理的。)

11. $\begin{Bmatrix} \text{In checking the goods against your in voice} \\ \text{Upon examination} \end{Bmatrix}$, we $\begin{Bmatrix} \text{discovered} \\ \text{found} \end{Bmatrix}$
→ a considerable shortage in the number of toy animals and toy pianos.

　　* case was in good shape:箱的形狀良好

　　* does not appear to have been tampered with:無動過手脚的跡象

　　* surmise:認爲　　　　* shortshipped:短裝

12. The damage was apparently caused by $\begin{Bmatrix} \text{poor} \\ \text{improper} \\ \text{insufficient} \end{Bmatrix}$ packing.

A machine of this size and weight should be blocked in position inside the export case.

　　(這種笨重的機器，應在出口用木箱內部加以固定起來。)

13. The casks were not apparently strong enough for the purpose they

were used for; the result was that several casks sustained a leakage while in transit and the contents had all run out when delivered.

＊run out＝leak out＝be lost

14. We are compelled, therefore, to request you to make up for the loss of $…which we have sustained by the damage to the goods.

15. Please investigate this matter and adjust it $\begin{cases} \text{without delay.} \\ \text{as soon as possible.} \end{cases}$

二、答覆買賣索賠的有用例句

1. Immediately upon receipt of your $\begin{cases} \text{letter} \\ \text{cable} \end{cases}$ of...inquring about →

   → your order...,we consulted our files and records.

   ＊ consult our...records:查卷

2. Thank you for $\begin{cases} \text{calling our attention to...} \\ \text{notifying us so promptly of...} \end{cases}$

   ＊ call our attention to:提醒我們注意到...

3. As soon as we received your letter, we got in touch with the →

   → $\begin{cases} \text{packers} \\ \text{shipping agents} \\ \text{manufacturers} \end{cases}$ and asked them to look into the matter.

4. We have looked into the matter and find that your claim is perfectly justified.

   ＊ perfectly justified:確有道理

5. After a careful investigation, we have discovered that by some unaccountable carelessness your order was misplaced and was not attended to. We are wholly to blame for the delay.

   ＊ unaccountable carelessness:無法說明的疏忽;

   ＊ misplaced:誤置

   ＊ not attended to:未予處理;

＊ we...for the delay:對此遲延自應負全責

6. We deeply regret to find that the wrong goods have been shipped thru a mistake on the part of our shipping clerk. We are temporarily understaffed and had to hire new hands, but all this is no excuse.

　　＊ shipping clerk:發貨員　　　＊ understaffed:人手不足

　　＊ to hire new hands:雇用新手；＊ all this is no excuse:這些都不是藉口

7. We frankly admit that delivery was delayed, but it was really beyond our control since it was caused by a fire in our works. We asked for your understanding then and you kindly made allowance by extending the relative L／C.

　　＊ fire in works:工廠火災

　　＊ asked for your understanding:徵求你的諒解

　　＊ you kindly made allowance:蒙你惠允；蒙你原諒

8. The shortage of 4 out of the 60 cases of No.18, we understand, must have been caused thru unskilled packing for which please accept our profound apologies for the inconvenience you have been put to by this irregularity.

　　＊ unskilled packing:不熟練的包裝

9. As the articles were packed with the utmost care, we can only conclude that the damaged cases has been stored or handled carelessly. We have reported your claim to our insurance company.

　　＊ stored or handled carelessly:保管或搬動不小心

10. We regret these faulty sets were sent to you, and have today sent a replacement of 21 sets. We hope you will be pleased with the new lot.

　　＊ faulty sets:不良的貨品（成套的貨品）　　　＊ new lot:新品

11. We are, therefore, taking the matter up with the agents of the... Lines. In the meantime we are sending you a replacement today and

hope it will arrive in good order.

＊ take the matter up with:將本案向…提交

＊ Lines:輪船公司或航空公司。

12. If you will meet us and keep the goods, we shall offer you an allowance of 10% off the invoice value.

＊ meet us＝meet our request

13. We are very sorry this mistake occurred. You may be sure that →

→ we $\left\{ \begin{array}{l} \text{will make every effort} \\ \text{shall do everything in our power} \end{array} \right\}$ to see that such →

→ mistake does not happen again.

＊ you may be sure that:請相信　　＊ to see that:設法使…

14. We believe that the matter is now settled to $\left\{ \begin{array}{l} \text{your satisfaction.} \\ \text{our mutual satisfaction.} \end{array} \right.$

＊ mutual satisfaction:（使）雙方滿意

15. We appreciate the leniency you have shown in keeping the wrong goods and trust that you will give us an opportunity to supply you with further goods.

＊ leniency:寬容　　＊ keep the wrong goods:收下不良貨品

16. As requested, we agree to pay five cents per yard for the quantity specified above to compensate you for the loss.

17. We apologize for the inconvenience this transaction has caused you and assure you of our better attention to your future orders.

三、過失在買方或不在賣方

1. We have closely examined the sample taken from our last consignment and find it is no way different in quality from the TP-123 that we have here in stock. We can only surmise that there be a mistake

somewhere.

2. We have made a careful investigation but have failed to find that your letter was ever received by us. It is possible, of course, that it went astray in the mail.

　　* go astray in the mail:郵途中遺失

3. We are enclosing the report to show you that your order has been despatched from this end as promised. The shipping agents are now tracing it; they will check at the port of transhipment also.

4. We are, however, unable to explain the damage, since we took the best possible care in packing the cases and the shipping company received the whole lot in perfect condition as can be seen from the clean B／L we obtained.

5. The goods were in good condition when they were shipped on board the vessel as shown in the clean B／L issued by shipping company.

6. The shipping company is to be responsible for the shortage referred to in your letter. We suggested that you take up the matter with the shipping company.

7. We regret that we cannot see our way clear to make as exception in that case. The time limit for the claim expired a fortnight ago.

　　* see our way clear:認為適當（可能）＝regard as suitable or possible （主要用於否定句或疑問句）

8. Much as we would like to be of service to you, we are unable, in this instance, to accept your claim.

9. As you consider our proposals $\left\{ \begin{array}{l} \text{unacceptable,} \\ \text{unsatisfactory,} \end{array} \right\}$ we suggest that →

　　→ the matter be submitted to arbitration.

10. We wish to settle this dispute in a friendly way, and we, therefore, suggest that we submit it to arbitration and agree to abide by the

arbitrators' decision.

11. Under these circumstances, we trust you will agree that we cannot be expected to grant a free replacement of these damage machines. However, we are ready to compromise by replacing them at a reduced price of US$500 each.

12. Granting that you suffered a certain loss, we do not think it fair that we have to bear the whole responsibility. We fully realise, however, the awkward position in which you are placed, and in view of our friendly relations, we will offer you an allowance of 3%.

13. We hope the above explanation has convinced you that we have been abiding by our agreement with you.

14. Under the circumstances, we cannot see our way clear to →

→ $\begin{cases} \text{fall in with} \\ \text{agree to} \\ \text{yield to} \end{cases}$ your view in respect of the present claim.

　＊ fall in with＝agree to　　＊ yield to＝agree to:順從

四、待查明後再議

1. In order to make a detailed report, however, certain investigation will have to be made. These will require a week to conclude.

2. We propose to have the goods inspected immediately. If the inspection confirms the accuracy of your estimate, generous compensations will be allowed at once.

　＊ generous compensations:大幅補償　　＊ allowed:予以…

3. In order to settle the matter, we should be pleased to send a representative to conduct an inspection.

4. We are immediately investigating the matter and just as soon as we have anything definite to report, we shall write you again.

習　題

一、將下列中文譯成英文。

1. 此次延誤已令本公司深感不便(has caused us much inconvenience)，目前(at present)本公司已無這些存貨，無法滿足客戶之要求(cannot meet the demands of our customers)，以致許多客戶已改向他處訂貨 (placed their orders elsewhere)。

2. 對于延期裝運貴公司十月一日所訂貨品一事，本公司深表歉意，此實由於(due to)本公司機器發生嚴重故障(serious breakdown in our machinery)而停工數日所致(brought all our work to a stand still for several days)。

3. 謹向貴公司保證本公司必在履行貴公司今後之訂貨時(in filling your future orders)加倍留意(doubly careful)避免引起此款困擾(not to cause such trouble)。

4. 我們的檢驗顯示(shows)顏色不能令人滿意。

5. 由於損害非同小可(not negligible)，我們請你立卽補足短少之數(make up the shortage)。

二、將本書 No.137 的英文信譯成中文。

第二十七章　挽回客戶

(Regaining Lost Customers)

第一節　挽回客戶信的寫法

爭取新客戶固然重要，但是保住老客戶更重要。做生意如能一方面與老客戶保持繼續往來，他方面又能積極爭取新客戶，是最理想的事。然而往往事與願違。在交易往來過程中，或由於誤會、不滿、齟齬，或其他種種原因，往往交易了一兩次之後，音訊就不繼。精明的生意人，一方面開拓新客戶，他方面則絕不慢待老客戶，經常與老客戶保持連繫，對老客戶表示關心，並隨時詢問有什麼不週到或需要改進的地方。這樣才能保住客戶。

當發覺客戶中止訂購或訂單越來越少時，應即檢討爲什麼會發生這種現象，並進而採取適當的措施，以挽回失去了的或將失去的客戶。挽回客戶的最經濟的方法就是寫一封令他感動的信。這種信的內容包括下列各項：

1. 直率地或婉轉地提醒客戶很久沒有接到訂單。
2. 誠懇地詢問客戶爲什麼不再惠賜訂單，以便檢討、改進。
3. 誠懇地表示願意恢復以往愉快、密切的往來關係。
4. 假如有什麼使對方不滿的事，即告訴客戶，以後不致再發生類似情事，並要求客戶給予機會，以便能提供最佳服務。
5. 最後，表示謝意並祈惠覆。

第二節　拉回客戶信函

No.159　久未收到訂單，查詢何故

Dear Sirs,

We have not had either your order or your letter for almost six months. It seems to us that you have gone away from us. Surely, there must be some reasons for your having done so, and it is our sensitive assumption that possibly we have failed to satisfy you in some way.

As the saying goes: "Customer is always right." Your frank comments will be a help for us to renew and promote the good relationship between our two firms. Would you please let us know what is in your mind?

We shall appreciate receiving your kind reply.

Yours faithfully,

【註】

1. go away from us:離開我們。

2. sensitive assumption:敏感的假定。

3. in some way:在某一點（方面）cf. in some ways:在某些方面。

4. as the saying goes＝at the saying is:俗語說得好。

No.160　問客戶久未訂貨的原因

Dear Sirs,

We have not received your order quite a long time. We are eager to find out what the trouble has been. Did we do something wrong to your firm? Was our service satisfactory to you? Whatever it may be, could you kindly let us know?

It is our policy to render the best service to customers. You have always been considered as one of our regular customers, for you have given us remarkable patronage. We need you and we do not want to lose your business. We hope to have the pleasure of serving you again soon.

Your kind reply will be much appreciative.

Yours sincerely,

【註】

1. to render the best service:提供最佳服務。

2. remarkable patronage:很多惠顧。

No.161　問老客戶爲何已很久沒有訂貨

Dear Mr. Johnson,

I wish it were possible for me to step into your place and have a heart-to-heart talk with you. But as this is not possible, I am using this letter to keep up the contact between us.

The most unpleasant thing I have discovered in a long time is that we have not had an order from you for more than 100 days. There is a reason, of course. Won't you tell me what it is? If anything is wrong—whatever it is—I will fix it.

As you know, there isn't more pleasant feeling than to know you have a pleased customer. A satisfied customer is the best asset any business can have. Errors cannot always be helped, but they can be corrected. That's what I am here for. If you're not satisfied with our service from any angle, I am not doing my job right.

Your account is always appreciated. We consider it so valuable that we will do whatever is necessary to keep it alive.

An addressed envelope is enclosed to make it easy for you to tell if anything is wrong—and if you will send an order along at the same time, I will personally see that you get special attention.

Yours sincerely,

【註】

1. step into your place:到你那裡。

2. heart-to-heart talk:坦率的洽談; 促膝相談。

3. keep up the contact:保持聯繫。

4. fix:處理。

5. as you know,...a pleased customer:你知道，沒有比知道你有一個滿意的客戶能更使人感到愉快了。

6. business:指「事業」。

7. errors cannot always be helped:錯誤是無可避免的，"help"當「避免」解釋時總與"can't"或"can"連用。

8. That's what I am here for:那就是我的職責所在。

9. from any angle:從任何角度看。

10. I am not doing my job right:我沒有盡到責任; 我沒有做好工作。

11. Your account is always appreciated:我們經常感激你的照顧。

12. keep it alive:保持住 (你的照顧)。

13. I will personally...attention:我將親自加以特別的留意。

No.162 寫了一封信沒有反應, 再接再勵

Dear Sirs,

We are still wondering why we seem to have lost an old friend in you. A short time ago, we wrote to ask why we're not receiving your orders recently, but as yet we have had no reply.

It is possible that something we may have done—or did not do—has disturbed you. This could be a misunderstanding.

The matter is of great importance to us. Frankly, we want to keep the business of old customers like you, and we are sure it can't be too late to restore the pleasant relations that we formerly enjoyed. In this old world, we can't get anywhere without old friends.

So once again we ask: won't you please take a minute to jot down on the back of this letter the reason we haven't been hearing from you.

Yours sincerely,

【註】

1. we are still...in you:我們還在想: 爲什麼我們會失去了一個像閣下的老朋友。

2. as yet＝so far:到目前爲止。

3. disturbed you:得罪了你。

4. restore:恢復。

5. In this old world,......without old friends:在這古老的世界中，我們如果沒有老朋友，一定不會有什麼成就。

6. jot down＝write briefly 摘記下來。

△寫了一封信沒有反應，應該再接再勵，鍥而不舍。

No.163　老朋友才是好朋友

Dear Sirs,

You will agree that old friends are the best friends. If an old friend suddenly seems to forget you, you want to know the reason.

That's just the situation we are up against. Several months have passed since that last order. Of course, we would like to know the reason.

A frank discussion of any problem is the quickest and best way to a satisfactory solution. It is possible that something we may have done—or did not do—has annoyed you. Often it is just a misunderstanding.

In this old world we can't get anywhere by ourselves. We've got to have someone else's help, so won't you tell us just why we haven't heard from you? Be absolutely frank.

We very much appreciate your business and it is our sincere hope that we may renew a relationship with you that has been most pleasant.

May we hear from you soon?

$$\boxed{\text{Yours sincerely,}}$$

【註】

1. that's just the situation...up against:我們正面臨這種情形。

2. in this old world...by ourselves:在此古老的世界裡，我們不能獨善其身。如因我方錯誤致遭到客戶指責，以後，再也沒有收到他們的訂單時，我們如想拉回生意，則寫信的方式有二，一為向他們認錯，請他們原諒，二為裝迷糊，不知有何錯誤，上面一封信就是裝迷糊的寫法。

No.164　拉回因糾紛而失去的客戶

Dear Sirs,

Your orders have discontinued since the settlement of a claim you approved for the damaged goods of your order No.123.

We still feel sorry for the trouble that has caused you much inconvenience. We reiterate that we will make every effort to avoid similar mistake in our future transactions.

If one of your good friends once did something wrong in spite of his paying best attention when handling some case for you, but he apologized to you for that and expressed sincerely to do a better job for you in the future, would you forgive him? Now we are of the same situation. As the saying goes: "To err is human, to forgive divine." therefore, could you give us opportunities to provide you with our better service?

Welcome your challenge! We are sure that we are in

a position to do outstanding service for you. We shall not fail to meet your satisfaction in executing any of your orders in the future.

We are expecting your kind reply.

Yours sincerely,

【註】

1. reiterate:重申。

2. to err is human, to forgive divine: men are apt to make mistakes and to forgive mistakes is a divine character.意指「犯錯是人的常情，寬恕錯誤才是超凡的」。

3. challenge:挑戰。

4. shall not fail to:必定。

No.165 詢問客戶有何需改進之處

Dear Mr. Charleston,

In looking over accounts, we are very glad to note the nice business you have given us, and want to assure you of our appreciation.

However, in view of the somewhat larger order received from you last year, we are just wondering if any difficulty has arisen which is within our power to remedy.

If you have not received full satisfaction from us in every way, either in product or service, we would deeply appreciate your letting us know about it; not only because of our earnest desire to serve you well individually, but to help us avoid disappointing other customers.

Please understand that we appreciate your continued patronage, but we want to be sure that we are rendering the kind of service you deserve, and will appreciate a word from you as to whether or not anything may have gone wrong.

Sincerely yours,

【註】

1. 當客戶訂單減少時, 固然要寫信爭取訂單, 但是訂單很多時, 也要寫信表示謝意, 並詢問對我們有何意見及缺點, 以便改進, 這樣才能保住客戶。上面的一封信, 就是一個例子。

2. look over accounts:看了帳目。

3. nice business you have given us:你給我們的好生意（意指很多生意）。

4. within our power to remedy:我們能力範圍內所能補救的…。

5. in every way:在各方面; 在任何方面。

6. render the kind of service you deserve:給你應得的服務。

7. go wrong:差錯。

習 題

一、將下列中文譯成英文。

1. 我們想知道發生了什麼, 不但是因爲我們很想收到你的訂單, 而且是因爲我們很關心(are very much concerned about)你這位老客戶。

2. 如俗語所說:「顧客永遠是對的」。

3. 中國有一句俗語說(A Chinese saying goes)「過而能改, 善莫大焉」(A fault confessed is half redressed), 所以希望你能依照這種精神(in the same

spirit)原諒我們。

4. 服務是我們的責任(Service is our business)，所以請告訴我們你需要什麼。

5. 我們發覺貴公司已經一年多沒有向我們購買了，我們為此深感遺憾。所以不繼續光顧敝公司(discontinued your patronage of our firm)，內中必有重大原因(There must be some important reason)，假使確有使貴公司不再和我們往來的重大原因(if it is important enough to cause you stay away from us)，那麼我們盡一切努力去查出原因，自然也夠重要的。

二、將本書 No.163 的英文信譯成中文。

第二十八章　代理關係的建立
(Establishment of Agencyship)

產品的銷售需要靠人去推銷，尤其在國際行銷時爲然。當廠商着手選擇產品銷售路徑之前，首先要考慮的是應經那一類商人去推銷最佳。廠商將其產品推銷到國際市場的通路固然很多，但在很多場合，以透過代理商 (Agent) 優點較多。

反之，進口業者爲確保貨源以及採購的順利，也往往在出口地指定採購代理 (Buying Agent) 代辦採購業務。

以下就較常見的售貨代理 (Selling　Agency)，及採購代理 (Buying Agency) 有關的書信往來分別擧例說明。

第一節　售貨代理信的寫法

一、申請充任售貨代理函件的撰寫要領

要求一廠家指定本公司爲售貨代理商的函件，其撰寫原則與買賣客戶的招攬函件撰寫原則類似。所以這種 Agency Application 的內容大致如下：

1. 如何知悉對方 (如係初次認識)。
2. 具體說明本公司的優點，例如組織、資本、經驗、能力以及商業關係等。
3. 表示願充任其售貨代理。
4. 請其提示代理條件 (或主動提出代理條件)。

5.表示願衷誠合作，共同開拓業務。

No.166 希望指定本公司爲售貨代理商

Dear Sirs,

We owe your name to CETRA, through whom we have learned that you are seeking an agent here. Therefore, we write to ask if you are interested in extending your export business to our country by appointing us agents for the sale of your products.

We are sure that you know very well the advantages of representations. This is especially necessary when foreign exporters want to do large volume of business with our country, because most of our government, military and industrial purchases are done through open tenders. It is only through a regular representative that will have a chance to offer or bid in time and effectively.

We, the Taiwan Trading Company, are a well-established firm with 20 years of history and integrity in Taiwan foreign trade. We not only have very good connections with the government procurement agencies and government enterprises, but also have close business relations with the domestic private enterprises. We know how to meet tender requirements and could handle them effectively and successfully. We feel confident that if you give us any opportunity to deal in your products, the result will be entirely satisfactory to both of us.

So, we shall be very much pleased to act as your sole agents in Taiwan for your products. If our request is accepted by you, we shall thank you to let us know your terms and conditions under such agency agreement, and forward quotations, samples and other helpful literature, with a view to getting into business in the near future.

For any information you may desire regarding our standing, we are pleased to refer you to the following bank:

Bank of Taiwan

Chungking S. Road, Sec. 1

Taipei

We hope that this letter will be a forerunner to many years of profitable business to both of us, and look forward to the pleasure of hearing from you.

Yours faithfully,

【註】

1. we are sure that:我們相信。類似的用語有:

   we have the confidence that...

   we firmly trust that...

   we feel confident that...

   we are confident that...

2. representation:代表; 代理。

3. open tender:公開招標。

4. bid:投標; 報價。

5. government procurement agencies:政府採購代理機構，在我國爲中央信託局 (Central Trust of China)。

6. government enterprises:政府經營的企業。

7. private enterprises:私人經營的企業。

8. act as:充任。

9. forerunner:先驅；預兆。

No.167 同意指定對方爲售貨代理商

Dear Sirs,

Thank you for your letter of October 11 in which you indicate your desire to act as our sole agents for our products in Taiwan.

After completion of our credit files and careful consideration of your proposal, we have now made our decision to accept your proposal and to appoint you our sole agents in Taiwan district for a period of one year.

For the guidance and to help towards pleasant relations between us in our transactions, we have prepared an Agency Agreement, which is enclosed for your approval. Any suggestions you may have to make on it would be welcome. If you have no objection to any of its articles please return us the duplicate duly signed.

It is our firm belief that this agency relationship would prove mutually beneficial, and we trust that you will be prepared to co-operate with us closely.

Yours faithfully,

【註】

1. sole agent:獨家代理，又稱爲 exclusive agent。
2. completion of credit files:完成信用檔卷，即指已調查過代理商的信用情況。
3. Agency Agreement:代理契約
4. objection to its articles:對（代理契約的）條款有異議。
5. firm belief:堅信。
6. mutually beneficial:互相有利。cf. mutual benefit:互相的利益。

　　當雙方同意建立代理以後，即可簽立代理契約（Agency　Contract, Agency Agreement 或 Agency Arrangement）規定雙方的權利義務。至於代理契約的內容，視情形而定。有的規定得很簡單，有的則規定得很詳細。

No.168　獨家售貨代理契約

Suitable for exclusive and sole agents

(representing manufacturers overseas)

An Agreement made this...day of...19...between...whose registered office is situated at...(hereinafter called "the Principal") of the one part and...(hereinafter called "the Agent") of the other part.

Whereby it is agreed as follows:

1. The Principal appoints the Agent as and from the... to be its Sole Agent in...(hereinafter called "the area") for the sale of...manufactured by the Principal and such other goods and merchandise (all of which are hereinafter referred to as "the goods") as may hereafter by mutually agreed between them.

2. The Agent will during the term of...years (and

thereafter until determined by either party giving three months' previous notice in writing) diligently and faithfully serve the Principal as its Agent and will endeavour to extend the sale of the goods of the principal within the area and will not do anything that may prevent such sale or interfere with the development of the principal's trade in the area.

3. The Principal will from time to time furnish the Agent with a statement of the minimum prices at which the goods are respectively to be sold and the Agent shall not sell below such minimum price but shall endeavour in each case to obtain the best price obtainable.

4. The Agent shall not sell any of the goods to any person, company, or firm residing outside the area, nor shall he knowingly sell any of the goods to any person, company, or firm residing within the area with a view to their exportation to any other country or area without consent in writing of the Principal.

5. The Agent shall not during the continuance of the Agency constituted sell goods of a similar class or such as would or might compete or interfere with the sale of the Principal's goods either on his own accout or on behalf of any other person.

6. Upon receipt by the Agent of any order for the goods the agent will immediately transmit such

order to the Principal who (if such order is accepted by the Principal) will execute the same by supplying the goods direct to the customer.

7. Upon the execution of any such order the Principal shall forward to the Agent a duplicate copy of the invoice sent with the goods to the customer and in like manner shall from time to time inform the Agent when payment is made by the customer to the Principal.

8. The Agent shall duly keep an account of all orders obtained by him and shall every three months send in a copy of such account to the Principal.

9. The Principal shall allow the Agent the following commissions (based on F. O. B. United Kingdom values)...in respect of all orders obtained direct by the Agent in the area which have been accepted and executed by the Principal. The said commission shall be payable every three months on the amounts actually received by the Principal from the customers.

10. The Agent shall be entitled to commission on the terms and conditions mentioned in the last preceding clause on all export orders for the goods received by the Principal through Export Merchants, Indent Houses, Branch Buying offices of customers, and Head Offices of customers situate in the United

Kingdom of Great Britain and Northern Ireland and the Irish Free State for export into the area. Export orders in this clause mentioned shall not include orders for the goods received by the Principal from and sold delivered to customers' principal place of business outside the area although such goods may subsequently be exported by such customers into the area, excepting where there is conclusive evidence that such orders which may actually be transmitted via the Head Office in England are resultant from work done by the Agent with the customers.

11. Should any dispute arise as to the amount of commission payable by the Principal to the Agent the same shall be settled by the Auditors for the time being of the Principal whose certificate shall be final and binding on both the Principal and the Agent.

12. The Agent shall not in any way pledge the credit of the Principal.

13. The Agent shall not give any warranty in respect of the goods without the authority in writing of the Principal.

14. The Agent shall not without the authority of the Principal collect any moneys from customers.

15. The Agent shall not give credit to or deal with any person, company, or firm which the Principal shall from time to time direct him not to give credit to or

deal with.

16. The Principal shall have the right to refuse to execute or accept any order obtained by the Agent or any part thereof and the Agent shall not be entitled to any commission in respect of any such refused order or part thereof so refused.

17. All questions of difference whatsoever which may at any time hereafter arise between the parties hereto or their respective representatives touching these presents or the subject matter thereof or arising out of or in relation thereto respectively and whether as to construction or otherwise shall be referred to arbitration in England in accordance with the provision of the Arbitration Act 1950 or any re-enactment or statutory modification thereof for the time being in force.

18. This Agreement shall in all respects be interpreted in accordance with the Law of England.

As Witness the hands of the Parties hereto the day and year first hereinbefore written.

(Signatures)

（中譯）

本契約係於一九＿＿年＿＿月＿＿日由當事人一方＿＿＿＿＿＿＿＿

＿＿其已登記之辦公地址在＿＿＿＿＿＿＿＿＿＿（以下簡稱本人）與他

方當事人＿＿＿＿＿＿＿＿＿（以下簡稱代理商）所簽訂。茲約定事項如下：

1. 本人茲任命代理商自＿＿＿＿＿＿＿＿＿＿起為＿＿＿＿＿＿＿＿＿ ＿（以下簡稱代理地區）之獨家代理以推銷本人所製造之＿＿＿＿＿＿＿ ＿＿＿＿＿＿以及以後雙方互相同意之其他貨物及商品（以下簡稱商 品）。

2. 代理商在其＿＿年任期內（以及其後由當事人一方在三個月以前書面 通知他方所決定之期間內）將竭誠為本人服務並努力在代理地區內推 銷本人之商品，並將不作任何有礙此一推銷之行為或干擾本人在代理 地區內所作之貿易推廣。

3. 本人將經常向代理商提供每種商品最低售價表，代理商不得以低於表 定價格出售，並應設法在每一場合爭取最好之價格。

4. 代理商未經本人書面同意，不得將商品之任何部分售與代理地區以外 之任何個人、公司或行號，亦不得故意將商品售於代理地區內之個人、 公司或行號以便其再行輸往其他國家或地區。

5. 代理商在本代理契約有效期間內，不得以自己名義或代表任何其他個 人、公司或行號銷售與本人商品類似之商品，因而干擾本人商品之銷 售或與之競爭。

6. 代理商接到商品訂單後，應立即將訂單寄交本人，本人如果接受，則 將商品直接運交顧客。

7. 本人履行每一訂單後，即將開往顧客之發票副本寄交代理商，並當顧 客向本人付款時，以同樣方式隨時通知代理商。

8. 代理商所獲得之一切訂單均應記帳並每隔三個月向本人寄送一份帳 單。

9. 代理商在代理地區所直接獲得之一切訂單經本人接受並履行後，由本 人給予代理商佣金＿＿＿＿＿＿＿＿＿＿（按 F.O.B.英國計算）。佣金 每三個月付一次，按本人收自顧客實際金額計算。

10. 本人經由出口商、購買代理商、顧客之採購辦事處，以及顧客在英國、北愛爾蘭與愛爾蘭自由邦之總公司，輸往代理地區之一切出口訂單，代理商均得依照上條所述之條件請求支付佣金。惟顧客在代理地區以外之營業處所寄往本人之商品訂單縱然此等商品爾後再由顧客從該營業處所運往代理區，亦不包括在本條所稱之出口訂單以內。但如有確切證明，此等實際上可以經由英國總公司之訂單，係由於代理商向顧客爭取所致者，不在此限。

11. 如因本人付給代理商之佣金數目發生爭執時，將以本人之稽核員所決定之數目爲準，稽核員之證明對本人及代理商均有最後的拘束力。

12. 代理商不得對本人之信用能力作任何保證。

13. 代理商未獲本人之書面授權，不得對商品作任何保證。

14. 代理商未獲本人授權不得向顧客收取任何金錢。

15. 曾經本人指示代理商不得對某些個人、公司或行號授予信用或與之交易者，代理商不得對此類客戶授予信用或與之交易。

16. 本人有權拒絕履行或接受代理商所獲得之訂單或訂單之一部分，而代理商對被拒之訂單或其中之一部分無任何佣金請求權。

17. 當事人間，或其各別代表人間，今後有關本契約所指之事項或有關解釋等所引起之一切異議，均應在英國依照 1950 年仲裁條例或任何新制定或法定修正本，以仲裁解決之。

18. 一切有關本契約之解釋，以英國法律爲準。茲證明本契約係雙方當事人於前述之年月日簽訂。

(簽字)

No.169　拒絕指定對方爲售貨代理商

Dear Sirs,

Thank you very much for your proposal of October 20 offering your services to act as our agent for sale of our products in Nigeria. Although we greatly appreciate your offer, we regret to inform you that we have already made our decision to appoint Rammy Trading Company, Lagos, our agent in your area prior to receiving your proposal.

As it has been our policy of setting, one agent for one country up to the present time, we are obliged to decline your kind offer. We will, however, keep your name and address in our file so that we may communicate with you when our activity becomes more extensive in your country.

Thank you again for your proposal and hope a friendly business relationship between us will be possible at some future time.

Yours faithfully,

【註】

1. prior to＝before

2. keep in file:存卷、歸檔，同樣的用法尚有"keep on file"或"keep in a file" cf. put this letter in the file:將此信件放入檔卷中。

3. at some future time:將來有一天。

　　cf. in the future:在將來。

　　　　in future:下一次；嗣後。

No.170　謀任佣金代理

Dear Sirs,

We have a well-developed sales organization in Tanzania and are represented by a large staff in various parts of the country. From their reports it seems clear that there is a good demand for nylon textiles and as we believe you are not directly represented in Tanzania we are writing to offer our service as commision agent.

We have numerous connections throughout the country and there are good prospects of a very profitable market for your manufactures. Provided satisfactory terms could be arranged, we would be prepared to guarantee payment of all amounts due on orders placed through us.

You will naturally wish to have information about us. For this we refer you to the Barminster Bank, Dares Salaam and to Electrical Household Appliances Ltd., Taipei, with whom we have held the sole agency in Tanzania for the past three years.

We hope to hear favorably from you and feel sure that we could come to a satisfactory arrangements as to terms.

Yours faithfully,

【註】

1. well-developed sales organization:指「很好的推銷網」。

2. large staff:很多職員（指推銷員）。

3. there is a good demand for nylon textiles:這裡需要大量的尼龍紡織品。

4. not directly represented:無直接代表。

5. commission agent:佣金代理商。

6. good prospects:有很好的前途; 前途樂觀。

7. manufactures:製品。

8. arrangements as to terms:有關代理的條件。

　　要求增加佣金不是容易的事，特別是契約中已有規定時爲然。因此，任何要求增加佣金的信，應將其所以要求增加的理由具體地說明，而且要技巧地提出，以博得受信人的諒解與同情。

No.171　代理商要求增加佣金

Dear Sirs,

We should be glad if you would consider some revision in our present rate of commission.　The request may strike you as unusual since the increase in sales last year resulted in a corresponding increase in our total commission.

Marketing your goods has proved to be more difficult than could have been expected when we undertook to represent you.　Since then, German and American competitors have entered the market and firmly established themselves.　Consequently, we have been able to hold our own only by putting pressure on our salesmen and increasing our expenditure on advertising.

We are quite willing to incur the additional expense and even to increase it still further because we firmly believe there is business to be had if the required effort is made;

but we do not think we should be expected to bear the whole of the additional cost without some form of compensation. You may feel that this could most conveniently take an increase in the rate of commission by, say, 2%. We suggest this figure after carefully calculating the increase in our selling costs.

You have always been considerate in your dealings with us and we know we can rely on you to consider our present request with understanding and sympathy.

Yours faithfully,

【註】

1. strike you as unusual:使你認為不尋常。

2. corresponding increase:相應的增加; 相稱的增加。

3. more difficult than could have been expected:遠較我們預期為難。

4. undertake to represent you:承受代理你。接下代理你的任務。

5. hold our own＝maintain our position in the market:維持我們的地位。

6. put pressure on:加以壓力。

7. incur＝to be responsible for:負責。

8. there is business to be had:有生意可作。

9. you may feel:你可能認為。

10. say:譬如說。

11. considerate in your dealings with us:與我們交易時你們顧慮週到。

12. rely on:靠（你）; 信賴（你）。

No.172　對於代理商要求增加佣金的覆函

Dear Sirs,

Thank you for your letter of August 28.

We note the unexpected problems presented by our competitors and wish to say at once that we appreciate the extra efforts you have made with such satisfactory results.

In the long run we feel that the high quality of our products and the very competitive prices at which we offer them will ensure steadily increasing sales despite the competition from other manufacturers. At the same time we realize that in the short term this competition must be met by more active advertising and agree that it would not be reasonable to expect you to bear the full cost. To increase commission would be difficult as our prices leave us with only a very small profit. Instead, we propose to allow you an advertising credit of US$500 in the current year towards your additional costs, this amount to be reviewed in six months' time and adjusted according to circumstances.

We hope you will be happy with this arrangement.

Yours faithfully,

【註】

1. unexpected problems presented by:由…引起的預料之外的問題。
2. extra efforts:額外的努力。
3. in the long run＝in the end＝eventually:最後。
4. ensure:保證。
5. steadily:穩定地。

6. in the short term:短期內。

7. be met by:以…應付。

8. bear the full cost:負擔全部費用。

9. to allow:給予。

10. in the current year:今年內。

11. adjust according to circumstances:隨情況而調整。

　　供應商想提出降低佣金率的意見時必須慎審，否則將引起惡感而破壞感情。這種信應具體說出所以要降低的原因，藉以說服代理商。

No.173　提議降低佣金

Dear Sirs,

　　We write this letter with the utmost regret. It is to ask you to accept a temporary reduction in the agreed rate of commission. We make the request because of an increase in manufacturing costs due to additional duties on our imported raw materials and to our inability either to absorb these higher costs or, in the present state of the market, to pass them on to consumers. In the event, our profits have been reduced to a level that no longer justifies continued production.

　　We hope that in this disturbing, but we trust purely temporary, situation you will accept a small reduction of, say, $1\frac{1}{2}\%$ in the agreed rate of commission, with our promise that as soon as trade improves sufficiently, we shall return to the rate originally agreed.

　　　　　　　　　　　　　　　　Yours faithfully,

【註】

1. we write...utmost regret:以最大的歉意寫這封信。

2. to absorb these higher costs＝to accept without raising prices:負擔這些
   增加了的成本而不漲價。

3. pass them on to: 轉嫁給 (用戶)。

4. in the event, our profits...continued production:結果，我們的利潤已經低
   到無法繼續生產下去。

5. disturbing situation:令人困擾的情況。

第二節　採購代理信的寫法

進口商委託出口地的商號就地代辦採購的交易，稱爲委託購買
(Indent)。在委託購買交易中，委託商稱爲"Indentor"，受委託商稱爲
"Indentee"，一般又稱爲採購代理商(Buying Agent)。

No.174　申請擔任採購代理

Dear Sirs,

Your name has been recommended to us by Bank of
America, San Francisco, as one of the leading importers
of handicrafts.　If you are not already represented in
this market, we venture to inquire whether you are disposed to accept us as Buying Agents for Taiwan.

We have been established here in a fairly large way for
the last ten years, and from our intimate knowledge of
the business we believe that we can entirely meet your
requirements.　As our business has extended considerably of late, we have decided to devote our principal

attention to electronics and consider that the reputation we enjoy here and the connections we have all over this island, justify us at this stage in enlarging our operations and in shipping to U.S. importers.

We, therefore, feel confident that if you entertain our proposal, the result will be perfectly satisfactory to both sides.

We shall be glad to learn your views in the matter, and, to assist you in coming to a decision, we subjoin the name of several firms in U.S. whom we are regularly supplying what they require. We are sure that they will be pleased to reply to any inquiries from you concerning us.

For the guidance and to help towards pleasant relation between us in our transactions we encloses a copy of Agency Agreement for your approval, and, as we are from the first putting you on a very advantageous footing, we doubt not that you decidedly accede to our proposal.

Yours faithfully,

【註】

1. venture:冒昧。

2. are disposed to:有意。

3. in a fair large way:相當大規模地。

4. intimate knowledge of:熟悉。

5. of late＝recently

6. devote to:專心致志於…；專營。

7. enlarge our operations:擴大營業。

8. in shipping to U.S.:運銷美國。

9. entertain our proposal:接受我們提議。

10. assist you in coming to a decision:協助你們裁決。

11. subjoin:附上。

12. putting...footing:把你放在一種極有利的地位。

No.175 採購代理契約

Agreement for Buying Agency

THIS AGREEMENT, made and entered into this 30th day of June, 19…, by and between The American Trading Company, Inc., of New York, NY, U.S.A., party of the first, and Taiwan Trading Co.,Ltd., Taipei, Taiwan, party of the second part, witnesseth:

1. The party of the first part (hereinafter called the Principal) here by grants to the party of the second part (hereinafter called the Agent) the exclusive right to buy and ship its requirements in Taiwanese electronics for the period of three years from the date hereof unless sooner terminated as hereinafter provided.

2. The Agent covenants that it will use every reasonable diligence in executing all orders placed with it by the Principal and will not, without the consent of the Principal, deviate from any instructions given by

the Principal in regard to the purchase and shipment of its orders.

3. On all orders executed by the Agent during the continuance of this agreement, the Agent shall be paid the following commission:

on an order amounting to US$10,000 or less:3%.

on an order amounting to more than US$10,000:2.5%.

4. The amount of an invoice rendered by the Agent, including its commission and its disbursements except postages and petties, shall be paid by the Principal under Irrevocabe L/C against the delivery of shipping documents.

5. This agreement may terminated by either party upon three months written notice delivered or sent by registered mail to the other, and may be terminated at any time, without such notice, upon breach of any its terms and conditions.

IN WITNESS WHEREOF, the parties hereto have caused this agreement to be executed in quadruplicate, two for each party, as of the date first above written by their duly authorized representative.

The American Trading Company, Inc.
(signed)

Taiwan Trading Co., Ltd.
(signed)

【註】

1. made and entered into:締結。"made"與"entered into"重覆，是法律用語。

2. party of the first:第一當事人。

3. witnesseth:證明，爲"witnesses"的古語用法，但法律文件卻喜用此語，難怪(no wonder)法律文件都晦澀難懂。

4. exclusive right:獨占權利。

5. terminate: (契約) 終止。

6. deviate from:逸出; 違反。

7. disbursements:開支。

8. petties:零星支出。

9. written notice:書面通知。

10. in witness whereof:茲證明……

No.176　同意指定對方爲採購代理

Gentlemen:

We have received your letter of August 12 together with its enclosures and have taken your proposal to act as our Buying Agents for Taiwan into serious consideration.

You will readily understand that we wish to proceed without undue haste. We, therefore, suggest that the proposed Agreement for Buying Agency shall hold good at first for a period of twelve months only and be terminated on 30 days' notice from either side.

If, as we have no reason to doubt, the results are such as may justly regarded as satisfactory, we shall, after the lapse of that period, be prepared to confirm the agreement for a longer time.

> We trust we may both have reason to congratulate our-
> selves upon this departure.
>
> Very truly yours,

【註】

1. have taken...into serious consideration:對…做了審愼的考慮。

2. readily＝easily

3. wish to proceed without undue haste:願意不過於草率的進行。

4. hold good:有效。

5. justly:公正地；理由充份。

6. after the lapse of that period:滿期之後。

7. congratulate...upon:恭喜。

8. this departure:這個新的發軔；這個新行動的開始。

第三節　有關代理的有用例句

1. We should be glad if you would consider our application to act as agents for the sale of your...(products)...

2. We are acting as agents for a number of American publishers and are wondering whether your firm is represented in...(country)...

   ＊ your firm is represented in...你的行號（是否）有代表在…

3. We have received your letter of...(date)...and shall be glad to offer you a sole agency for the sale of our products in...(country)...

4. We thank you for your letter of...(date)...and are favorably impressed by your proposal for a sole agency.

5. We hope our handling of this first order will lead to a permanent agency.

6. I hope to hear favorably from you and feel sure we should have no

difficulty in arranging terms.

　＊ arranging terms:指代理條件的安排

7. If you give us the agency we should spare no effort to further your interests.

　＊ spare no effort:盡力

　＊ to further your interests:增進你的利益

8. If you are interested we can give you first-class references.

　＊ first-class references:最佳的備詢人 (或信用資料)

9. We already represent several other manufacturers and trust you will allow us to give you similar service.

10. We accept your terms and conditions set out in the draft agency agreement and look forward to a happy and successful working relationship with you.

習　　題

一、售貨代理與採購代理有何區別? 與佣金代理又有何不同?

二、將下列中文譯成英文。

1. 我們敢說(venture to say)以本公司對本地情形的熟悉(knowledge of local conditions)、本公司的聲譽(reputation)以及我們的關係(connections)將對貴公司有很大的用處(service to you)。

2. 鑒於起初階段(initial stages)將遭遇到相當的困難(be attended with considerable difficulties)，我們認爲貴公司將同意一成佣金並非不合理。

3. 本公司已代表許多日本廠家多年，因此我們相信可對貴公司同樣效勞(render you similar services)。

4. 由於市場有限(limited)，對貴公司銷售(marketing)我們的競爭對手產品必須予以若干限制(some restrictions must be placed)。

5. 我們發覺過去數月貴公司銷貨量(sales)銳減(a sharp decrease)，甚感遺憾。

三、試述訂立代理契約時須注意那些事項?

第二十九章　招標與投標

(Invitations & Bids)

第一節　招標與投標的意義

　　國際貿易的交易方式買賣雙方除採協議(Negotiation)方式進行外，也可以公開招標(Open tender)的辦法。所謂公開招標乃買方或賣方利用公開競爭方式，使多數的商號出價(Bid)，從而以最理想的條件買進或售出貨物的行為，上述多數商號的出價行為稱為投標(Bidding)。

　　報紙上常有"Invitation-to-bid"的啓事，這個"Invitation-to-bid"詞，照字面解釋，是「邀請出價」之意，而實際上即為「招標」之意。「標」即是「標價」(Bid or Price Quotation)，也即「報價」(Offer or Quotation)之意。所以，"Invitation-to-bid"又可寫成"Invitation-to-offer"或"Invitation-to-tender"。因其公開邀請多數商號競標，所以稱為「公開招標」，但自參加競標商號的立場而言，這種交易方式，又稱為「公開投標」(Open bidding)。

　　如果買方以招標方式購入貨物，則就買方立場而言，可稱為公開標購(To purchase through open tender)；如賣方以招標方式出售貨物，則就賣方而言，可稱為公開標售(To sell through open tender)。

　　招標在法律上乃為一種向不特定人的買賣要約引誘行為。在招標之前，招標商首先要擬定招標單。所謂招標單乃規定標購或標售貨物名稱、規格、數量、交貨、付款、驗收、投標商資格、開標時地、押標金、決標方法者。

擬定招標單之後, 即可公告(Announcement)。公告除張貼於招標機構佈告欄外, 通常都在報紙上刊登廣告。

第二節　有關招標與投標的函件

No.177　招標單(Form of Invitation-to-Bid)

CENTRAL TRUST OF CHINA
TRADING DEPARTMENT
INVITATION-BID-CONTRACT

Invitation No. CTCTD 123　　　　Date: Sept 20, 19…

Contract No.＿＿＿＿＿＿

Sealed bids in triplicate, which shall be valid for at least six (6) hours and subject to the terms and conditions of the attached sheets, will be received at CTC Trading Dept. 4 th floor, 49, Wu-Chang Street, Taipei, not later than 3:00 P.M., Oct. 9, (Tuesday) 19… (Taipei Time), and then they will be publicly opened in the presence of bidders. The successful bidder shall be informed individually by CTC for negotiation, if necessary, within the validity of the offer, CTC will disclose the result of bidding to the unsuccessful bidders. CTC reserves the right either to award or to reject any or all bids, or to negotiate with the lowest acceptable bidder/s.

1. Commodity:

U.S. Yellow Soybeans.

2. Quality & Specifications:

U.S. Grade No. 2, of 19··· crop

Oil contents: 18.5% min.　　Bushel weight: 54 lbs. min.

Moisture: 14.0% max.　　　Foreign materiai: 2.0% max.

Splits: 20.0% max.　　　　Damaged total (including

　　　　　　　　　　　　heat damaged): 3.0% max.

Heat damaged: 0.5% max.　Brown & balck kernels:

　　　　　　　　　　　　2.0% max.

3.Quantity: 10,000 metric tons, 5% more or less in bulk, up to 10% in bags for stowage purpose if required, gross for net, at ship's option.

4. Price:

Price to be offered either on FOB ex-spout basis (stowing & trimmings ship's account) or on C&FFO Kaohsiung basis at CTC's choice.

5. FOB Award:

CTC will arrange the vessel (s) at the following terms and conditions: (All detailed shipping conditions shall be settled before awarding by bidder & shipowner directly, CTCTD serves only as a witness, without any responsibility on CTCTD's part)

a)Loading rate: CQD

b)Carrying charges: In the event that the vessel contracted could not arrive at the loading spouts within the shipping period as stipulated above, any carrying charges and interest thus incurred shall be for shipowner's account. However, in case that the vessel arrives at the loading port in accordance with the shipping period as stipulated, but could not load soybeans due to port congestion or other causes which are beyond the control of the vessel, no carrying charges and interest shall be paid by the shipowner. Carrying charges, if any, shall be borne by the seller, irrespective of force majeure clause under Article 25 of conditions.

c)The shipowner or his agent shall notify the seller of the ship's name, her ETA to loading port, at least seven days prior to the arrival of the vessel. In counterpoint, the seller shall make a penalty payment at the rate as stipulated in Article 18 of general conditions, should the seller fail to load the commodities within the stipulated shipping period after receipt of Notice of Readiness forwarded by the shipowner or his agent.

d)Other terms & conditions: According to Gencon Charter Party with overtime clause.

6. C&F Award:

(I)Terms of carrier (s)

a)Bidder must specify vessel(s) name, age, speed, tonnage, number of hatches, draft etc. otherwise the bid is not acceptable.

b)Vessel(s) over 20 years of age and below 12 knots per hour are not acceptable. Vessel (s) should be sufficiently and efficiently equipped with derricks／winches to handle the loading and discharge of cargo for each and every hatch.

c)Bidder／owner must guarantee that the carrier(s) used must be able to berth alongside the Kaohsiung wharf for discharge alongside the wharf, otherwise all losses incurred should be for Bidder's／owner's account.

(II)Discharging rate:

Cargo to be discharged at the rate of 200 MT per hatch per weather working day, Sundays and Holidays excepted, even it used, provided that each hatch is equipped with one set of efficient derrick with adequate power. However, C.Q.D. discharge is preferred.

(III)Discharging laytime:

Laytime to commence at 1p.m. if Notice of Readiness to discharge is given to the consignee before Noon and at 8 a.m. next working day if Notice of Readiness is given during office hours after Noon, whether in berth or not. If work commenced earlier than the time stipulated above, laytime still to be counted from the time stipulated not counted earlier.

(Ⅳ)Demurrage & Despatch at discharging Port:

Discharging US$200／US$100 per hatch day or pro rata. Others according GENCON Charter party, with BFC Saturday clause & overtime clause.

(Ⅴ)Vessel's overages insurance premium shall be for seller's account.

7. Shipment:

Loading port: US Gulf port, one safe berth port only.

Discharging port: Kaohsiung.

Shipping period: November, 1989 or earlier.

8. Payment:

By Irrevocable Letter of Credit in full amount available against at sight draft and documents, to be established before shipment in favor of seller.

9. Bid Bond:

One percent Bid Bond shall be posted before 17:00 of the day previous to the bidding date and shall be valid until 60 days after the bidding date.

10. Performance Bond:

Five percent performance bond shall be posted with CTC within ten working days after awarding. The porformance Bond shall be valid until 90 days after the contracted shipment date and shall be extended accordingly in case the shipment is delayed or the contracted shipment date is extended.

11. Insurance:

To be covered by CTC, but seller shall noticfy CTC by telex of all necessary data before loading for insurance coverages.

12. Documentation:

a)Official grade certificate issued at time of loading certifying that all soybean loaded are strictly in conformity with the specifications as stipulated in article 2 of this Invitation, is required, and will be considered final as to quality of the soybean shipped.

b)Weight Certificate issued by a licensed weighmaster at the point of loading on vessel is required and will be considered final as to quantity of the soybean shipped.

c)Official Phytosanitary Export Certificate is required.

d)Official Certificate of Origin is required.

e)First & Second Originals with copies of clean-on-board Bill of Lading, ten copies of seller's detailed Invoice, and stowage plan are required.

f)As soon as available, the seller shall telex CTC Trading Dept.

(Telex No. 11-377 "Centrust" Attn: "Trustrade") the following information:

Invitation & Contract numbers, brief description of quantity shipped, ship's name, port of departure,

ETD, ETA, Kao-hsiung, Invoice value, a copy or this telex shall constitute a part of the repuired documents for payment.

g)At the time of loading, the seller shall forward by registered airmail two complete sets of duplicates of documents stipulated as Articles a) to f), to CTC Trading Dept. immediately.

h)All expenses of quality／weight inspection and analysis, together with the fee of all documents required in the L／C, etc, except that expenses paid to the independent surveyor will be for CTC's account, shall be for the seller's account.

13. CTC reserves the rights to award the contract to any bid or reject all bids. In case the terms quoted by the bidders cannot fully meet with CTC's requirements, CTC shall request all bidders to submit their firm offer(s) according to the same terms and conditions of this Invitation-Bid-Contract to CTC, in sealed envelop marketd "Firm Offer" day by day consecutively or any day at their choice, within the next ten days after the tendering date subject to CTC's prior acceptance, if acceptable, of such Firm Offer as stated above.

14. General Instructions & Conditions:

(I)Attached "General Instructions & Conditions" are incorporated hereto to form a part of this

> Invitation-Bid-Contract, if any contradiction in-between, the latter shall govern.
>
> (II) CTC's award of this tender is valid subject to the approval of BOFT.

【註】

1. sealed bid:密封標單。

2. bidders:投標商。

3. successful bidder:得標商，又稱爲 awarded bidder。

4. negotiation:議價。

5. disclose:公開。

6. reject any bids:否決任何「標」。

7. lowest acceptable bidder:可接的最低標投標商。

8. foreign materials:夾雜物；雜質，又稱 foreign matter。

9. in bulk:散裝。

10. for stowage purpose:做爲積載之用。

11. gross for net:以毛重代替淨重，意指按毛重計價。

12. at ship's option:意指多裝或少裝 5%由船方決定。

13. FOB ex-spout:利用噴注管的船上交貨價，"spout"爲用以將穀類注入船艙的噴注管。

14. stowing & trimmings:堆積及平艙費。

15. FOB award:按 FOB 決標。

16. carrying charges:加計費用或持有費用。按 FOB 成交時由於買方未能在約定期間內裝船時，買方應負擔的各種費用：包括(1)儲存費用(2)利息

17. shipowner:船東。

18. in counterpoint:對應。

19. overtime clause:逾期條款。

20. hatch:艙口。

21. draft: (船的) 吃水 (＝draught) :cf. a ship of 10 feet draft 吃水十呎的船。

22. knot:海哩。

23. derrick／winch: derrick:起重機, winch:絞盤。

24. overage insurance premium:逾齡保險費。

25. bid bond:押標金。

26. posted:繳存。

27. performance bond:履約保證金。

28. documentattion:單證的提供, 本條規定賣方押滙時應提供的單證。

29. official grade certificate:官方出具的等級證明書。

30. save for＝except

31. licensed weighmaster:特許 (指有執照的人) 重量檢定人。

32. Official Phytosanitary Export Certificate:官方出具的植物檢疫證明書。

33. stowage plan:積載計劃。

No.178 招標佈告(Announcement for open tender)

INVITATION-TO-BID

The Central Trust of China, Trading Department, intends to purchase the following listed commodity through open tender. Any registered and qualified firms or manufacturers, if interested in this tender may obtain from this office Invitation-to-Bid and submit their bids to this office on or before the date specified hereunder:

Inv. No.	Description	Area of source	Invitation issuing date	Tender opening date

CTCTD 123	US Grade No.2 Yellow Soy-bean 19⋯ crop	U.S.A.	Sept.20 19⋯	10:00 a.m. Oct. 9, 19⋯

【註】

1. purchase...through open tender:以公開招標方式購買⋯。

2. registered and qualified:登記有案的合格（廠商）。

3. invitation-to-bid:招標單。

4. area of source:貨物來源地區，即指採購地區。

5. invitation issuing date:招標日期，即印發標單日期。

6. tender opening date:開標日期。

No.179　Agent　通知供應商招標消息

Gentelmen:

We confirm we have sent you the following telex today:

CTC TRADING DEPT WILL OPEN TENDER 10:00 A.M. OCT 9 FOR 10,000 MT US GRADE NO.2 YEL-LOW SOYBEANS 19⋯ CROP NOV SHIPMENT PLS SEND US YOUR FIRM BID IN FOB EX SPOUT AND CNF KAOHSIUNG DETAILS AIRMAILING

We are pleased to inform you that CTC Trading Depart-ment intends to pruchase 10,000 metric tons US grade No.2 yellow soybean, 19⋯ crop through open tender at 10:00 a.m. on October 9, 19⋯, Invitation-to-Bid for which is enclosing for your information and study. If you are in a position to make the supply, please send us your firm bid on FOB ex spout Gulf Port and C&FFO Kaohsiung. You are requested to get your offer here before October 8, our time.

In addition to the particulars carried in the Invitation-to-Bid, we want to call your attention to the following:

1. Bid Bond: one percent (1%) bid bond should be deposited by you either in cash, L/C or Bank guarantee before 17:00 of the day previous to the bidding date and shall valid until 60 days after the bidding date.

2. Performance Bond: Five percent (5%) performance bond should be deposited by you in cash, L/C or Bank guarantee within 10 working days after the award of the contract.

3. Penalty for delay shipment: the buyer has made it known that suppliers must make a penalty payment to buyer on the basis of 0.1% of the contract value of the delayed portion for each one day's delay.

Please note that your offer should include one percent (1%) commission for us on FOB ex spout value or 0.8% on C&FFO value.

We are hoping to receive your most competitive offer with details in due course.

Yours faithfully,

【註】

1. 電文中的 Telexese:

   CNF: C&F

2. U. S. Grade No. 2 yellow soybeans:美國二級黃豆。

3. 19… crop: 19…年生產的作物。

4. Gulf port:這裡是指美國的 Gulf port,包括 Mobile, New Orleans, Beaumont, Houston, Galveston 等港口。

5. get your offer here:將你的報價寄達此地。

6. our time:意指此地時間(October 8).

7. Bid Bond:為投標人因投標而向招標人繳存的保證金。投標人於得標後如拒絕簽約時, 招標人即可沒收該保證金。參加國際投標時多以 Bank Letter of Guarantee 或 Stand-by L/C 代替現款或票據。

8. Performance Bond: Performance Bond 為賣方與買方簽約時, 為防賣方無力履行契約, 而要求賣方繳存的保證金。在國內買賣, 多用現款或票據繳存, 但在國際貿易則多以 Bank letter of guarantee 或 Stand-by L/C 代替, 如賣方不履約, 則買方可沒收 Performance Bond。

9. bidding date:指開標日而言。

No.180　供應商通知 Agent 應標

Dear Sirs,

Re: open tender of Yellow Soybeans

Thank you very much for your telex and your letter of September 20 and Invitation-to-Bid with which you called our attention to the forthcoming bid of the CTC Trading Deptartment for 10,000 M/T of yellow soybeans.

After due consideration, we have decided to authorize

you to participate in this tender on behalf of ourselves in accordance with the terms and conditions stipulated for subject tender.

As you will note from the attached enclosures, the price is left open. We will telex you one day before the tender opening date.

In the meantime, we shall ask our bankers, Chemical Bank, New York, to issue a letter of Guarantee favoring CTC, Trading Department through CTC, Banking and Trust Department as a bid bond for the subject tender.

We trust that our offer will enable you to secure this bid for us and thank you for your kind cooperation.

Yours faithfully,

【註】

1. participate in this tender:應標。

2. price is left open:價格未決定。

3. secure this bid:奪得此標。

No.181 供應商以 Telex 通知 Agent 價格

RYL SEPL 20 OUR LETTER SEPT 30 CONCERNING CTCTD'S INVITATION NO 123 FOR YELLOW SOYBEAN WE OFFER FIRM VALID UNTIL 4:00 PM OCT 9 USD 220 PER MT FOB EX SPOUT GULF PORT AND USD 236 PER MT CANDFFO KAOHSIUNG INCLUDING YOUR COMMISSION STOP PLS TELEX US RESULT SOONEST

【註】

1. RYL: refer to your letter
2. USD: U.S. Dollar

No.182　Agent　向招標商報價

Central Trust of China

Trading Department

Dear Sirs,

<u>Re: Invitation No. CTCTD 123</u>

In response to the captioned Invitation and subject to all the terms and conditions thereof, we are pleased to offer frim 10,000 metric tons, 5% more or less, of US grade No.2 yellow soy-beans for account of Cargill Inc., New York valid until 4:00 p.m. October 9 at price quoted and on conditions as stipulated in our bid. Enclosed please find our bid in duplicate for your evaluation. Under separate cover, we are submitting samples of yellow soy-beans which were received from our principals.

We shall appreciate it very much if you will give our bid your favorable consideration.

Faithfully yours,

Taiwan Trading Co., Ltd.

No.183　Agent 通知 Principal 得標，並請繳履約保證金

Cargill Inc.

New York, NY.

Dear Sirs,   Re: Your Bid No. NY-123

                        CTC's Invitation No. CTCTD 123

This confirm our telex dispatched to you today regarding to the captioned bid.

We are pleased to inform you that we have, on your behalf, submitted your bid to CTC. Trading Department for 10,000 M/Ts (±5%) yellow soybeans with a total value of US$2,200,000, and the contract has been awarded to us.

According to CTC's regulations, the suppliers to whom a contract is awarded should post performance bond equivalent to 5% of the contract value with them by cash, L/C, or L/G. Therefore, you are requested to post the bond with them on or before October 19. Meanwhile, please let us be informed as soon as the performance bond has been posted.

One copy of the original contract has been airmailed to you today for your reference and perusal.

                        Faithfully yours,

                        Taiwan Trading Co., Ltd.

【註】

1. supplier:供應商。

2. post:繳存。

No.184　Principal 通知 Agent 已繳履約保證函

Taiwan Trading Co., Ltd

Dear Sirs,　　Re: CTC's Invitation No. CTCTD 123

We are very much pleased to learn from your telex and
your letter both dated October 9 that the contract for
10,000 M／Ts yellow soybeans has been awarded to us
and appreciate your efforts to obtain it.　We have also
received the original copy of the contract to which we
have given our careful attention.

As requested, we asked the Chemical Bank, New York,
on October 16, to issue a letter of guarantee for US$110,
000 through CTC, Banking & Trust Department as our
performance bond to guarantee our performance under
the contract.　Enclosed please find one copy of letter of
guarantee for your reference.　Please pass this information on to CTC, Trading Department for their reference.
We are now going to start preparing the commodity and
trust that we can deliver them within the period as
stipulated in the contract.

　　　　　　　　　　　　Yours faithfully,

　　　　　　　　　　　　Cargill Inc.

　　由招標而投標以至簽約，文書作業(Paper Work)甚繁雜，爲簡化起
見，可以將招標(Invitation)、投標(Bid)以及契約(Contract)合併成一定
的格式，稱爲"Invitation, Bid and Contract"，將有關各種條件或規定，
印妥備用，甚爲方便。

茲將中信局購料處所用的格式列示於 No.178:

No.185　Agent 要求招標商發還履約保證函

CTC, Trading Dept.

Dear Sirs,　　　　　Re:Contract No. 321

<u>Letter of Guarantee</u>

We refer to our letter to you of October 21, 19⋯ informing you that our principals, Cargill Inc., have posted the performance bond in letter of guarantee form as required under the subject contract.

Now, the yellow soybeans have arrived and been inspected to the fullest satisfaction of yourselves.　In view of the fact that the supplier have performed satisfactorily the obligations under the contract, you are requested to release the performance bond as early as possible.

Your compliance with our request will be appreciated.

Yours faithfully,

Taiwan Trading Co., Ltd.

No.186 Invitation, Bid and Contract.

FORM B
Bid Bond is required

INVITATION, BID & CONTRACT

INVITATION NO._____   CONTRACT NO._____

CENTRAL TRUST OF CHINA, PROCUREMENT DEPARTMENT. 49, WU CHANG ST. SEC. 1, TAIPEI. 100, REP. OF CHINA, CABLE ADDRESS: TRUSTPRO TAIPEI, TELEX: 11377 CENTRUST, ON BEHALF OF THE CLIENT

ADDRESS:_____

INVITATION   DATE_____

Sealed bids subject to the instructions and conditions on the attached sheets will be received at this office until_____ o'clock,_____ and then publicly opened for furnishing the following supplies from_____ _____ for shipment to_____ on or before_____

CENTRAL TRUST OF CHINA, PROCUREMENT DEPARTMENT

Bids to be given on_____ basis valid for_____ days.   **MANAGER**

ITEM/ CODE NO. (1)	DESCRIPTION OF SUPPLIES (2)	QUAN- TITY/ UNIT (3)	UNIT COST FOB/ FAS (4)	TOTAL COST FOB/ FAS (5)	UNIT OCEAN FREIGHT (6)	TOTAL OCEAN FREIGHT (7)	INSU- RANCE PRE- MIUM (8)	TOTAL C&F/ CIF (9)
	GRAND TOTAL:							

Insurance _____
Inspection by _____
L/C Opening Bank _____

BID   DATE_____

In response to the above invitation and subject to the instructions and conditions thereof, the undersigned offers and agrees, if this bid be accepted within _____ days from the date of the opening, to furnish any or all of the items, upon which prices are quoted, at the price set opposite each item and deliver at the point(s) specified in accordance with the delivery schedule as shown below or in the attached continuation sheet.

Shipment on or before _____
To be shipped by Liner Vessel/Tramp Steamer/Parcel Post/Airlift _____
Port of Shipment_____ Source of Origin (indicate country)_____
Name of L/C Beneficiary_____
Address _____
Supplier _____   Address_____
Manufacturer,_____   Address_____
Bidder_____
Address _____
Telephone No._____ Telex or Cable address_____ By_____

(Signature and title of person authorized to sign)

Opening of bids witnessed by   CTC   Auditor   Client

CONTRACT

ACCEPTED BY CENTRAL TRUST OF CHINA   DATE_____

1. Accepted as to items numbered:_____
       Total Price_____ _____
2. Shipment on or before _____
       to_____
3. Inspection by _____
4. Markings to include "_____"
   in addition to those required in Article 13 of attached Conditions.
5. "On deck" B/L _____ acceptable. Transhipment _____ allowed.
6. Partial shipments _____ allowed, but not more than _____ shipments.

CENTRAL TRUST OF CHINA PROCUREMENT DEPARTMENT

MANAGER

【註】

1. have performed satisfactorily the obligation·under the contract＝have fulfilled satisfactorily the contractual obligations.
2. release:解除；"release the performance bond"爲解除履約保證之意，也即發還履約保證函之意。

習　　題

一、將下列用語譯成英文。

1.公開招標	2.公開標購
3.公開標售	4.決標
5.得標商	6.押標金
7.開標	8.底價
9.決標單	10.圍標

二、將下列英文譯成中文

1. Bidder (hereinafter called Seller, if successful and awarded a contract) shall submit his bid to the Central Trust of China, Procurement Department (hereinafter called CTC) in five complete sets each consisting of quotation on Invitation Bid and Contract form B, specifications and all other necessary information. The bid after being accepted by CTC will constitute the contract covering the item(s) accepted and one set will be returned to the seller. In case Invitation and Bid form C is used. A separate contract will be concluded after the award is made. Quotations, specifications and all necessary information which form a part of the bid shall be in the English language.

2. Bids shall be submitted to CTC by or through qualified local firms whose registration cards and tex-paying certificates will be checked

before the opening of bids. Bids submitted direct by foreign companies, if they themselves are not manufacturers or producers, must be certified by the Chinese consul in that country or the manufacturer whose products are being offered. Otherwise the bids will not be considered. Cable bid may be considered provided written bid has been airmailed CTC before the bidding deadline.

3. Bidder must state FOB cost, insurance premium, if any, ocean freight and independent surveyor's or laboratory's charges separately. FOB cost, unless otherwise specified, shall include ex-factory cost export packing cost, factory inspection charges, inland freight, forwarding fee, certificate charges, export taxes and duties and all other export expenses up to the point of the supplier's obtaining a clean-on-board ocean Bill of Lading. Ocean freight, unless otherwise specified, shall include the handling, loading, stowing aboard and trimming ocean vessel at the port of shipment and discharge at the port of destination.

第三十章　貿易電報與電報交換

(Business Cable and Telex)

　　做貿易，利用電報(Cable)、電報交換(Telex)或電話傳眞機(FAX)交易的情形相當普遍。諸如詢價、報價、還價、簽約，乃至交涉索賠等等，都可以利用電報、電報交換或電話傳眞機進行。然而國際電報或電報交換，費用昂貴，假如對于電文的撰寫不得要領，則不但交易不能順利達成，而且還浪費金錢。因此，要成爲一個勝任的貿易從業員，必需對於電報及電報交換，尤其電報文體(Cablese)及電報交換文體(Telexese)應具有相當的知識。

第一節　國際電報

　　「電報」英文寫成"Telegram"或寫成"Cablegram"或簡稱爲"Cable"。而國與國之間的電報往來則稱爲國際電報(International Telegram)。從事貿易的人士，對於國際電報的有關知識，如能諳熟，對於業務的發展，必有助益。

一、國際電報的用語

　　國際電報的用語可分爲明語(Plain Words, Plain Language)和密語(Code Words, Secret Language)兩種。二者可單獨使用，也可混合使用。

　　1.明語：乃以國際電報通訊所准用的各國文字繕成而且每字及每一辭句均保存其所屬文字的原來意義者。凡電文及署名完全用明語書寫的，稱爲明語電報。明語電報內有下列各種情形時，亦作明語論：

①用字母或阿拉伯數字書寫的數目，而無秘密意義的。

②電報掛號字樣。

③商務通訊上習用的名詞及編號而無秘密意義的。

④電文第一字母所用的對號（查對、押號）字或數目字。

2.密語：是由文字或數字加以組合而成，每個字都有特定的意義，但已失去文字的原來意義。

二、國際電報的種類

國際電報按處理緩急，可分爲下列三種：

1.尋常電報(Ordinary or Full-rate Telegrams)是一種快遞電報，電報局收到拍發電稿後，必須立即遞往國外目的地。這種電報優先拍發的權利，僅在加急電(Urgent Telegram)之後。尋常電報須照全費率(Full-rate)計算，所以無納費標識。我國規定報費最少以七個字起算。經營國際貿易的商人常在日間利用這種快電拍往同一時區(Time Belt)的外國顧客，使對方在營業時間內收到電報，並能在當日營業時間內獲得覆電。尋常電文可用明語、密語或兩種混合者。

2.加急電報(Urgent Telegrams)加急電爲尋常電的加速，享有優先傳遞投送之權，報費按尋常電加倍收，以七個字起算，在收報人姓名地址之前加"Urgent"標識，並作一字計算。但須注意有些地區及國家不適用加急電。

3.書信電報(Letter Telegrams)凡電文冗長而不急於傳遞的電報，多用書信電。這種電報投遞較遲，但費用較廉，故在商業通訊上應用最廣。書信電報最爲經濟，在我國按尋常電減半收費，並規定最少以二十二字起算，包括地址二字納費標識一字。須在收報人姓名地址前加註"LT"標識，並作一字計算。

茲將三種電報收費比較如下:

電　報　種　類	電報種類標識	費　　用　　比	最低收費字數	用　語　限　制
加　緊　電　報	URGENT	2	7	明・密
尋　常　電　報	—	1	7	明・密
書　信　電　報	L／T	½	22	明

三、電報字數計算方法

　　1.業務標識、收報人姓名地址、收報局名、電文及署名, 概按十個字碼以內作一個計費字數計算, 每逾一至十個字碼, 加算一個字。

　　2.明語電報的電文各字, 應按照各種明語文字的標準字典所載單字書寫, 其他各種不規則的縮寫或湊合字, 應隔開分別計字。

四、電報實例說明

ZCZC JAF846 PTM672 LHB18 ①

CHIA HL JATKO23 ②

TOKYO ③ 27／26 ④ 12 ⑤ 1530 ⑥

LT ⑦　　　　　　　　　　19… JUNE ⑨ 12PM3 40

TAICO ⑧

TAIPEI ⑩

PLEASE OFFER⑪ BEST C&F YOKOHAMA 10000DOZ
LADIES　UMBRELLAS　AS　PER　YOUR　SAMPLE
MAY10 PAYMENT SIGHT LC JULY AUGUST SHIP-
MENT CABLE REPLY

JATRA⑫

COLL 10000⑬

茲依編號順序說明於下:

①電報經由路線

②呼號

③發報地

④電報字數。實際數字為 26 字，
收費數字為 27 字，因為
10,000 Doz 應算二個字

⑤發報日期: 本月 12 日

⑥發報時間: 下午 3 點 40 分

⑦電報種類: LT

⑧收報人電報掛號

⑨收報時間: 收報局收到電報時
間，通常用橡皮戳表示

⑩收報地

⑪電文

⑫發報人電報掛號

⑬電文複核: 凡電文中涉及阿拉
伯數字或發電可能錯誤的文字，
再行核對一遍。

五、撰寫電報文體 Cablese 的要領

1. 第一、二人稱代名詞可省略。

書　　信　　體	電　　報　　體
We confirm having accepted your order No. 10 dated October 5.	YOURS 5 ORDER NO 10 ACCEPTED

2. 省略標點符號: 非絕對必要時，標點符號不必使用，尤其地址中的符號為然。

書　　信　　體	電　　報　　體
454 Market Street, San Francisco, California, USA	454 MARKETSTREET SANFRANCISCOCALIFOR-NIA

3. 省略縮寫字及 Apostrophe 的符號: 縮寫字及 Apostrophe 的符號是單獨計算為一字，因此應省略。

書　　信　　體	電　　報　　體

Mr.	MR
Mrs.	MRS
Don't	DONT
Can't	CANT

4. 省略冠詞: 冠詞如 a, an, the 均可省略。

書　　信　　體	電　　報　　體
send us a sample	SEND US SAMPLE

5. 省略介系詞。

書　　信　　體	電　　報　　體
In compliance with your cable of 12th, we have opened L／C today.	YOURS 12TH OPENED LC TODAY

6. 省略助詞: 如 will, shall, can, have been 等可予省略。

書　　信　　體	電　　報　　體
We shall send you the letter of confirmation.	CONFIRMATION LETTER FOLLOWS

7. 以"un-", "in-", "dis-", 等接頭詞(Prefix)簡化否定詞。

書　　信　　體	電　　報　　體
(1)We are not interested in it.	(1) UNINTERESTED
(2) We cannot obtain offers.	(2) OFFERS UNOBTAIN-ABLE

8. 以"-able"代替可能性, 許多表示可能性的字, 如"can", "possible" 等字, 都可在適當動詞之後加"-able", 而有同樣意義。

書　　信　　體	電　　報　　體
(1) We can accept your terms.	(1) TERMS ACCEPTABLE

(2) We can obtain orders.　　　(2) ORDERS OBTAINABLE

9. 用命令式。

書　　信　　體	電　　報　　體
Please inform us by cable whether you have opened L/C.	CABLE WHETHER LC OPENED

10. 以現在分詞表達未來的行動。

書　　信　　體	電　　報　　體
(1) We shall effect shipment on July 10.	(1) SHIPPING JULY 10
(2) We shall write for the result.	(2) RESULT WRITING

11. 以被動代替主動。

書　　信　　體	電　　報　　體
We have obtained import licence.	IL OBTAINED

12. 以單字代替數個字。

書　　信　　體	電　　報　　體
Do not agree	DISAGREE
Do your best	ENDEAVOUR
At your earliest convenience	SOONEST
Send by air mail	AIRMAIL
Per piece	APIECE
As soon as possible	SOONEST
In the middle of May	MIDMAY
At the end of April	ENDAPRIL

Will send you a letter　WRITING

13. 將過去分詞放在句末以省略動詞。

書　　信　　體	電　　報　　體
⑴ We would like to ask you to effect shipment promptly.	⑴ PROMPT SHIPMENT RE-QUESTED

14. 地址使用電報掛號: 如"Central Trust of China, 49, Wuchang Street. Section 1, Taipei"共有 10 個字, 如使用電報掛號 "CENTRUST TAIPEI"則只有二字。

15. 數字(Figure)宜予拼出(Spell Out)。

書　　信　　體	電　　報　　體
We offer 100 dozen	OFFER ONEHUNDRED DOZ

六、貿易電報英文實例

1. 詢價

letter form	cablese
We are informed that you are the sole agent for Ningta Bicycles and we shall be pleased if you will send us your price list and state your best terms.	INFORMED YOU ARE AGENT FOR NINGTA BICYCLES STOP SEND PRICELIST WITH BEST TERMS

2. 寄型錄及價目表

letter form	cablese

We have sent you today our illustrated catalog of the bicycles suitable for your area together with the price list.

SENT TODAY BICYCLES CATALOG SUITING YOU WITH PRICELIST

3. 報穩固價

letter form	cablese
With reference to your letter of June 5, we are pleased to offer firm for your reply reaches here by August 1,1,000 dozen of pullovers at US$15 per dozen C&F New York, for December shipment.	YOURS FIFTH OFFER FIRM SUBJECT REPLY HERE AUGUST 1 ONETHOUSAND DOZ PULLOVERS FIFTEEN USDOLLARS DOZ CANDF NEWYORK DECEMBER SHIPMENT

4. 發出訂單

letter form	cablese
With reference to your offer of July 20 we are pleased to send you our order for 500 dozen of sport shirts style A, at US$20.34 per dozen CIF New York.	YOURS TWENTIETH ORDER 500 DOZ SPORT SHIRTS STYLE A USD 20.34 PER DOZ CIF NEWYORK

5. 通知裝運

letter form	cablese

We have shipped today 200 cases of camphor tablets per s.s. "Pacific Transport" which is scheduled to sail from Keelung July 10 and to arrive at Kobe July 15.

SHIPPED CAMPHOR TWO HUNDERED TODAYPACIFIC TRANSPORT ETD KEE LUNG JULY 10 ETA KOBE JULY 15

第二節　國際電報交換

貿易廠商爲了節省時間、費用，避免到電信局拍發電報的麻煩，可在自己辦公室裡利用預先裝就的電傳打字機(Teletypewriter)和國外裝有同樣設備的客戶，直接交換信息。這種國際間利用電傳打字機的交換電信，稱爲電報交換，俗稱電傳，英文爲 Teletypewriter Exchange，簡稱爲 Telex 或 TLX.

一、Telex 的優點

1. 自動記錄：憑電傳打字收發電信、電文內容可自動紀錄下來，並可產生副本。

2. 直接筆談：可以打字方式和對方直接交談，與電話通話相似。

3. 可行不在現場通訊：如電傳打字機無人看管時，也可自動收錄電文，對於時差較大的國際通訊，尤屬便利。

4. 費用低廉：Telex 按通訊時間計費，較電報按字計費者，低廉甚多。冗長電文使用 Telex 尤爲經濟。

5. 通信文體自由：明語、密語、簡體字均可使用。

6. 可在辦公室內直接收發。

二、Telex 呼叫接線的種類

Telex 呼叫的接線種類，視各被呼叫用戶所在國(或所在地)的交換設備情形，可分爲下列二類：

1. 人工交換接線(Manual Operation)：即呼叫用戶先將被呼用戶的 Telex 呼叫號碼(Call Number)及所在地名向國際電臺 Telex 席值機員掛號後，由電臺值機員請被呼叫用戶的交換值機員將被呼叫用戶接出後，相互通報。

2. 全自動交換接線(Fully Automatic Connection)：呼叫用戶可直接選撥國外被呼叫用戶的 Telex, 呼叫號碼後，即可直接將被呼叫用戶接出，直接通報。

三、Telex 費率的計算方式

Telex 的計費方式有二種。

1. 三分一分制：與未開放全自動交換作業系統區域通報時採用。每次基本計費時間爲三分鐘，未滿三分鐘也以三分鐘計算。如超過三分鐘，則超過部分以一分鐘爲計費單位，不足一分鐘也以一分鐘計費。

2. 六秒六秒制：與已開放全自動交換作業系統區域的通報時採用。每次基本計費時間爲一分鐘，未滿六秒鐘者以六秒鐘計算，超過六秒鐘者，超過部分也以六秒鐘爲計費單位，不足六秒鐘，以六秒鐘計費。

四、撰寫電報交換文體(Telexese)的要領

Telex 係按時間計費，因此 Telex 在文體方面，應力求簡潔。因此，乃有 Telexese 的產生。茲將撰寫 Telexese 的要領。列舉於下：

1. 省略主詞、助詞、介系詞、冠詞。

2. 以現在分詞代替未來式。

3. 以形容詞"-able"代替可能動詞。

4. 將主動改爲被動。

5. 以"un-"，"in-"，"mis-"，"dis-"表示否定。

以上可參閱"Cablese"部分。

6. 使用 Telex 用略語，例如:

As soon as possible → ASAP

Refer to your telex → ROTLX

Refer to our letter → ROL

7. 單字簡縮。

因 Telex 係按時計費，所以應力求「字要簡短」，「字數要少」，以下就單字簡縮要領加以說明。

⑴略去母音字母，保留重要子音字母，以及最後一個子音字母。但第一個字母一律保留。

accept	ACPT
dollars	DLS
freight	FRT
manager	MGR

⑵保留第一音節

　①保留第一音節及第二音節第一個子音字母。

answer	ANS
document	DOC

　②保留第一音節及第二音節

memorandum	MEMO
negotiation	NEGO

　③保留第一音節及第二、三音節第一個子音字母

approximate	APPROX
immediately	IMMED

　④保留第一音節及後面重要子音字母

airfreight	AIRFRT
consignment	CONSGT

government	GOVT

(3)保留最後一個音節全部，其他保留重要子音字母或全部子音字母

（重音在最後音節者，通常按此法簡寫）

addressee	ADRSEE
cancel	CCEL

(4)保留首尾兩字母

bank	BK
from	FM

(5)音譯

are	R
new	NU
you	U

(6)過去分詞字尾"ed"以"d"代替

arrived	ARVD
confirmed	CFMD
received	RCVD

(7)以"g"代替字尾的"ing"

airmailing	AIRG
manufacturing	MFG

六、貿易電報交換英文實例

1. message

Please open L／C immediately at contract price, otherwise we cannot apply for export license.

telexese:

PLS OPN LC IMMDLY AT CONT PRC OZWS EL UNAVL-BL

    * OZWS=Otherwise

2. message

We have not yet received the shipping documents. Please investigate and reply.

telexese:

DOC UNRCVD PLS CHCK N RPL

    * CHCK=check; N=and

3. message

Please inform us by cable whether import license is obtained or not.

telexese:

TLX WHZ IL OBTND

    * WHZ=whether; OBTND=obtained

4. message

We are pleased to inform you that we have accepted your order of April 16th. Therefore, please open letter of credit immediately and airmail the design sample to us.

telexese:

YS 16TH ACPT PLS OPN LC IMMDLY Y AIR DSN SMPL

YS=yours; Y=and; DSN=design; SMPL=sample

5. message

Please refer to our letter of October 12, concerning difficulty of allowing discount as requested. However, we have obtained a five percent reduction from the maker although the market here is advancing They emphasized such reduction would not apply to the shipments made after November.

telexese:

OURLET OCT 12 ABT DISCOUNT DIFFICULTY MAKER
ALLOWED 6% BUT UNAPLICBL SHIPTS AFTER NOV

七、常用 Telexese

About	ABT	Acknowledge(d)	ACK(D)
Accept(ed)	ACPT (D)	Application	APLCTN
Accordingly	ACDGLY	Applicable	APLICBL
Additional	ADDL	Advise	ADV
Addition	ADDN	Already	ALRDY
Agree	OK	Approximate	APPROX
All right	OK	Approve	APPR
Amount	AMT	Airmail	AIR
And	N	Airmailed	AIRD
Arrange	ARRNG	Average	AVRG
Arrive(d)	ARV(D)	Answer	ANS
Attention	ATTN	As soon as possible	ASAP
Balance	BALCE	Beginning	BEG
Bleached	BLCHED	Bill of Lading	B/L
Between	BTWN		
Cancellation	CANCELN	Cancelling	CCELG
Cancel(led)	CCEL(D)	Check	CK
Can not	CANT	Charter Party	C/P
Carton	CTN	Commission	COMM
Charge(d)	CHRG(D)	Counter Offer	C/OFFER

Colors	COLS	Credit	CR
Confirm(ed)	CFM(D)	Correct	CRT
Contract	CONT		
Confirmation	CFMTN		
Days	DS	Delivered	DELVD
Debit	DR	Destination	DEST
Deliver	DLV	Double	DBL
Delivery	DLVY	Dollars	DLS
Difference	DIFFRNE		
Document	DOC		
Each	EA	Estimated time of arrival	ETA
End October	ENDOCT	Estimated time of departure	ETD
Enquiry	ENQRY	Encluding	ENCLDG
Exchange	EXCHG	Export licence	EL
Export	EXP		
Factory	FCTRY	Following	FOLG
Figure(s)	FIG(S)	Forwarded	WDDFOR
Flight	FLT		
Forward	FORWD, FWD		
Freight	FRT		
Gallon	GAL	Guarantee	GURANTE

General	GENRL	Government	GOVT
Good	GD	Greasy	GRSY
Highway	HIWAY	How	HW
Hour	HR	However	HWEVR
Heavy	HVY		
Immediate(ly)	IMMD(LY)	Including	INCLDG
Import	IMPT	Information	INFMTN
Include(d)	INCLD(D)	Instructions	INSTRCTNS
Inform(ed)	INFM(D)	Instead of	I／O
Instead	INSTD	Invoice	INV
Interest	INTRST	Irrevocable	IRREV
Letter	LTR	Light	LITE
Liter	LIT		
Manager	MGR	Month	MO
Message	MSG	Measurement	MEASMT
Meter	MTR, M	Middle	MID
Negotiate	NEGO	New	NU
Negotiation	NEGN	Next	NXT
Night	NITE	Number	NR NO
Negotiating	NEGOTG	Nothing	NIL
Open	OPN	Original	ORGNL
Ordinary	ORD	Otherwise	OZWS
Origin	OGN	Our Cable	OC
Order	OD, ODR	Our telex number 100	OX-100
Payment	PYMT	Private(ly)	PRVT(LY)

Possible	POSSBL	Price	PRC
Purchase	PURCHS	Please	PLS
Quality	QLTY	Quantity	QTY QNTY
Quotation	QUOT QUTN	Refer your telex	RYTX
Refer our letter	ROL	Refer your telegram	RYT
Refer your letter	RYL	Regarding	REGRDG
Refer your cable	RYC	Remarks	RMKS
Refer our cable	ROC	Repeat	RPT
Request (ed)	REQST(D)	Receipt	RCPT
Respectively	REPCTVLY	Reference	REF
Receive(d)	RCV(D)	Reported	RPTD
Refer our telegram	ROT		
Sample	SMPL	Shipment	SHIPT
Second	SEC	Station	STN
Service	SERV	Sorry	SRY
Shipped	SHIPD	Stop	STP
Specification	SPEC		
Telex	TLX	Transfer	TRNSFR
Telegraph	TEL	Thanks	TKS
Text	TXT	Tomorrow	TMW

Though	THO	Today	TDY
Through	THRU		
Unacceptable	UNACPTBL	Understand	UNDSTND
Unknown	UKWN	Urgent	URGT
You	U	Your	UR
Your telex	YX, YTLX	Your letter	YL
Yours	URS, YS	Your cable	YC

習　　題

一、國際電報的用語有幾種?

二、試述國際電報的種類, 其計費方法如何?

三、Telex 有那些優點? 其計費方法如何?

四、將下列文字改寫電報文體 (cablese)。

1. Please quote us your lowest CIF New York price with best discount and delivery for 1000 dozen ladies umbrellas No. 123.

2. The market here is advancing considerably. Please cable us when you can ship the goods.

3. We have obtained export licence, please inform us you can open L／C.

4. Please increase the quantity of the order to 200 doz.

5. We shall write for the result.

6. Referring to your order for camphor tablets, we have not yet received your letter of credit. Please open it urgently. Otherwise, we can not ship the goods in time, as the steamer is leaving within a week.

7. In comliance with your request of May 15, we have sent you today our samples under separate airmail cover. Shipment will be effected within 30 days after receipt of your L／C. Other terms and conditions

remain unchanged.

五、將下列電報交換文體譯成普通文體。

1. OURLET OCT 12 ABT DISCOUNT DIFFICULTY MAKER ALLOWED 6% BUT UNAPLICBL SHPTS AFTER NOV

2. ODR 105 BUYERS JONES N CO COMPLAIN FINISH DULL N COLS SLIGHLY DIFFERENT WHAT CAN U DO JONES BEING MUTUALLY IMPTNT CUSTOMERS

3. ODR 123 SMPL UNRCVD UNLESS IT ARR BIENDMAY JUNE SHPMT IMPSBL PLS APPR JULY Y OPN LC ACDLY

4. YS 15TH FOR RBER SHOES 1000 DOZ PAIRS CIF $3.50 NY SHPMT OCT REPLY HERE BEFORE25TH

第三十一章　通　函

(Circular Letters)

　　公司行號有時候爲某種目的，需將內容千篇一律的函件寄發給許多客戶或可能的客戶(Prospective-customers)，在這種情形，可利用通函(Circular Letters)或定型函(Form Letters)的方式。

　　有些人以爲通函的措詞最容易撰寫，其實一不小心就流於刻板、陳腐、平淡無味。因此撰寫這種信件時，應特別注意，使其具有個性，如果能跳出千篇一律的通函的窠臼，而寫出清新動人的通函，其效果一定很大，即使事實的本身很單純，無法以別出心裁的格調寫出，也應避免使用陳腐的成套，而應以明確、簡潔、直截了當的方式寫出來。

第一節　宣佈開業

　　因爲一般人不輕信他們所不知的事情，所以撰寫開業通函時，必須多用腦筋，以求其具有吸引力與說服力，一封開業通函應包括：

　　1.商號創立日期、地址、經營項目。2.有充分的資金。3.幹部有豐富的業務經驗，可提供更佳的服務。4.能提供客戶所需要的各種貨品，而且價格具有競爭性。5.希望惠顧。

　　有時爲便於對方的徵信，也可將自己的往來銀行名稱寫明。

No.187　宣佈開業

Gentlemen:

　　Establishment of Taiwan Trading Co., Ltd.

We have the pleasure of informing you that on April 1, 19···,We established ourseleves as an export house of SUNDRY GOODS under the style of TAIWAN TRADING Co., LTD. at the following address:

　　48 Wuchang Street, Sec. 1, Taipei, Taiwan

Our staff members were trained in such prominent firms as Central Trading Co Ltd. Mitsui Shoji K. K. and China Trade and Development Corp., and because of our close connections with leading manufacturers of various kind of general merchandise in Taiwan, we are very well-placed to supply you with high-grade goods at most competitive prices.

Upon hearing from you about your requirements, we shall be happy to send you illustrated catalogs, which will give you a good idea of the kinds of merchandise we handle.

It will be deeply appreciated if you will give us a chance to serve you.

　　　　　　　　　　　　　　　　Yours truly,

【註】

1. We have the pleasure of informing you that:可以"we are pleased to

inform you that"代替。

2. We have estabished ourseleves as:我們業已創設（以…爲業的…商號）

3. an export house:出口商號，其他如"general importers and exporters; manufacturers; agents; representative; dealers; buying agents; selling agents; commission agents, commission merchant; sole agests, whole-salers;detailers 等均說明商號業務特性。

4. under the style of...＝under the name of...: 以…的名稱（義）。

5. prominent：著名的。也可以"outstanding"（卓越的）形容。

6. well-placed:很有利的地位；方便的；順手的。

第二節　宣佈成立分支機構或部門

成立分支機構本身就是一件很好的宣傳。在撰寫設立分支機構或新部門的通函時，應：1.說明（強調）因業務的蒸蒸日上(steady growth of the business)，不得不成立新分支機構或新部門。2.強調由於新分支機構或新部門的成立，能給客戶提供更佳、更有效率的服務。3.寫明新分支機構（或新部門）的地址，開業日期。4.介紹新分支機構或新部門的負責人。5.告知嗣後有關通信可逕致該分支機構或部門。

No.188　宣佈成立分公司

Dear Sirs,

In view of the rapid development of our business in Southeast Asia, we have now decided to open a branch at the following address:

Taiwan Trading Co., Ltd;Singapore Branch

900 Upper Cross Street, Singapore,1

Mr. Yang, manager of the new branch, has been

holding a responsible position of a sub-manager of the export department at our Head Office in Taipei for the past three years. His ample experience and unlimited resources will be of more service to you.

Since the branch office will open on 1st of August, we shall be pleased if you will send your future inquiries and orders direct to our Singapore office.

Yours faithfully,

【註】

1. in view of＝considering　　2. responsible positien＝負有責任的職位。

3. ample experience:豐富的經驗。　4. unlimited resources:多謀才略。

5. send direct to:逕寄。

第三節　通知遷移新址

通知遷移新址的通函，其內容只要將何時遷移至何地的事實加以說明即可，這種通知的形式有兩種，一爲通告方式，一爲私信性質(personal touch)方式，採取後一方式時，最好以下列文句結束:

1. Your continual confidence in us is hereby solicited.

2. Your continual patronage will be appreciated.

3. We wish you to continue patronizing us in the years to come.

No.189　通告遷移地址

NOTICE OF REMOVAL

Taiwan Trading Co., Ltd. would like to announce that

as from July 1, 19··· its office will be moved to the following address:

    33 Hung, Yang Road, Taipei, Taiwan

    Telephone:321-1234 (10 lines)

Cable Address "TAITRA, TAIPEI", TELEX "TP 3211" remain unchanged.

【註】

Notice of Removal＝Removal Notice

No.190　通知遷移新地址

Dear Sirs,

NOTICE OF REMOVAL

We are pleased to inform you that owing to steady growth of our firm in the past years and in view of facilitating business expansion, we have decided to move our office to the following address:

    TAIWAN TRADING CO., LTD.

    Huai Ning Building

    Room 201

    29-31 Hsiang Yang Road

    Taipei, Taiwan

Our calbe address, telex and telephone numbers remain the same, as shown above.

Your continual patronage will be appreciated.

                    Yours faithfully,

【註】

本例係採 personal touch 方式，一般而言，效果較佳。

第四節　通知人事異動

No.191　通知重要幹部人事異動

Dear Sirs,

We are pleased to inform you that our Board of Directors has announced the changes of responsible officers of our company. The purpose of such change is to streamline the organization as well as the operations of our company.

With a view to facilitating your contacting persons in charge of the divisions, the following information may be helpful to you.

Mr. Charles Chang has been transferred from Domestic Sales Division to the Foreign Division, and Mr. William Ling has been promoted to take Mr. Chang's place. Mr. George Wu, who has just returned from West Germany, has been appointed as Production Manager of our factory. Mr. Wu is an expert in the manufacture of products of our company, and his wide experience and knowledge will surely enable him to improve the quality of our products and reduce the production costs.

With the reshuffle of our personnel, we can assure

you that you will, from now on, receive our products of much better quality and at much reduced prices. In order to avail yourselves of our better services, we request you to write us for more specific information.

You are cordially invited to send us your enquiries.

Faithfully yours,

【註】

1. board of directors:董事會。

2. streamline:使現代化；改進。

3. reshuffle:改組。cf. a reshuffle of cobinet:內閣改組。

4. avail yourselves of:利用。

cf. You should avail yourself of the books in the library.

（你應利用圖書館的書）

I will avail myself of your kind invitation and come this evening.

（承您好意相約，今晚定往奉陪）

第五節　通知漲價

No.192　通知漲價

Gentlemen:

Owing to the rapid and unprecedented rise in cost of raw materials, as a result of the oil crisis, from which our products are manufactured, we are compelled to announce that from and after October 20 our previous price list will be cancelled and replaced by the new one which we enclose for your reference.

All orders received up to October 20 will still be accepted by us at old prices, after which time the new quotations will come into force.

The ever-increasing demand for our products and the continued strong-tone in the market for the raw materials make us bound to look forward to still higher rates within a relative short time. We, therefore, advise you to lose no time in covering your season's requirements.

Yours sincerely,

【註】

1. rapid and unprecedented rise: (原料價格的) 上漲迅速及空前。

2. oil crisis:石油危機 cf. energy crisis:能源危機。

   3. from and after＝on and after＝as from

4. new quotation:新定價。

5. come into force:生效。

6. ever-increasing demand:日益增加的需要。

7. continued strong-tone in the market for the raw materials:原料在市場上漲風繼續熾烈。market-tone:市況。

8. relative short time:相當短的時期內。

9. lose no time:勿失機會。

10. covering:置辦；購儲。

第六節　有關通函的有用例句

1. We are pleased
   We have the pleasure $\Big\}$ to inform you that we have established →
   We have the honor

   → ourselves in $\left\{\begin{array}{l}\text{this city}\\\text{New York}\\\text{Taipei}\\\text{this district}\end{array}\right\}$ as $\left\{\begin{array}{l}\text{commission merchants}\\\text{general importers \& exporters}\\\text{general agents}\\\text{shipping agents}\end{array}\right\}$ →

   → under $\left\{\begin{array}{l}\text{the name of}\\\text{the style of}\\\text{the title of}\\\text{firm name of}\end{array}\right\}$ ...

2. The new firm will devote its attention principally to commission business, in which the shipping of textiles will form an important feature.

3. Feeling confident of our ability to conduct any transactions and to execute any orders committed to my charge in a speedy, economical, and satisfactory manner, we solicit the favor of your commands.

4. The thorough knowledge and trade experience which I have gained in this branch of business during a ten years' engagement as an employee and manager of prominent firms in this branch both at home and abroad, will enable me to cope with all reasonable requirements.

5. The excellent reputation which your firm enjoys here, renders us extremely desireous of entering, if possible, into business relations

with you, and we therefore offer you our services for any purchases you may have to make in this area.

6. We can promise you immediate attention to all orders no matter how small, and it is our firm intention to do our utmost to meet our clients' wishes with regard to delivery as on all other points.

7. As the volume of our trade with Singapore constantly increasing, we have this day opened a new branch in Singapore.

8. For the convenience of our customers, we have decided to open a new branch in New York, and have appointed Mr. Kao as Manager.

9. The rapid development of our business in Taichung has compelled us to open a branch at the following address:

10. We have entrusted the management of our new branch to $\left.\begin{matrix} \\ \\ \end{matrix}\right\} \rightarrow$
   We have placed the management in the hands of
   → Mr...., who has, for years past, held a responsible →
   → position in our firm.

11. We have appointed Mr.... $\left\{\begin{matrix} \text{to our manager} \\ \text{as manager} \end{matrix}\right\} \rightarrow$ Mr.... has been
   $\left\{\begin{matrix} \text{connected with our firm} \\ \text{with us} \end{matrix}\right\}$ for →
   → many years, and, as he is thoroughly conservant →
   → with the manufacture of our goods, our clients can →
   → rely upon the exact and prompt execution of any →
   → orders placed in his hands.

12. We are pleased to inform you that in consequence of the increase of our business, we have been obliged to occupy larger premises, and that we have moved our office to:

13. In consequence of the rapid expansion of our business, we have found it necessary to take larger and more convenient quarters,

which we will occupy early next month. We would therefore ask you to address will future correspondence to our new office at:

14. Unfortunately, this growing expense had made it necessary for us to raise our selling price. But in order to facilitate office arrangement we have decided to adjust the matter by withdraw the discount on certain orders and not by changing catalog prices.

15. We trust that you will realize that this step was taken under pressure, and that no alternative course was open to us.

習　　題

一、將下列中文譯成英文。

1. 本公司將於五月六日以臺灣貿易公司名義，在臺北市成立，從事雜貨的出口，特此奉達(be pleased to inform you)。

2. 鑒於本公司在東南亞的業務迅速發展(Rapid development of business)，經決定於下列地址成立分公司。

3. 因爲敝公司出品的原料價格上漲劇烈，空前所未有，我們不得不宣告敝公司從前的價目表自十月廿日取消而代以所附的新價目表。

4. 謹通知本公司自 1989 年 1 月 1 日起遷往下列新址。

二、將本書 No.192 的英文信譯成中文。

三、用下列片語造句。

1. avail oneself of

2. under the style of

3. come into force

4. loss no time

第三十二章　求才與求職

(Position Vacant & Position Wanted)

　　由於對外貿易的發展，經營貿易的公司行號越來越多。外國廠商也紛紛前來設立分支公司行號。這些公司行號在在需要諳熟英文的人才。他們招聘人才，在很多場合都是利用報章雜誌刊登「求才」廣告，藉以選聘適當人才。而且這些公司為配合需要，多以英文刊登啓事。應徵求職的人也需要以英文應徵。因此，本書特闢一章討論如何撰寫求才廣告(want ads)以及如何撰擬應徵信(letter of application for job)。

第一節　求才廣告的寫法

　　求才廣告(Want ads; Advertisement for position vacant)是廣告的一種，所以求才廣告的文字必須能引人注意(attract attention)，由而激起適當的人才能前來應徵。一般而言，求才廣告的內容，應包括下列各項：

一、徵求人才的公司行號名稱：通常以求才的公司行號為主詞，並以醒目的字體放在最上方。但如不願將其名稱標明時可以類如："A leading trading Company"，"A newly-established German Electronic Company"的詞句代替，以免推薦、保舉、關說麻煩，同時可向報館暫租一信箱，在廣告上說明回信請寄某報第幾號信箱，或請應徵人將信逕寄第幾號郵政信箱(P. O. Box)。

二、徵求那一種人才：說明徵求那一種人才，有時將此一項以醒目的字體放在最上方，以引起讀者的注意。

三、說明應徵人(applicant, candidate)應具備的資格，例如性別、年齡、學歷、經驗、技術等。

四、必要時，說明待遇優厚、福利好、工作環境佳、升遷機會多等。

五、說明應徵方法：例如應檢送履歷表、相片、希望待遇。

六、寄信地址以及截止期限等。

No.193　徵求經理女秘書

WANTED——SECRETARY TO MANAGER

A leading American Trading Company in Taipei requires female secretary to the manager. All applicants must be college graduates with a bachelor degree and have complete understanding of spoken and written English, and accounting. Applicants are requested to apply in their own handwriting giving personal data including past experience and salary expected and including one recent photo to P. O. Box 1234, Taipei.

　　All applications will be treated confidentially.

【註】

1. wanted:「被需要」之意，也即「求才」之意，這是求才廣告中常用的開頭字。求才、求職的標題，常以下面的句子表現：

POSITION VACANT
SITUATION VACANT ｝求才; 招聘; 遺缺需人。

POSITION WANTED
SITUATION WANTED ｝求職。

2. a leading American Trading Company:一規模宏大的美國貿易公司。

3. female secretary:女秘書。　cf. male secretary:男秘書。

junior secretary, assistant secretary, senior secretary, executive secretary, confidential secretary（機要秘書）

4. with a bachelor degree:具有學士學位。

5. have complete understanding of spoken and written English, and accounting:具有完全瞭解說、寫的英文能力及諳會計。

6. applicants are requested:應徵者必需⋯。

7. to apply in their own handwriting:以親筆寫應徵信。

8. personal data:個人資料，類似的有：résumé（簡歷），personal history, curriculum vitae, autobiography 等。

9. salary expected:希望待遇。

10. recent photo:近照。　cf. bust photo:半身照。"photo"為"photograph"的略語。

11. will be treated confidentially:將予秘密處理。

cf. will be kept confidential:將予保密。

will be held strictly confidential:將嚴予保密。

will be kept in strict confidence:將嚴予保密。

No.194　徵求行銷經理

A billion dollar multi-national West German corporation in cosmetic product field is activating its operations in Taiwan and requires a:

MARKETING MANAGER

Responsibilities

The selected candidate will be responsible overall for attaining the corporate marketing objectives in Taiwan. He will formulate and supervise the execution of marketing plans and also be responsible for sales and budget allocation.

Qualifications

- Chinese, male, between 30 and 40 years of age
- Degree in Business Administration, Marketing or other related studies preferred
- Minimum 4 years experience in marketing consumer goods preferably with cosmetics sales experience
- A good knowledge of retail trade, finance, and logistics
- Good command of English, both spoken and written
- Dynamic, confident and vigorous personality

The successful candidate will receive overseas training on appointment. Interested candidates are invited to submit detailed resume in English (name and address in Chinese) marked "CONFIDENTIAL-MM" to:

華成企業管理服務有限公司

SGV-SOONG & CO.

P. O. Box 1539, Taipei

【註】

1. a billion dollar multinational corporation:一個億萬元的多國性公司。

2. cosmetic product:化粧品。

3. activate its operations:拓展業務。

4. marketing manager:行銷經理。

5. selected candidate:入選的候選人（應徵人）。

6. overall:全盤的。

7. attaining the corporate marketing objectives:達成公司的行銷目標。

8. formulate:規劃；設計。

9. execution of marketing plans:執行行銷計劃。

10. budget allocation:預算分配。

11. degree in Business Administration:企業管理學位。

12. preferred:較好；最好…;更佳。

13. retail trade:零售業。

14. logistics:後勤。

15. dynamic, confident and vigorous personality:有幹勁、有自信、活潑的個性。

16. successful candidate:入選的（成功的、錄用的）候選人（應徵人）。

17. receive overseas training:接受國外訓練。

18. interested candidate:感興趣的候選人，意指「對此職位有意的人」。

No.195　徵求進出口助理

IMPORT／EXPORT
ASSISTANT WANTED

Energetic young man preferably with 1 or 2 years work-ing experience in Import／Export field is required. Good command of English and typing ability are necessary. Challenging and excellent opportunity to develop with the company. Please send application in English (with name and address in Chinese & English) with full details and recent photo marked "CONFIDENTIAL" to: General Manager, P. O. Box 19—279, Taipei.

【註】

1. assistant:助理；助手。

2. energetic:精力充沛的；有幹勁的。

3. preferably:最好是。

4. are necessary＝are essential。

5. challenging:挑戰性的; 考驗的。

6. marked "CONFIDENTIAL":註明「密」字樣。

No.196　徵求打字員

SITUATION VACANT
Typist／Clerk

Applications are invited for newly created position of female Typist／Clerk in the Export Department of Taiwan Trading Co., Ltd. Applicants must be proficient in speaking, reading and typing English. Preference will be given to applicants with English-language stenography qualifications. Starting salary in the vicinity of NT$8,000 monthly. With attractive fringe benefits. Send application in English, in own handwriting, with full details of experience and qualifications and enclose recent bust photo to Export Manager, 1 Shing Shen South Rd., Sec. 1, Taipei. Applications close Saturday 15th October,19…

【註】

1. newly created position:新空缺。

2. proficient in:精通於…。

3. preference:優先選擇; 優先考慮。

4. stenography＝shorthand 速記, stenographer:速記員。

5. starting salary＝commencing salary

6. in the vicinity of:左右; 大約。

7. fringe benefits:福利（如壽險、醫療保險等）。

8. applications close Saturday:星期六截止申請（收件）。

第二節　應徵求職信的寫法

求職信(Employment Application Letter)是私人信函中最重要的一種，一封良好的就業應徵信可以決定一個人的一生事業。因此，對於如何撰寫求職信，對於求職的人而言，甚為重要。

本來，求職信就是一封推銷信——推銷自己。不用說，以英文應徵某一職位時，求才的公司，必以應徵人所寫英文的好壞作為決定取採的重要標準。求職信既然是推銷信的一種，那麼求職信應包括推銷信的四要素，即

1. Attention:開頭部分，包括如何得悉求才的事表示願意應徵。使用精緻的信紙，繕打清楚，遣詞用字適當，拼字正確，標點無誤。

2. Interest:說明自己的學歷、經驗、訓練、特殊技能，表示自己對求才公司能有貢獻，但不可過份誇張。

3. Desire:敍述應徵的理由，使求才公司覺得你有前途、潛力，而引起採用的欲望。

4. Action:最後，打動求才公司的心，而採取行動。為此，表示希望能會晤面談，並將電話號碼寫上。

求職信可分為兩種，一種是對於公開求才的應徵信(Solicited letter of application)；另一種是得自他人的消息或毛遂自薦的應徵信(Unsolicited letter of application)。無論屬於那一種，應徵信都應特別強調 "your attitude"，在求職信中雖然需將自己描寫一番，但應從求才公司的立場考慮。換句話說，不僅說明你的能力(Qualifications)可適合該職位所要求的條件，而且也應表明對求才公司的業務有濃厚的興趣。綜上所述，

一封完善的 Solicited Letter of Application，應包括下列各項：

1. 首先敍述應徵的由來（看到求才廣告或經人介紹）以及表明寫此信的目的。

2. 次之，說明自己的學歷、經歷、訓練、特殊才能。

3. 強調自己的學驗、才能符合該職位。

4. 表明對求才公司的業務有濃厚興趣，希望能爲求才公司貢獻自己的才能。

5. 提供履歷表、畢業證書、照片、推薦信。

6. 必要時提供備詢人（reference）。

7. 表示希望賜予面談的機會。

以上各項，可視情形而增減，學經歷、訓練等應力求具體明確，可在信中說明，也可另附履歷表或自傳，尤宜將自己的特殊才能具體說明。

假如求職者是剛自學校畢業，初次應徵工作，又無工作經驗，那麼學歷特別重要。因此，在應徵信中宜含蓄地，但有自信地說明，能將理論靈活地應用到實際工作，爲該求才公司貢獻一身之長。

對於報紙「求才廣告的無機關或公司名稱者」（Blind Advertisement），應徵信稱呼用 "Gentlemen" 或 "Dear Sirs"。

以下 No.197 至 No.203 爲 Solicited Letter of Application.

No.197　應徵經理女秘書職位

Gentlemen:

In reply to your advertisement in today's China Times regarding a secretary to manager, I wish to apply for the position.

I am confident that I can meet your special requirements indicating that the candidate must have a good knowledge of English and accounting, for I graduated from Business Administration Department of Fujen Catholic University last year.

In addition to my study of Business English and Accounting while in the university, I also have had the secretarial experience for two years in Formosa Trading Co., Ltd. The main reason for changing my job is to gain more experience with a superior trading firm like yours. I believe that my education and experience will prove useful for the work in your office.

Enclosed please find my curriculum vitae, certificate of graduation, letter of recommendation from the Dean of Business Administration of the University and one recent photo. With respect to salary, although it is difficult for me to estimate how much my services would be worth to your company, I should think NT$20,000 a month, which is my present monthly salary, a satisfactory beginning salary. I shall be much obliged if you will give me an opportunity for a personal interview.

Sincerely yours,

【註】

1. in reply to:敬覆。

2. apply for the position:應徵這一職位。

3. meet your special requirements:符合您的特別需要。

4. candidate:候選人，也可以 applicant 代替。

5. change job:換工作。

6. a superior trading firm:優越的貿易公司。

7. certificate of graduation＝diploma

8. letter of recommendation:推介書信。

9. dean:系主任。

10. personal interview:面談，面試。

No.198　履歷表

CURRICULUM VITAE

Name in Full:　　Wang Ta-wei

Date of Birth:　　December 7, 1964

Family Relation:　　Eldest daughter

Permanent Domicile:58 Tung San Street, Taipei

Present Address:　　Same as Permanent Domicile

Educational Background:

> Entered Provincial Hsin Chu girl middle School in September, 1976, finished in July, 1982.
>
> Entered Fujen Catholic University Business Administration Department September 1982, finished in September 1986.

Working experience: As stated in the letter of application.

Rewards:Won a prize for three years' regular attendance at the Provincial Hsin Chu Girl Middle School. Won the second place in the intercollegiate

English Speech Contest sponsored by the International Rotary Club Taipei Branch in 1983.

Personal details:　Age:23 Height: 165 cm.

Weight: 54 kgs.

Health: first class

Marriage: single

Hobbies: reading, stamp collection and sports.

I hereby declare upon my honor the above to be a true and correct statement.

【註】

1. in full:全名, 中國人之姓名不必將姓氏(surname)放在名字後面, 但如果是英文名字, 則應將姓氏放於名後, 如: Charles Chang, George Liu 等。另中國人之名字, 第二個字之拼字必須小寫如: Chang Chin-yuan。

2. Permanent Domicile:永久通訊處(domicile 有戶籍之意義)。

3. Rewards:獎勵。

4. three years regular attendance:三年全勤上學(attendance 出席上課)。

5. second place:第二名。

6. intercollegiate English Speech Contest sponsored by the International Rotary Club Taipei Branch:由國際扶輪臺北分社主辦之大專學校校際英語演講比賽。

7. personal details:個人的資料。

8. declare upon my honor:以個人的榮譽保證。

No.199　個人資料

PERSONAL DATA

Name: Elizabeth M. Chang

Address: 1 Wuchang Street, Sec. 1, Taipei

Tel: 321-1234

Age: 24 years　　　　　　　Date of Birth: Aug. 8, 1964

Place of Birth: Tainan　　Marital Status: Single

Height: 5 feet 6 inches　Weight: 120 lbs.

Health: Excellent　　　　Dependents: two; parents

Education:

　College: National Taiwan University, B. A. in
　　　　　　Economics, 1983~1987

High School: North Gate High School, Tainan,
1980~1983

Technical Skills:

Stenography: 100 WPM

Typing:60 WPM

Office machine:Memeograph　　Adding and calculat-
　　　　　　　　　　　　　　　　ing

　　　　　　Offset duplicator　Photocopy

　　　　　　　　　Telex　　　　Filing Systems

Working Experience:

　July, 19—to present　　Taiwan Trading Co; Ltd. Sec-
　　　　　　　　　　　　　retary to President

January, 19— to June, 19— Ark Company Steno-typist

Special interests and activities:

　　Golf: active amateur

　　Swimming

Piano

Reference:

Dr. Peter Pan, Professor of National Taiwan University

sity

【註】

1. marital status:婚姻狀況。"Single"爲未婚，"Married"爲已婚。

2. Dependent:家屬。

3. WPM:爲"Words per minuste"的縮寫。

Stenography:100 WPM 爲速記每分鐘100字之意，Education 也可以 "Studies Pursued"或"School attended"代替，"entered"爲入學，"(was) graduated from"爲畢業於……，如係「結業」則用"finish"。

No.200　應徵行銷經理職位

SGV-SOONG & CO.

P.O. Box No. 1539

Taipei

Sir,

Your advertisement in the United Daily News for the position of one Marketing Manager prompts me to offer you my qualifications for this vacancy.

As required, I am enclosing one copy of my resume from which. I am sure, you will be able to evaluate my qualifications. As to English, I can write and speak fluently and am quite confident I have the capabilities required of the position so far as this language is concerned.

With regard to the required experience, I am pleased to

inform you that I have had experience in selling consumer goods for more than six years. I am now in charge of the Sales Section of Taiwan Cosmetics Company which is one of the leading manufacturers of cosmetics in Taiwan.

Should my qualifications meet with your approval and an interview be arranged for me, I can be reached by telephone 321-1234 between the hours of 9: 00 a.m. and 5: 30 p. m., Monday through Saturday.

Sincerely yours,

No.201　簡歷表

RESUME

Name: Charles C. Y. Wang (王昌陽)

Sex: Male

Date of Birth: October 8, 1950

Address:1 Hangchow S. Road, Sec. 1, Taipei (臺北市杭州南路一段一號)

Marital Status: Married

Education: Graduated from the Department of Business Administration, Fujen Catholic University in 1974 with B. A. degree.

Experience: Salesman, Sales Section, Taiwan Cosmetics Company, 1975—1981;Chief of Sales Section, Taiwan Cosmetics Company since 1981.

Business skills: Typing 60 wpm

Market research

Credit analysis

【註】

1. resume 也可寫成 r′esum′e:簡歷表之意。

2. market research:市場研究。

3. credit analysis:信用分析。

No.202　應徵進出口助理職位

General Manager

P. O. Box 19—279

Taipei

Dear Sir,

Your advertisement in the China Post for an Import／Export assistant, indicates that your company is in need of a competent man, preferably with 1 or 2 years' working experience in Import ／Export field. Please consider me an applicant for the position because I possess all the qualifications you required.

In 1984, I was graduated, the second place from the Department of International Trade, Tung Hai University. Right after completion of my military service in 1985, I joined a firm which has been engaged in international trade for many years. I am now still working in this firm and my main responsibility is to handle and process all the paper work concerning import ／export.

In addition, I also draft routine letters, cablese, telexese and operate telex machine.

I am quite confident that both my education and experience would fit your requirements fairly well. Should my application receive your favorable consideration, please grant me an interview. You may write me at the above address or reach me by phone at 321—1234 in care of Mr. H. H. Yang. I am sure an interview will convince you I am the right man for the job. Enclosed please find my recent photo.

Sincerely yours,

【註】

1. competent man:適任的人。

2. second place:第二名。

3. be engaged in:從事於……。

4. operate telex machine:操作電報交換機。

5. convince you:使你相信。

6. right man:適當的人。

7. cablese:電報文。

8. telexese:電報交換文。

No.203 應徵打字員工作

Manager

Taiwan Trading Co., Ltd.

1 Shing Shen South Rd., Sec. 1

Taipei

Dear Sir,

I read with immense interest your advertisement in the China Post of May 4 that there is a newly created position of female Typist／Clerk in your Export Department. Considering my education, training and experience, I am quite confident that I would be the right person for the job.

Right after my graduation from the Department of Commercial Correspondence of Providence College of Arts and Science in 1975, I was employed through very keen competition among 100—odd applicants by the Formosa Enterprises Ltd. as Typist／Clerk and have been working in the firm for the past two years. During this period, I have acquainted myself with all the routine clerical work of foreign trade.

So far as my capability is concerned, I can type at 70 WPM and am proficient in speaking and reading English. Besides typing, I can operate telex machine.

I shall be much obliged if you would extend my application your favorable consideration and, should my qualifications be satisfactory to you, arrange an interview with me at any time that is convenient to you. As required, I am enclosing one recent photo of mine.

Please keep my application in strict confidence.

Sincerely yours,

【註】

1. Providence College of Arts and Science:靜宜女子文理學院。

2. acquainted myself with:熟悉；諳熟。

3. routine clerical work:例行的書記工作。

4. capability:與"ability"互通，指智力（或體力）的「能力」，後面接"of"或"for"。"capability"常指天生或潛在的能力，在此意義上，與"ability"不同，後者常可指學到的能力。

5. be proficient in:精通；諳熟。

6. arrange an interview with me:爲我安排面談。

第三節　自薦求職信的寫法

自學校畢業後，爲了求職，常常需寫自薦信。這種自薦信並非應求才的廣告而寫，但如寫得充實而有力，往往比起憑求才廣告的應徵更有希望。因爲應求才廣告的應徵信，數目較多，不易引起收信人的注意；而非憑求才廣告的自薦信，在同一時期內，不至於很多，所以容易引起收信人的注意。這種自薦信，除開頭更應着重引人注意之外，其他部分，與根據求才廣告而寫的應徵信大同小異。

No.204　自薦求職信之一 ──貿易人員

Gentlemen:

On my graduation from college this fall, I am desirous of securing a position that will offer me opportunity in the field of foreign trade. Knowing something of the scope and enterprise of your huge export department, I thought perhaps you would keep me in mind for a possible opening.

I am strong and alert, and shall be twenty years of age in July next year. At present I am a student in the college of一, but I shall graduate from the college this coming July, finishing the requirements in three years. I have had no business experience, but my college record has been good, a copy of my antecedents is enclosed for your reference.

Dr. A. B. Chien, President of the College of一, will be glad to tell you more about my character and ability. I shall be glad to call at any time for an interview.

Very truly yours,

【註】

1. secure a position:謀取一職。"position"也可以"job"代替。

2. keep me in mind＝have me in mind＝bear me in mind＝remember：記住我；把我記在心裡。

3. finish the requirements:修完必修課程。

4. college record:大學成績。

5. antecedents:經歷；履歷表。

6. President:校長。

No.205　自薦求職之二——秘書

Dear Mr. Williams:

Someday in the future you may have need for a new secretary.

Here is why I should like to offer myself for the job, and here is why I am so much interested in obtaining it.

For one thing, I know that you do an enormous variety of work very fast and very well. This offers a real challenge to whoever works for you. It is the kind of challenge I like to meet because, with all due modesty, I have so trained myself in secretarial work that only exacting problems are interesting to me.

As to my mechanical abilities, I can take dictation at the rate of 180 words a minute, and type at the rate of 70 words per minute.

I cannot think of any job in which I would be so useful as that of secretary to you, since, in addition to my business training and experience, I could put to work for you and your organization the know-how of practical, everyday business handling.

Yours very truly,

【註】

1. some day in the future：將來有一天。
2. for one thing:首先，一則（在申述理由時用之）。

   for one thing I haven't money, for another....

   一則我沒錢，二則…。
3. an enormous variety of work:各式各樣的工作。
4. a real challenge:真正的考驗。
5. it is the kind of challenge I like to meet:我願意嘗試這考驗。
6. with all due modesty:我敢說（客氣地說）。
7. exacting problems:需要特別注意的問題；費力的問題；難於處理的問題。
8. mechanical abilities:技術上的本領。

9. I cannot think...to you:我想不出有任何的職位可以比做你的秘書, 更能顯出我的本事。

10. buiness training:商業訓練。

11. I could put to work...business handing:我能把我的技巧和知識來替貴公司處理每天的業務。

No.206　在報紙上刊登「求職」廣告

POSITION WANTED

Very capable secretary-stenographer, many years' experience, can hold executive position, open for engagement.
Best references.
Replies to Box 123, China Post.

要別人來聘, 就該將自己的長處寫出來, 例如上面的廣告待聘的人是一位極能幹的秘書兼速記員, 而且有多年的經驗, 並且能夠擔任行政事務。

【註】

1. position wanted:待聘; 求職。

2. can hold executive position:能夠擔任行政事務。

3. open for engagement＝is open for engagement:徵求職務。
$$＝seeks\ position$$
$$＝seeks\ employment$$

4. best references:有最佳備詢人之意。

第四節　推薦信的寫法

推薦信(Letter of Recommendation)是推薦人向求職人的未來雇主(Prospective employers)所寫有關求職人能力、品德及為人的信, 也就是介紹信的一種。

撰寫推薦信時，應兼顧到被推薦人及收信人（雇主）雙方。換言之，推薦人固應寬大爲懷，爲被推薦人美言，但也應有是非感，不可因太熱心，導致收信人誤會(misleading)。須知因推薦人的陳述不實或誇大，往往會引起嚴重的後果。因爲雇主可能因誤信推薦人的陳述，而將主要的職位給與求職人，以致雇主受到損失。

推薦信可分爲不指名的一般推薦信(General Recommedation)與指名的特別推薦信(Special Recommendation)兩種。前者是寫給未指定收信人的(To who it may concern)，這種推薦信比較正式，持信人(卽被推薦人)可視情形向任何人提出，指名的推薦信則是寫給某特定人的信。

不指名的一般推薦信(General letter of recommendation)，不指名誰是收信人，通常以 "To Whom It May Concern:" 或 "To whom it may concern:" 做爲 Salutation，且其後面通常用 Colon(:)，信的結尾沒有 Complimentary close。這種推薦信類似證明書，效果較差，它的第一句往往是 "This is to certify that...", 或 "This is to testify that..."，整個信中不應有 "you" 的字眼。

一封完善的推薦信的內容應包含下列各項：

1. 被推薦者的全名(full name)：僅寫 Mr. Hang, Miss Wu 是不可以的。
2. 說明認識多久了(How long have you known him?)
3. 認識程度(How well do you know him?)
4. 說明與被推薦人的關係(Relationship)：同事、主管、師生關係。
5. 被推薦人的表現(Performance)：工作表現、學習表現。
6. 結論：無保留地或有保留的推薦、普通推薦或極力推薦。

No.207 不指名一般推薦信──推薦翻譯人員

To whom it may concern:

　　This is to certify that Mr. George M. Cheng worked under me as a translator for three years. He left me two years ago on account of his family's removal to Taipei. I was very sorry to lose him.

　　Mr. Cheng did fine work in both English-Chinese and Chinese-English translation, and I find that he has improved a great deal since he left me. He possesses a keen intellect and aims at perfection in doing any piece of work. He is industrious, painstaking, and above all, punctual.

　　I recommend him without reserve for translation work.

William Taylor

【註】

1. to whom it may concern:逕啓者。

2. English-Chinese and Chinese-English translation:由英譯中和由中譯英。

3. any piece of work:任何一件工作。

4. painstaking:刻苦耐勞的。

5. above all:尤其 (是 adverbial phrase)。

6. punctual:守時。

7. without reserve:沒有保留地; 完全地。

　　下面是一封不用 "This is to certify that..." 開頭的一般推薦信。

No.208 不指名一般推薦信——推薦速記打字員

To whom it may concern:

The bearer, Miss Wang I-min was in our employ as steno-typist from May 20, 1984 to June 30, 1987, during which period she rendered satisfactory services. On account of her efficiency her salary was raised in March 1985, to NT$20,000 a month. She left our services at her own request.

She is painstaking and conscientious worker. Her character and habits are entirely appreciable.

We can confidently recommend her to any one who desires her services.

Taiwan Trading Co., Ltd.

K. Kang, President

【註】

1. the bearer:持信人。

2. in our employ:受雇於我們。

3. render satisfactory services:服務成績令人滿意。

4. at her own request＝of her own accord＝by her own wish:自願地；出於 自願；出於本意。

5. conscientious worker:盡職的工作者。

6. her character and habits are entirely appreciable:她的品性與習慣也甚可 貴。

至於指名的特別推薦信,其形式與普通信一樣Salutation用"Dear…",
Commplimentary Close 則用"Very truly yours,"或"Yours very

truly," 等詞句，而且大都封口逕寄，間亦有交本人面投者。書寫特別推薦信的動機，或爲應本人的請求，或爲應公司的詢問（這種情形卽爲後述的 Letters of Reference），求才公司對予特別推薦信較爲重視，因爲其所述較爲切實。

No.209　指名的特別推薦信

Dear Mr. Ward:

　　Mr. Richard Lin has requested that I write you regarding his work during the three years he has been in our employ.

　　During this time he has performed his work in a highly satisfactory manner. He has been punctual for all appointments and has executed his assignment efficiently. I can truthfully say he is conscientious and ambitious. In closing I am glad to recommend him unqualifiedly for a position with your company.

　　　　　　　　　　　　　　　Very truly yours,

【註】
1. truthfully:誠實地。
2. unqualifiedly:無條件地。
   形容被推薦人的用語有：

honest	誠實	dynamic	有幹勁
sincere	誠懇	responsible	負責
faithful	忠實	aggressive	有衝勁
enthusiastic	熱情	patient	有耐心
cooperative	合作	conscientious	盡責
cheeful	樂觀	initiative	主動
reliable	可靠	amicable	和藹
		diligent	勤勉

第五節　有關求職、推薦的有用例句

一、開頭句

1.
$$\left.\begin{matrix} \left.\begin{matrix} \text{In} \\ \text{With} \end{matrix}\right\} \text{reference} \\ \text{In} \left\{\begin{matrix} \text{reply} \\ \text{response} \\ \text{answer} \end{matrix}\right. \end{matrix}\right\} \text{to your advertisement in} \left\{\begin{matrix} \text{today's} \\ \text{yesterday's} \\ \text{April 10 th's} \end{matrix}\right\} \rightarrow$$

$$\rightarrow \left\{\begin{matrix} \text{Central Daily News} \\ \text{United Daily News} \\ \text{China Times} \\ \text{China Daily News} \\ \text{China Post} \\ \text{China News} \end{matrix}\right\} \text{for a(n)} \left\{\begin{matrix} \text{Manager of Export Department} \\ \text{assistant to export manager} \\ \text{purchasing manager} \\ \text{executive secretary} \\ \text{steno-typist} \\ \text{sales manager} \\ \text{production manager} \end{matrix}\right\} \rightarrow$$

$$\rightarrow \text{I} \left\{\begin{matrix} \left.\begin{matrix} \text{wish to offer} \\ \text{hope to offer} \end{matrix}\right\} \left\{\begin{matrix} \text{my qualifications} \\ \text{myself} \end{matrix}\right. \\ \left\{\begin{matrix} \text{should like to apply} \\ \text{wish to apply} \end{matrix}\right. \\ \left\{\begin{matrix} \text{wish to offer you my services} \\ \text{am pleased to offer myself as a candidate} \end{matrix}\right. \end{matrix}\right\} \rightarrow$$

$$\rightarrow \text{for} \left\{\begin{matrix} \text{the job.} \\ \text{the post.} \\ \text{the vacancy.} \\ \text{the position.} \end{matrix}\right.$$

＊ job:工作；＊ post:職務；＊ vacancy:空缺；＊ position:職位

2. I have learned from $\begin{cases} \text{Mr. A} \\ \text{one of my} \end{cases}$ $\begin{cases} \text{friends} \\ \text{classmates} \\ \text{schoolmates} \end{cases}$ → that you are →

→ $\begin{cases} \text{in want of} \\ \text{in need of} \\ \text{looking for} \end{cases}$ a(n)...

3. I shall be grateful if you would consider my qualification for...advertised in today's China Post.

4. I am writing this letter to you because of your reputation in business circles. I am hoping that there might be a suitable place for me on your staff.

5. Please consider this letter my application for the position of book-keeper in your Accounting Department which was advertised in April 10 th's China Times.

6. I have seen your advertisement in the China News of March 6, and I should like to apply for the job of secretary now open in your company.

7. Your advertisement in this morning's China Times for a(n)... prompts me to offer you my qualifications for this position.

8. I understand that there may soon be an opening for a(n)... in your company and I should like to apply for the job.

9. I am 23 years of age and an accounting major at the...college.
   I have also had training in typing, bookpeeking and shorthand.
   As I shall graduate next June, I should like to apply for the position of assistant accountant in your company for which job, I understand, there may be an opening soon.

10. I am writing to inquire whether you will need the services of a young woman with educational training in college (and some part-

time experience.)

二、關於年齡或學歷的說明

1. I am...years $\begin{Bmatrix} \text{of age} \\ \text{old} \end{Bmatrix}$, and $\begin{Bmatrix} \text{graduated from} \\ \text{was graduated from} \\ \text{a graduate of} \\ \text{shall graduate from} \end{Bmatrix}$...

2. I shall graduate $\begin{Bmatrix} \text{in June} \\ \text{this coming summer} \end{Bmatrix}$ from $\begin{Bmatrix} \text{college.} \\ \text{university.} \end{Bmatrix}$

3. I $\begin{Bmatrix} \text{graduated} \\ \text{was graduated} \end{Bmatrix}$ $\begin{Bmatrix} \text{last summer.} \\ \text{two months ago.} \end{Bmatrix}$

4. $\begin{Bmatrix} \text{I majored in} \\ \text{My major was} \\ \text{I took up} \end{Bmatrix}$ $\begin{Bmatrix} \text{Foreign Language and Literature.} \\ \text{International Trade.} \\ \text{Business Administration.} \\ \text{Secretarial Science.} \end{Bmatrix}$

5. I have a degree of $\begin{Bmatrix} \text{Bachelor of Business Administration (B.B.A.)} \\ \text{Bachelor of Arts (B.A.)} \\ \text{Master of Arts (M.A.)} \\ \text{Master of Science (M.S.)} \end{Bmatrix}$

6. I attended...Primary School,....Junior Middle School,...Senior Middle School, and...College.

7. I entered...School in 1985, changed to...School in 1986, and finished in 1987.

三、關於經歷及專長的說明

1. I
   - have a
     - good（充分）
     - fair（相當）
     - slight（約略）
     - sound（徹底）
     - thorough（完全）
   - knowledge of（瞭解）
     - typewriting.
     - shorthand.
     - telex.
     - export/import procedures.
     - English.

   am well acquainted with（熟諳）

2. I
   - can
     - fulfil the ordinary duites of
       - office work.
       - accounting.
     - take dictation in
       - English.
       - Spanish.
   - am able to
     - take dictation
     - write shorthand
     - type
     at 80 WPM.

3. For the past...
   - months
   - years
   →

   → I
   - have been
     - in the office of
     - in the employment of
     - employed by
     - with
     ...Co., as
     - secretary.
     - accountant.
     - typist.
   - have
     - served as an
       - administrative secretary
       - assistant secretary
       to Mr....
     - worked as...in...Company.

4. I have had...years' experience
   - in
     - my present post.
     - a trading company.
   - with a factory as an accountant.

5. Since my graduation from...,

$$\rightarrow \text{I} \begin{cases} \text{have been} \begin{cases} \text{in the employ of} \\ \text{employed in} \end{cases} \text{Co. as a(n)...} \\ \text{engaged in}(職務) \text{ under the employment of...Co.} \\ \text{have been working for a} \begin{cases} \text{government} \\ \text{stated-owned} \end{cases} \begin{array}{l} \text{enterprise as} \\ \text{a(n)...} \end{array} \end{cases}$$

四、學歷、經歷不在信中說明而以附件處理時

1. Enclosed are a $\begin{cases} \text{personal data sheet} \\ \text{resume} \\ \text{curriculum vitae} \\ \text{personal record sheet} \end{cases}$ and two copies of letters of

recommendation.

2. The details of my education and experience are given on the

enclosed $\begin{cases} \text{personal record.} \\ \text{resume sheet.} \\ \text{personal data sheet.} \end{cases}$

五、關於能適任的表示

1. I $\begin{cases} \text{believe} \\ \text{think} \\ \text{feel} \\ \text{am sure} \end{cases}$ that $\begin{cases} \text{I am competent to meet the raquirements} \\ \text{I can fill satisfactorily the position} \\ \text{my experience has been of the kind} \\ \text{my ability can meet the requirements} \end{cases} \rightarrow$

$\rightarrow \begin{cases} \text{as advertised.} \\ \text{which you have specified.} \\ \text{which you advertised.} \\ \text{called for by the advertisements.} \end{cases}$

2. I am confident that my experience and references will show you that I can fulfil the particular requirements of your secretary position.

3. I feel quite certain that as a result of the course in filing which I

completed at...College, I can install and operate efficiently a filing system for your organization.

六、關於待遇

1. Should consider NT$ 6,000 a month
$$\begin{cases} \text{a fair} \begin{cases} \text{starting salary.} \\ \text{initial salary} \end{cases} \\ \text{satisfactory} \begin{cases} \text{compensation.} \\ \text{remuneration.} \end{cases} \\ \text{satisfactory.} \end{cases}$$

2. The salary at which I should $\begin{cases} \text{desire to commence is NT\$15,000} \\ \text{require would be NT\$17,000} \end{cases} \rightarrow$

$\begin{cases} \text{a month.} \\ \text{per month.} \\ \text{monthly.} \end{cases}$

3. I should require $\begin{cases} \text{a starting} \\ \text{a commencing} \end{cases}$ salary of NT$20,000 per month.
$\text{a salary of NT\$20,000 a month to begin with.}$

4. $\begin{cases} \text{In regard to} \\ \text{With regard to} \\ \text{Regarding} \end{cases}$ salary, I think it is better to leave it to

→ you to fix after experience of my capacity.

（至於薪水，在考驗我的能力之後，聽由你裁奪）

5. Although it is difficult for me to say what compensation I should deserve, I should consider NT$… a month a fair starting salary.

6. As much as I should like to join your company, it would not be advisable for me to change my position for less than NT$…, which

is my present $\begin{cases} \text{monthly} \\ \text{yearly} \end{cases}$ salary.

7. At present, I am receiving NT$…monthly. The salary at which I

should desire to commence is NT$…per month.　If you give me a chance.　I will accept whatever salary you think reasonable.

8. I feel it is too bold for me to state what my salary should be.　My first consideration is to satisfy you completely.　However, while I am in probation, I should consider NT$…a month satisfactory compensation.

　　＊ too bold:太大膽

　　＊ in probation＝serving apprenticeship 試用期間

9. I hesitate to state a definite salary, but, as long as you have requested me to, I should consider NT$…a month satisfactory.

　　＊ hesitate to state:難於提出

　　＊ as long as:既然

七、關於面談

1. May I have $\begin{cases} \text{an interview?} \\ \text{the opportunity to discuss this matter further with} \\ \text{you?} \end{cases}$

2. I shall be pleased if you will be good enough to favor me with an opportunity for a personal interview.

3. If you $\begin{cases} \text{wish} \\ \text{ask} \\ \text{desire} \end{cases}$, I shall be $\begin{cases} \text{pleased} \\ \text{glad} \\ \text{happy} \end{cases}$ to →

→ $\begin{cases} \text{call for an interview} \\ \text{call in person} \\ \text{have an opportunity} \end{cases}$ →to $\begin{cases} \text{tell you more about myself and my qualifications.} \\ \text{demonstrate my ability.} \\ \text{prove my ability.} \\ \text{talk with you so that you may} \\ \text{judge my qualifications → further:} \end{cases}$

　　＊ call for an interview:接受會見。

* call in person: 親往拜訪。

* demonstrate my ability: 證明本人的能力。

4. I assure you that if
- appointed
- successful
- my application be successfull

Should you
- consider my application favorably
- entertain my application favorably
- think favorably my application
- kindly entertain my application
- give me a trial
- select me to fill the vacancy

, →

→ I would
- endeavor
- do my best
- do my utmost

to
- satisfy your requirements.
- give you satisfaction.
- afford you every satisfaction.
- gain your confidence.
- deserve your confidence.
- please you.

* if appointed; if successful; if my application be successful 均可譯爲「如蒙錄用」

* should you consider my application favorably; entertain my application favorably; think favaorably my application, Kindly entertain my application 均可譯爲「如蒙考慮」 * Should you give me a trial: 如蒙試用

5. You can
- reach me by phone at 321-1234
- reach me by dialing phone 321-1234
- call phone number 321-1234 and ask for me

between →

$$\rightarrow \begin{cases} \text{the hour of } 7 \text{-} 9 \text{ a.m. and } 6 \text{-}10 \text{ p.m.} \\ \text{eight and six during the day.} \end{cases}$$

6. I can be reached by telephone 123-3214 between the hours of $9:0\ 0$ am. and $5:0\ 0$ p.m., Monday through Saturday.

八、推薦信用語

I、開頭

1. It is $\begin{cases} \text{my} \\ \text{a real} \end{cases}$ pleasure to recommend Mr. George Liu....

2. I am pleased to recommend without reservation Miss...

3. It is a great pleasure to recommend to you Mr. H.K. Ling as a worthy candidate for...

4. This is to recommend Miss Nancy Liu for your favorable consideration of her application for...

5. Mr. H.H. Wang, my former student at National Taiwan University, has asked me to write in the interest of his application for...

6. $\begin{cases} \text{In my capacity as} \\ \text{In the capacity of} \end{cases} \begin{cases} \text{President of Taiwan University} \\ \text{Dean of Department of Commerce} \end{cases},$

I would like to recommend

$\begin{cases} \text{Mr. William H. Lin for...} \\ \text{most enthusiastically Mr. A who has applied for...} \end{cases}$

II、學業、能力、品格的描述

1. Mr. Lin graduated with honors and received a B. A. degree from this University.

2. During the academic year of 1975-1976, Miss Wang took my courge in Business Administration at Fujen Catholic University and achieved outstanding scores at the semester finals.

＊ at the semester finals.在期終考試。（注意：finals 爲多數）

3. He is qualified to conduct correspondence, and is expert and accurate at calculations.

4. He has a good command of the English language.

5. As to his English proficiency, he is able to speak, read and write the language efficiently.

6. He is honest, sincere and faithful, always adhering to high ethical standards.

　　＊ adhering to high ethical standard:遵守高度的倫理標準。

7. In his performance of duties, he has displayed a high sense of responsiblity and always helped whenever he could.

　　＊ always helped whenever he could：只要他能力所及經常幫助別人。

8. As to his character and personality, he is honest, reliable, responsible, and cooperative.

9. After his undergraduate study and military service, he worked for two years with an industrial firm, handling the quality control of electronics production.

10. In the performance of her duties, she displayed profound knowledge in market surveying, which proved greatly helpful to our sales programs.

習　　題

將下列中文譯成英文：

1. 頃自本日中國郵報閱悉貴公司刊登廣告徵求出口部經理，本人願意應徵該項職位。

2. 本人今年卅歲，於 1978 年畢業於臺灣大學，獲得商學士，主修國際貿易。

3. 我能每分鐘打 60 個字。

4. 我剛服役期滿。

5. 素仰貴公司在商界的聲譽，故寫此信希望能在貴公司謀一合適職位(a suitable place)。

6. 台北某外商公司徵求幹練(efficient)雜貨出口助理(export assistant)，須英文說寫流利並持有良好證件(possess good credentials)。

7. 職求人——女性速記員。

至少有二年之工作經驗；極良好機會；月薪新臺幣貳萬元起碼；條件適當者可望迅速晉級；每週工作五天；現代化辦公室，有空調設備。

8. 如蒙錄用(if appointed)，本人保證當盡力贏得您的信賴(gain your confidance)。

第三十三章　商業社交信

(Social Letters in Business)

第一節　商業社交信的特色

英文商業社交信雖不屬於本書範圍，但是做貿易之餘，人與人之間難免也有社交。因此，特闢一章介紹英文商業社交信(Social Letters in Business Social Correspondence in Busiuess)的寫法，諒對讀者不無益處。

社交書信的寫法，在遣詞用字或體裁方面，與商業書信略有不同，社交書信的內容及語氣方面，多較富感情、親切而不必太拘禮。

然而，社交書信與商業書信一樣，最起碼包括 Heading, Inside Address, Salutation, Body of the Letter 及 Complimentary Close 五大主要部分。這五大主要部分，原則上雖與商業書信的寫法沒有什麼不同，但仍須注意下列各點：

1. Heading:社交書信既是個人與個人間的通信，通常不用公司或機關的信紙。不過如係業務上認識的朋友，而其通信雖然是私人的社交性質，實質上則與公務多少有關係的，仍可使用公司或機關的信紙。

2. Inside Address:寫信給親友時，通常不必寫 Inside Address，如要寫上 Inside Address，則可仿商業書信的寫法。

3. Salutation:社交書信的 Salutation 比起商業書信，較非正式(In-

formal)。通常在"Dear"後面加上見面時所用的稱呼，例如見面時稱他爲 Mr. Chang，則社交書信的 Salutation 可以用"Dear Mr. Chang"或"My dear Mr. Chang"；如見面時稱他爲"Charles"，則 Salutation 可用"Dear Charles"或 Dear Dr. Chang。再者，在標點方面，社交書信的 Salutation 後面的標點，通常係用逗點（,）或驚嘆符號（!）而不用冒號（:），這與商業書信略有不同。

4. Body of the Letter:語氣及措詞宜親切(affectionate)、友善(friendly)、親密(intimate)。

5. Complimentary Close:商業書信中的 Complimentary Close 多用"Faithfully yours,"或"Yours faithfully"等詞句，但在社交書信則視寫信人與收信人關係的深淺，用不同的詞句。例如"Sincerely yours,"，"Yours sincerely,"，"Yours very sincerely,"，"Sincerely,"等。但"Sincerely yours,"，"Cordially yours,"，"Cordially,"等在社交信中，不管關係深淺都可通用。要好的朋友之間或親戚之間，則可用下面任何一種："Affectionately yours,""Lovingly yours,""With love"等字樣。大致說來，商業社交信有下列幾種：

1. Letters of Introduction　2. Letters of Invitation
3. Letters of Thanks　4. Letters of Congratulations
5. Letters of Sympathy　6. Letters of New Year's Greetings

第二節　介紹信的寫法

進出口廠商爲了拓展業務往往有機會出國旅行訪問，在出國之前，如能由相關的人士或公司行號，尤其由往來銀行寫信介紹，請求對方予以協助，那麼辦起事來方便的多，介紹信應 brief 而 sincere，介紹信的內容應包括：

1. 被介紹人的名字。　　　2. 介紹人與被介紹人的關係。

3. 介紹的理由（目的）。　4. 有關被介紹人的背景等。

5. 結束時表示如能惠予協助當感激不盡。

No.210　銀行爲其客戶寫介紹信

Mr. Charles C. Y. Chang　　　　　　　May 6,19…
Senior Vice President
Central Trust of China
4 Wu-Chang, Section 1
Taipei, Taiwan

Dear Mr. Chang,

We have the pleasure of introducing Mr. Archie Gann and Mr. Frank Ricciardi of Columbus, Ohio. Mr. Gann is President and Mr. Ricciardi is Merchandise Manager for Dorcy Cycle Corp., a valued client of our bank.

They will be visiting Taiwan from March 17 to March 20 on business. Although this letter of introduction is not to be construed as an authorization for extension of credit, any courtesies or assistance you may provide will be appreciated. They may desire to discuss export financing with someone at your bank. Mr. Gann and

Mr. Ricciardi will idendtify themselves with appropriate travel documents and a copy of this letter.

Thank you very much for your cooperation in this matter and we shall look forward to the opportunity to assist you in the future.

Sincerely,

A. Z. Sofia

Senior Vice President

AZS／sah

【註】

1. a valued client of our bank:敝行的重要客戶。

2. on business:為了業務; 因商務。

3. is to be construed as an authorization for extension of credit:被解釋為貸款的授權書。

4. any courtesies or assistance you may provide will be appreciated:如蒙照顧及協助, 將不勝感激。

   any courtesy you may show to him will be considered →

   $\rightarrow \left\{ \begin{array}{l} \text{a favor} \\ \text{as shown} \end{array} \right\}$ to myself:如蒙照顧, 將感同身受。

5. export financing:出口融資。

6. identify oneself with:以⋯⋯做為識別之用。"identify oneself with"另有「和⋯⋯提携」之意, 例如: He identified himself with the Labor party:他和工黨提携。

   注意: introduce 該把被介紹的人做 object, 不該把收信人(you)做 object,例如 introduce him to you 不可改作 introduce you to him, 又如 introduce to you my cousin 不可改作 introduce you to my cousin.

　　將介紹信正本交給被介紹人後, 另把介紹信副本寄給介紹信的收信人，是一種既合乎禮貌，又使大家方便的做法。

No.211　寄介紹信副本

Dear Mr. Yano,

We attach a copy of a letter of introduction to you we are handing to Mr. C. Y. Chang, who will be leaving early next week for a business trip to Far East.　He expects to be in Tokyo for a short time only, but while there will be interested in contacting woolen, cotton and silk piece goods manufacturers and exporters.

Mr. Chang and the company bearing his name have been well and favorably known to us for a number of years.　He is a man of substantial means.　We bespeak for him the courtesies of your office.　Mr. Chang will be amply supplied with funds and this letter is not to be construed as an authority to extend credit.

Yours faithfully,

【註】

1. business trip＝business tour:商務旅行。

2. a man of substantial means:很富有的人。

3. bespeak for him:爲他請求。

No.2|2 對於介紹人的覆信

Dear Mr. Wang,

We have recevied your letter of September 1, enclodsing a copy of a letter of introduction which you have given to Mr. Andrew F. Parker of the American Merchandise Company.

It will be a great pleasure for us to meet Mr. Parker upon his arrival and to assist him in any way we can in order to make his trip enjoyable and successful.

Yours sincerely,

【註】

1. to meet:接見。

2. enjoyable and successful 也可以"pleasant and fruitful"代替。

No.2|3 通知訪問

Dear Mr. Liu,

I am pleased to inform you that I am scheduled to leave on January 20 and arrive in Taipei on January 25 by PAA flight No. 2.

I look forward with pleasure to seeing you and all your colleagues at your company and will contact you upon arrival in your city. With kindest personal regards,

Yours sincerely,

【註】

1. PAA: Pan American Airline:泛美航空公司。

No.214　歡迎來訪

Dear Mr. Mu,

Thank you for your letter of January 2, 19…. It's certainly good to know that you will be visit us in the latter part of this month.

We are very happy to have you with us. Please let us know your travel plan, so that arrangements can be made to make your stay here both enjoyable and prodductive.

Our greatest sage, Confucius said, "It's always a pleasure to greet a friend from afar." As a faithful supporter of our company, you are certainly our friend and a very good friend and will be accorded our very warm hospitality.

You must have heard of the beautiful scenery and delicious food here. Come and enjoy them yourself. We will take good care of you.

　　　　　　　　　　　　　　　　Yours sincerely,

【註】

1. happy to have you with us:能跟你在一起很高興; 承蒙光臨不勝愉快。

2. so that arrangements can be made:以便能安排一切。

3. productive:有收穫; 不虛此行。

4. It's always…from afar:「有朋自遠方來，不亦樂乎?」

5. come and enjoy them yourself:請您自己親自來享受吧!

6. We will take good care of you:我們會好好照顧您。

第三節　感謝信的寫法

出國考察業務如承國外廠商接待，那麼回國後，應由本人或公司出面，向接待廠商寫信致謝，這種感謝信的內容應包括：

1. 平安地回到本國。　2. 表示訪問對方時承蒙接待。

3. 表示收穫良多。　　4. 希望貴我雙方關係能更加密切。

No.215　由本人出面致謝

Dear Mr. Brown,

Our visit to your country was most enjoyable, thanks to you and your colleague.　You were all so generous with your time, and so patient with us and our endless questions!

Bill and I want you to know how very much we appreciate all you did to make our stay interesting and enjoyable.

We hope that when you visit Taiwan, you will plan on visiting us.　We'd welcome the opportunity to return your hospitality.

Thanks to all of you, for a truly unforgettable interlude in your country.

Yours sincerely,

【註】

1. you were all so generous with your time:那麼慷慨花那麼多時間陪我們。

2. so patient with us and our endless questions:不厭其詳的替我們解答問題。

3. how much...and enjoyable:多麼地感激你，使我們停留的期間，感到愉快而有趣。

4. We'd welcome...hospitality:希望有機會來報答你的款待。

5. Thanks to...in your country:謝謝你們大家，給我們在貴國種種無法忘懷的一切。

No.216　由公司出面致謝

Dear Sirs,

This is to express our sincere appreciation for the courtesies and assistance you have kindly extended to our Mr. Ma when he recently stopped over at your port on his way from Indonesia.

We learn with a great pleasure that during his brief stay in your port he was able to discuss with you various aspects of our business relationship with you and at the same time obtain valuable information on the current business conditions in your market. We have no doubt that the personal contact he made with you will contribute not a little to our mutual understanding and promotion of business.

For our part we place a high value on these personal discussions.

On behalf of this Company, the writer would like to express our appreciation of your kind reception of our representative.

　　　　　　　　　Yours faithfully,

【註】

1. stop over:中途停留。

2. mutual understanding:雙方的瞭解。

第四節　邀請信的寫法

邀請信(Letter of Invitation)分爲二種，一爲正式的(Formal)，一爲非正式的(Informal)。正式的邀請信，與其說是信，不如說是請帖。邀請信的發出數量較多時，大多用正式的；規模小的則用非正式的。

現在先介紹正式的邀請信：

正式的邀請信如上所述實際上就是請帖。它沒有 Heading, Inside Address, Salutation, Complimentary Close 和 Signature。正像中文請帖的沒有什麼「鑒」和什麼「安」，形式和文字都很呆板，不過也有若干點應注意，一不當心，就弄出笑話。下面是邀請參加酒會的正式邀請信。

No.2l7　邀請參加酒會

Mr. Stewart H. Cole

Vice President and General Manager

of Chemical Bank, Taipei Branch

And Mrs. Cole

request the honor of your presence

at a reception

on Friday, October Eighth, 1989

from six to seven o'clock

at International Reception Hall, Grand Hotel, Taipei

to meet

> Mr. Donald C. Platten, Chairman of the Board
> Chemical Bank, New York
> and Mrs. Platten
> R. S. V. P. (Regret only)
> Tel: 321-2211, Miss Helen Wang

非正式的邀請信，猶如普通的友誼信 (friendly letter)，內容通常很簡單，但語氣要親切，例如：

No.218 邀請晚宴

> Dear Mr. Lin,
>
> Our president, Mr. W. L. Stones, is giving a diner party in honor of Mr. George L. Smith at the Grand Hotel on Wednesday evening June 18, six O'clock.
>
> We would be pleased to have you join us at that time and look forward to the pleasure of your company.
>
> Yours sincerely,

答覆正式的邀請信 (即請帖)，該用正式的；答覆非正式的邀請信，該用非正式的。不論答覆那一種，該說明出席或不出席，不可說「也許可以來」，「希望能來」，「或許不能來」，「恐怕不能來」等不確定的話，不能出席時，宜說明不能出席的理由。

下面是正式答覆的幾個例示。

No.219 覆接受邀請

> Mr. C. Y. Chang
> accepts with much pleasure
> the kind invitation of

Mr. and Mrs. Stewart H. Cole's

reception

on Friday, October eighth, 1989

from six to seven o'clock

at International Reception Hall, Grand Hotel

to meet

Mr. Donald C. Platten, Chairman of the Board

Chemical Bank, New York

and Mrs. Platten

No.220　謝絕邀請

Mr. C. Y. Chang

regrets that a previous engagement

prevents him from accepting

Mr. and Mrs. Steward H. Cole's

kind invitation to a reception

on Friday, October 8, 1989

from 6:00 p.m. to 7:00 p.m.

at International Reception Hall, Grand Hotel

to meet

Mr. Donald C. Platten, Chairman of the Board

Chemical Bank, New York

and Mrs. Platten

注意下列各點：

1. 接受邀請的答覆用"accepts with (much) pleasure"「(極) 樂意接

受（邀請）」或"is pleased to accept"謝絕邀請的答覆用"regrets that..."（當然如主詞是多數，則該用 accept, are 和 regret）。

2. 接受邀請的答覆依照請帖寫明星期幾，年月日和時刻和地點；但謝絕時可以將時刻省略。

3. 接受邀請的答覆裡的"is pleased to accept"不可改作"will be pleased to accept"。

4. 謝絕的答覆裡的"prevents"不可改作"will prevent"。

5. 謝絕的答覆裡的"prevents him from accepting"也可改寫成"prevents his accepting"。

6. 謝絕的答覆裡的"regrets that...from accepting"也可改寫成 "regrets that on account of a previous engagement he is unable to (cannot) accept"。

7. 答覆非正式邀請信，最簡單的便是把正式的覆信句子改成 first person 的語氣，例如：

No.221　接受邀請

Dear Mr. and Mrs.…

Thank you for your kind invitation to a reception on Friday, October 8, 1989 from 6:00p.m. to 7:00p.m., at International Reception Hall, Grand Hotel.

I shall be honored to attend.

　　　　　　　　　　　　　Sincerely yours,

【註】

I shall be honored to attend:本人將一定出席。

No.222　謝絕邀請

Dear Mr. and Mrs....

I deeply regret that I am unable to accept your kind invitation to a reception on Friday, October 8, 1989 as I have a previous engagement for that evening.

I am very sorry to miss the pleasure of meeting Mr. and Mrs. Cole of whom I have heard so much.

Thank you all the same for your invitation.

Sincerely yours,

【註】

1. miss the pleasure of meeting:無緣拜識。

2. thank you all the same for your invitation:雖無法參加，但承邀請仍深表感謝。all the same＝just the same:同樣地。

第五節　祝賀信的寫法

祝賀信須在祝賀之事發生後立即致送，遲則失去祝賀之旨，信中須表示衷心愉快之意。

No.223　祝賀就任新職

Dear Mr. Yu,

It gives us a great pleasure to learn that you have been appointed Manager of the Export Department of yeur highly esteemed firm.

On this occasion we should like to send your our sincerest congratulations on this honourable appointment wishing you every success in the future.

We hope that with your friendly cooperation the amicable relations we have had the pleasure to maintain with your firm will be further intensified in the future.

Yours sincerely.

【註】

1. highly esteemed firm:寶號

2. on this occasion＝at this time

3. send you our sincerest congratulutions:謹祝賀。"send"可以"extend"代替。

4. honourable appointment:光榮的任命；榮任。

5. wishing you every success:祝您事事順遂。

6. intensified:加強。"wishing you every success"也可以"and to extend every good wish for your success in your new responsibilities."代替。

No.224　覆謝祝賀

Dear Mr. Chambers,

I take great pleasure in receiving your congratulatory letter dated May 10, 1989 on the occasion of my new appointment.

I would like to take this opportunity to express my personal appreciation for the support and cooperation which you extended to me during the term of my past

> office. I sincerely hope that you will continue the same to us.
>
> With my personal best regards,
>
> Sincerely yours,

【註】

1. congratulatory letter:祝賀信。

2. 第一段句子可以下列各種句子代替

   。Thank you very much for the greeting which you extended to me on May 10, on the occasion of my new appointment.

   。Thank you for your letter of May 10 conveying congratulations on the occasion of my new appointment.

   。I am very much obliged to you for your cordial congratulations.

   。It was most kind of you to write me such a cordial letter...

   。Thank you so much for your kind letter of congratulations...

   。A thousand thanks for your congratulations on my new appointment ...

3. I sincerely hope that...to us 可以下面句子代替:

   I hope that you will continue to give me the same support as you have always done to my predecessor.

No.225　祝賀設立新公司

> Dear Sirs,
>
> We are very pleased to receive the announcement of your opening of new office in Tokyo.
>
> Please accept our warmest wishes on this memorable

occasion and convey our words of congratulations to your new office for further expansion of your Far East business.

　　With kindest regards,

　　　　　　　　　　　　　　　　Yours sincerely,

【註】

1. memorable occasion:值得紀念的日子。

第六節　慰問信(Letters of Sympathy)的寫法

No.226　慰問病人(Sympathy upon illness)

Dear Mr. Johnson,

　　We were very sorry to learn from Mr. Brown of your recent illness.　I do hope you will be well on the road to recovery by the time of this letter reaches you.

　　My colleagues join me in sending you our regards and best wishes.

　　　　　　　　　　　　　　　　Sincerly yours,

【註】 寫信慰問病人時，應力求簡單，不可太囉嗦(wordy)。

　　"I do hope you will be well on the road to recovery...reaches you"也

　　可以"we extend our sincere best wishes for your speedy recovery."

　　(祈禱早日康復) 代替。

No.227 答謝慰問

Dear Mr. Chang,

This is to thank you for your very kind letter of inquiry during my recent illness. I should have answered it earlier, but the doctor did not allow me to write anything till this morning.

It is at time like this that one really appreciates the kindness of firends. I am feeling very much better now, and hope to be coming home soon. Thank you once more for your kind letter.

Sincerely,

【註】

1. I should have answered it earlier:該早答覆。

2. It is at time...your kind letter:這一段的意思是「一個人唯有在這種時期才能眞正瞭解友誼的可貴，我現在已經好多了。可望於短期內回家，對於你親切的信，再申謝忱。」

No.228 弔慰信 (Letters of Condolence)

Dear Sirs,

We were deeply grieved to hear from you of the death of distinguished Chairman, Mr....

We fully realize how much the loss of Mr....will be felt by all of you and would like to extend our deepest sympathy on this sorrowful occasion.

> Yours sincerely,

【註】

1. 第一段可以"We were shocked to learn that you had lost your Chairman, Mr...."代替, 均可譯成「奉聞貴公司董事長××先生逝世噩耗, 不勝震驚(悲傷)」。

2. and would like to extend...occasion 可以"and please accept my sincere sympathy and that of my entire organization."（本人及公司全體人員敬表誠摯的同情致慰問之忱）代替。

3. We fully realize...felt by you:我們十分瞭解您們是如何地悲傷。

第七節　賀年信的寫法

No.229　賀新年贈檯曆

Dear Mr. Smith,

Hearty congratulations to you on the advent of a bright and prosperous New Year.

I am assure you of my deep appreciation of your patronage during the past year and solicit a continuance of your favors.

As a small token of my best wishes, I am sending you a desk calendar by separate airmail, which you will please accept.

Sincerely,

【註】

1. as a small token of my best wishes:聊表祝福之意。

2. which you will please accept:敬請笑納。

No.230 覆謝贈檯曆

Dear Mr. Nixon,

Thank you very much for desk calendar which you sent me. It's simply beautiful.

I very much appreciate your thinking of me. You can be sure that your sentiments will be reciprocated.

As I turn over each page of the calendar, I, too, will think of you everyday.

As a token of my appreciation, I'm sending you under separate cover some preserved Chinese beef. It's nothing very much, and I hope you will like it.

My very good wishes to you,

Sincerely yours,

【註】

1. It's simply beautiful:眞精美。

2. your thinking of me:想到我, 想念我。

3. you can be sure...reciprocated:你可以確信你的情意將會得到回報的。

4. preserved beef:經加工便於保存的牛肉（如牛肉乾，牛肉罐頭等）。

5. it's nothing very much:這是小小的一點東西。

6. my very good wishes to you:特向您祝福；祝您萬事如意。

第八節 有用的介紹信例句

一、介紹信開頭句

a
We have the pleasure of
We have much pleasue in
We take great pleasure in
It gives us pleasure to have
this opportunity of

} introducing to you Mr.— who

b
This is to
This will
This letter will
This letter will serve to
This letter may serve to
It is with real pleasure that we

} introduce to you Mr.—who

c
The bearer of this letter, Mr.—,
The bearer of these lines, Mr.—,
This letter will be delivered (handed) to you by Mr.—who

will be in your city shortly.

is visiting your city on business.

is making a business trip to your city.

is coming to London in the interests of his company.

will be in your city for a few days on his way to Europe.

will be leaving early next week for a business trip to the U. S.

二、被介紹人姓名、身份

Mr. K. Wang in an officer of The Bank of Taiwan,

Mr. T. Ong, export manager of Tai Sninning Co. Ltd., Taipei, $\Big\}\rightarrow$

$$\rightarrow \left\{\begin{array}{l} \text{—with which we are closely connected,} \\ \text{—with whom we enjoy a very close connection,} \\ \text{—who are our principal bankers for many year,} \end{array}\right\} \rightarrow$$

三、介紹的目的

$$\rightarrow \left\{\begin{array}{l} \text{is proceeding to} \\ \text{is visiting} \\ \text{is coming to} \end{array}\right\} \rightarrow$$

$$\rightarrow \left\{\begin{array}{l} \text{—your port to open up a branch office.} \\ \text{—your city in order to form fresh connections and open new mar-} \\ \quad \text{kets for his products.} \\ \text{—your country for the purpose of extending the commercial rela-} \\ \quad \text{tions of his firm with leading firms there.} \end{array}\right.$$

四、希望予以協助

1. $\left\{\begin{array}{l} \text{We shall appreciate} \\ \text{We shall much appreciate} \\ \text{We would appreciate} \end{array}\right\} \rightarrow$

$$\rightarrow \left\{\begin{array}{l} \text{—anything you may be able to do to assist him.} \\ \text{—any courtesies (and assistance) which you may extend to him.} \\ \text{—any assistance you may be able to accord Mr.}\cdots \\ \text{—any assistance you may render (extend) to Mr.}\cdots \\ \text{—any courtesies or facilities that may be extended to him in} \\ \quad \text{accomplishing his mission.} \end{array}\right.$$

2. Anything you may be able to do to assist him →

$$\rightarrow \begin{cases} \text{—will be very much appreciated.} \\ \text{—will be greatly appreciated by us.} \\ \text{—will be appreciated not only by his company but also by us.} \end{cases}$$

五、預誌謝意

1. You will favour us by giving him any advice or assistance which it may be in your power to render during his stay.

2. We should feel very much obliged if you would kindly furnish him with any advice of which he may stand in need, or do him any other service which you may have the opportunity to render.

3. We would appreciate your rendering him every assistance, and giving him every information which he may require, or which may seem appropriate to ensure the success of his journey.

4. We shall consider any attention shown on Mr. Goto as a personal favour, which we shall be pleased to reciprocate whenever you allow us the opportunity.

第九節　國外訪問歸國後感謝信例句

一、報告平安返回本國的開頭句

1. Having now returned to Japan, I would like to express my thanks...

2. I arrived back safely yesterday evening, my first thought is to again thank you for...

3. The first thing I wish to do, now that I have returned to Tokyo, is to express to you my most grateful thanks...

4. After having been on my way home through a lot of interesting countries for a considerably long time I am now back in my own country and wish to express my thanks for...

二、感謝承蒙接待、協助

> Example:I wish to express my sincere appreciation for your courtesies and assistance extended to me during my recent visit to your city.

1)

We wish to express to you our sincere／great appreciation

We wish to express our thanks／appreciation

We wish to express our very sincere thanks

We wish to express our heartfelt appreciation

We wish to take this opportunity of expressing to you our appreciation

We would like to take this opportunity of thanking you

We would like to avail ourselves of this opportunity to let you know how grateful we are for your kindness

We thank you for…

We wish to convey to you our thanks…

We deeply appreciate

The first thing I wish to do is to express our most grateful thanks

This letter is written for the express purpose to thank you

2)　(*preceded by "for" where necessary*)

the kindness／generosity

the kindness and hospitality

many kindness and courtesies

your very great personal kindness

    the courtesies

your courtesies and assistance

your courtesies and thoughtfulness

the hospitality and friendship

the hospitality courtesy, and assistance

the most lavish hospitality you bestowed on us

the most wonderful hospitality and cooperation

the most effective cooperation

extended to me

shown to me

accorded us

which you showed to me

you extended to me

which we received at the hands of all whom we met at your company

that were so freely extended to me

which has been bestowed on me

三、表示此次旅行獲益良多

1. My journey to Japan will remain an unforgettable memory.

2. My visit will be one which I shall long remember with the greatest pleasure.

3. I enjoyed very much indeed my visit to your country, but I am quite certain that we would not have seen nearly as much or enjoyed it nearly so well had it not been for the attention that your staff gave us and the arrangements they made for our entertainment.

4. Your personal kindness and help to us were instrumental in making our visit exceptionally pleasant for us both.

5. Your wonderful hospitality and many courtesies extended to me had made my trip most enjoyable and a memorable one.

6. With kindest regards and looking forward to a long future of constantly increasing business and very close relations between our two firms, I am.

7. Our kindest regards to yourself and to all with whom we were associated in Australia.

8. Should you ever come to Taiwan, I certainly hope that you will call me and arrange to visit our office.

9. I trust you are well and hope that in the not too distant future you will re-visit Taiwan when we can repay a little of your hospitality. My wife joins me in sending our kindest regards to you and we would ask that you convey our best wishes to Mr....

四、結尾句

1. Once again, I would like to thank you for the wonderful welcome extended to me.

2. With best personal regards to you and your colleagues.

3. With very many thanks and kind personal regards to yourself, and would you please also convey my regards to Mr....

4. Many thanks again for all that you and your Company did for me, and with warmest regards.

5. Once more, my sincere and heartiest thanks for all that you did for me during my visit in your country.

習　　題

將下列中文譯成英文。

1. 謹將林先生介紹給你。

2. 希望隨時賜予照顧，我將視同身受。

3. 如蒙惠允協助，將感同身受。

4. 我們希望有機會報答你的好意。

5. 本人已回到臺北，本人在貴國停留期間，對你們所給予的好意和協助，謹致謝忱。

6. 無意中接到你那精美的禮物，我眞不知道如何感謝你。

7. 假如您和嫂夫人能於 3 月 3 日（星期六）下午 6 時光臨敝舍共進晚餐，則我們將會很高興，那晚餐會是個小小宴會，我們只請王大任夫婦參加，我們希望你們能來，祝好！

8. 承寵召，謹如命出席奉陪。

9. 欣聞吾兄被任命爲貴公司總經理，謹祝賀意，並祝萬事如意(offer you our heartfelt congratulations and best wishes)。

10. 祝你早日康復，順頌好(with best regards)。

第三十四章　標點符號的用法

標點符號的使用，其目的在於使文字清楚、易讀、易解。以下將標點符號用法加以介述。

一、Period　(·)──句點

1. 用在每一敍述句(declarative sentence)或祈使句(imperative sentence)之後。

例：(1) We mailed you our check on Thursday. (declarative)

(2) Write us as soon as you reach New York. (imperative)

但句尾的字如是縮寫字，則不要再加句點。例如：

The meeting will be held on May 15, at 10 a.m.

注意：如果是驚嘆句或疑問句，則句尾字即使是縮寫字，仍應加驚嘆號或疑問號，例如：

May I come to see you at 2:00p.m.?

Get up, it is already 8:00 a.m.!

2. 用在縮寫字起首字母 (initial) 後面。

例如：Mr.為 Mister 的縮寫；October 的縮寫字為 Oct.

但有些縮寫字往往不加句點，例如 CIF 是 Cost Insurance Freight 的縮寫，但在"C""I""F"的後面往往不加句點。類似情形尚有 FOB, C&F, FAS, BOFT (國貿局)，FBI, USA 等等。

有些縮寫字已不再視為縮寫字，所以在後面不加句點。例如：

Exam → Examination

Ad → Advertisement

3. 引用句中, 中間部分省略時, 用三點句點, 末尾省略時, 用四個句點。

例: The letter read:"We can not...until after you have filled out the questionnaire."

The letter read:"We will visit your country...."

4. 用於元與角分之間。

例: $4.99,　　$.08

二、Comma（,）——逗點

1. 用於複合句(Compound Sentence)中接續詞之前。

例: I plan to visit your country early next month, and I hope at that time to have a talk with you at your office.

但複合句很短而意義明確時, 可省略逗點。

例: Miss Wang is a secretary but Mr. Wang is a manager.

2. 複合句的主詞爲同一字, 而以"and"接另一動詞時不加逗點, 但以 "but"連接時, 應加逗點。

例: (1) We received your cable of May 10 and have shipped the goods today.

(2) We received your cable of May 10, but were unable to effect the shipment in time.

3. 用於分隔一系列(三個以上)同等名詞(coordinate nouns)、形容詞、動詞或副詞、片語或子句。

例: (1) I spoke to the receptionist, the secretary, the clerk, and the manager.

(2) We plan, fabricate, erect, and sell our own portable building.

(3) We want a careful, intelligent, conscientious clerk.

但如名詞前有兩個以上的非同等形容詞時，不用逗點。

例：The secretary wore a large green eyeshade.(green 修飾 eyeshade,而 large 則修飾 green eyeshade)

4. Jr., etc.出現在句子中間時，其前後應加逗點。

例：Mr. James Johnson, Jr., is our saleman.

5. 用於敍述性子句（即非限制性的nonrestrictive）前或後。

例：(1) Mr. Chang, who had been a faithful employee, retired last week. （子句在句子中間）

(2) Always faithful in carrying out his duties, Mr. Chang retired last week. （子句在句首）

(3) Mr. Chang retired last week, even though he would have liked to serve the company for more years.（子句在句尾）

6. 敍述性子句用逗點，限制性(restrictive)子句不用逗點。

例：(1) We showed your sample to our maker, who submitted us this offer.（我們把你的樣品向我們的製造商出示，他們提出了本報價）——敍述性用法。

(2) We showed your sample to our maker who submitted us this offer.（我們把你的樣品出示給向我們提出報價的製造商）——限制性用法。

(3) Employees who are careless should be discharged.

7. 同位語(appositive)，如取消也無妨,則前後要加逗點；如取消則意義不明顯時，則前後不用逗點。

例：(1) Mr. Chang, a faithful fellow, has been with the company since 1930.

(2) The term letter of credit is used frequently in business

correspondence.

職位或學位等，其前後均加逗點。

Mr. Chang, Vice President of our company, will visit your city next month.

8. 獨立子句、片語，或放在文首的副詞子句，以逗點分隔。

例：(1) My fiance, much to my regret, will go abroad.

(2) Frankly speaking, to write a letter is not so diffcult as you think.

(3) If our proposals are acceptable, please write us before the end of this month.

(4) If this offer is accepted, we shall give you another order next month.

9. 一系列名詞（三個以上），最後兩個名詞雖由"and"連接，但不相關聯時，"and"前面用逗點隔開。

例：Please note the delivery dates for tabacco, cotton, corn, and wheat.（如寫成 corn and wheat，則變成三種交貨期）

10. 下列引導性措詞（introductory expressions），出現在句子開頭時，用逗點隔開。

Accordingly	Hence
Actually	However
After all	In any case
Again	In brief
Also	Indeed
As a matter of fact	In fact
As a rule	In other words
Besides	In the first place

By the way	In the meantime
Consequently	In short
First	In general
For example	Meanwhile
For instance	Moreover
Fortunately	Next
Further	No
Furthermore	Nevertheless
Of course	Frankly speaking
On the countrary	Then
On the other hand	To be sure
Otherwise	To say the truth
Perhaps	Unfortunately
Secondly	Well
So	Without doubt
Still	Yes
Strictly speaking	Yet

11.用於分隔表示對比的字詞。

例: The factory workers, not the office staff, are on strike.

12.下面過渡性(trasitional)的單字或片語, 出現句子中間時, 其前後均加
逗點。

also	too
therefore	again
as a rule	in the first place
as we see it	however
as it were	if any

as you know	in addition
by the way	in fact
for example	in brief
indeed	in turn
finally	namely
i.e.	say
e.g.	of course
viz.	on the other hand
still	that is

例: (1) Our business is very good, that is, in the light of present times.

(2) We have established a sizable market, i. e., in U. S. A., Japan, and South Africa.

13. 在引用他人談話時，引用句的前或後，應以逗點分隔。

例: (1) He said, "I must go home now."

(2) "Today," he cried, "I am a free man."

14. 用於分隔年與月份之後的日期。

例: July 8, 1978.

15. 用於分隔地址（省、市、鎮、街、巷等）。

例: 100 Wuchang Street, Section 1, Taipei, Taiwan.

16. 用於直接稱呼時。

例: Mr. Chang, please come to my office.

17. 用於分隔句子中的釋意短句。

例: Your last letter, although mailed on the 18th, did not reach us until the 28th.

18. 用於輕微的感嘆詞後面。

例: Well, I did'nt think we'd get such a large order.

19. 用於句子中省略字時。

例: (1) One of the officers has been with us for fifteen years; the other, for ten. ("has been" is omitted before "for ten")

(2) This girl is a secretary; the other, a bookkeeper. ("is" is omitted before "a bookkeeper")

20. 用於分隔數目字，以便閱讀。

例: NT$123,456　　　123,456,789 dozen

但年代、電話號碼、門牌號碼、頁數、L／C 號碼等不用逗點隔開。

例: 1982　　1234 Park Avenue　　　page 1234　　L／C NO. 1256

三、Semicolon　(；)——半支點

1. 同等子句(coordinate clause)不用連接詞時，以半支點分隔。

例: The last day of the month is always a busy day for bank employees; it often forces them to work overtime.

2. 以下面的連接性副詞連接複合句時，在這些連接性副詞前面應用半支點。

however	otherwise
accordingly	for
nevertheless	hence
besides	consequently
then	likewise
thus	notwithstanding
therefore	

例: (1) I saw no reason for moving; therefore I stayed still.

(2) Your letter requesting cancellation of your order 123 was received only this morning; otherwise this order would have been shipped today.

However 當做 conjunctive adverb 時，其後面須加逗點。

3. 用於分隔一連串的名辭。

The new members of the Board and their home town follow: Mr. C. H. Chen, Taipei Taiwan; Mr, K. H. Wang, Tainan, Taiwan; Mr. M. N. Chia, Seatle, U. S. A.

四、 Colon （:） ——冒號

1. 用於表示依序詳細說明或列舉時：

例: (1) Four articles are on the desk: a stamp pad, a bottle of ink, an abacus and fountain pen.

(2) We have sent you today the following shipping documents:

Commercial invoice in two copies

Bill of lading in two copies

Packing list in three copies

2. 用於直接引述(direct quotation)也可用逗點。

例: (1) Our manager said: "These data must be kept confidential.

(2) Our motto is: "First come, first served."

3. 用在 appositive phrase 或 clause 之前。

例: Our chief requirement is this: We need good quality and prompt shipment.

4. 用於分隔時和分。

例: Time: 10:18 p.m.

5. 用在 Salutation 之後。

　例: Dear Sirs:

　　　但也可用逗點。

　例: Dear Sirs,

6. 下面字辭前面, 可以半支點分隔

namely	for instance
as	i.e.
that is	e.g.
for example	viz.
that is to say	

　例: He should be given the job; that is, he has the necessary qualifications.

7. 引號後面如還有文字時, 在引號後面用半支點。

　例: The cable read: "CANNOT SHIP BEFORE OCTOBER 10"; consequently we had to disappoint our customer.

五、Apostrophe(')——縮寫記號, 所有格記號

1. 用於表示所有格。

　(1)單數名詞: 其所有格加"'s"

　　例: My father's hat

　(2)多數名詞語尾有"s"時; 其所有格, 只在"s"後面加" ' "。

　　例: the shareholders' appeals

　(3)多數名詞語尾無"s"時, 其所有格應加" 's"。

　　例: men's suits

　　　children's coats

　(4)單數固有名詞以-s, -x, -ch, 或-sh 爲語尾時, 其所有格加"'s"或只加(')。

例: Mr. Fields's territory Mr. Fields' territory

　　Miss Cox's records Miss Cox' record

但加"'s"的較通行。多數固有名詞時，其所有格加 apostrophe.

例: the Clarks' letters

(5)縮寫字的所有格，apostrophe 應加在 period 後面。

例: A. I. B.'s regulations

　　CTC's Policy

(6)表示聯合的所有格時，在最後一名詞加"'s"。

例: Smith and Jones's opinions.

(7)複合名詞的所有格，在最後一字加"'s"。

例: My father-in-law's overcoat.

(8)非生物的所有格以用"of"表示為佳。

Poor: the shipment's date

Better: the date of shipment

Poor: the contract's signing

Better: the signing of the contract

但與時間、度量衡或人格化有關的習慣用語，不在此限。

a month's stay　　　　　For pity's sake

a day's work　　　　　a dime's worth

a mile's length

(9)在公司或社團名稱裡，常將所有格符號省略。

例: The Farmers Bank of China.

2.用於表示省略了字母或數字。

例: Let's (=Let us) have a drink

　　can't (=can not)

　　don't (=do not)

I was born in U. S. in the spring of '42. (＝1942)

It's(＝It is) just eight o'clock. (＝of the clock)

3.用於表示字母單字或數字的多數。

例：⑴ There are four o's and three m's in this sentence.

⑵ Do not use too many we's in a business letter.

⑶ Mind your P's and Q's. (當心說話或舉動)

六、Quotation mark (" ")──引號

1.雙引號用於直接引用語。

例：Our manager said, "Everybody must come to office at 8: 00 a.m."

2.雙引號用於表示強調或意義特殊的字。

The term "Super" is used freely today.

3.引用語中有引用語時，直接引用語用雙引號，引用語中的引用語用單引號。

例：Our manager always quote the words of our chairman, "Never forget that Confucius said, 'Do not do to others what you do not want others to do you'."

4.雙引號用以引用書中章節、雜誌中的論文名稱。

例："Writing effective collection letters" is a very informative chapter.

5.用以船名。

例：S. S. "Hupeh"

七、Hyphen (-)──連字號

1.用於複合單字。

例：father-in-law

well-known firm

2.數字由 21 至 99 的數目，用文字寫出而當做形容詞用時，用連字號。

例：twenty-one orders

eighty-eight invoices

3.用於接頭語。

例：Pro-America

re-elect

self-educated

4.當分數用文字寫出時。

例：five-sixths

5.當一個字分做兩行寫的時候在第一行的末尾加連字號。

6.當兩複合形容詞形容同一名詞時，第一形容詞後面只加連字號。

例：This molasses is sold in one-and two-quart bottles.

7.在多數場合，ex, vice 的接頭語都加連字號。

例：ex-president

vice-chairman

八、Question mark (?) ——問號

1.用於任何直接問句(direct question)之後。

例：What is your opinion?

但形式上是疑問句，而實際上卻是禮貌的請求句，則不用問號。

例：Will you please send us a catalog.

2.對某一敍述無把握或有所疑惑時，在其後面以 (?) 表示。

The check was for US$1,928.31(?).

九、Exclamation mark(!) ——驚嘆號

1.用在感嘆字句之後。

例：Wait! Don't release that shipment now!

2.用於呼告格之後。

例: O Liberty! O Liberty! What crimes are committed in thy name.(啊，自由啊！啊，自由啊！在你的名義下，不知犯了多少罪呀)

3. 用於形式上爲問句而內容爲感嘆句的句子。

例: How can you believe such a wild assertion!

十、Dash (—)──破折號

1. 用於表示思路或結構 (structure) 的突然破裂 (sudden break)。

例: ABC Company ordered ten of our mahogany coffee tables—I forget the style numbers—for one of their hotel accounts.

2. 用在短同位格片語之前，以加強語氣。

例: We called on the Fargo Company—the principal dealers of electrical appliances in Taiwan.

3. 用於分隔含有逗點的同位格措詞。

例: Three of our trucks—the '81 Chevrolet, the '81 Ford, and the '82 G. M. C.—are in excellent conditions.

4. 用於分隔重複字或重複片語，以加強語氣。

例: We must have your check within ten days—ten days, not a day later!

5. 當 as, namely, that is 等詞用於介紹一系列項目時，在這些詞後面加破折號。

例: There are three steps in our sales training program, namely—

① A six-month period in the factory

② A six-month period in the offices

③ A six-month period on the road working with an experienced salesman.

6. 用在限制或彙述前面一系列辭語的同位格子句前面。

例: Courtesy, reliability, and knowledge—these qualities we demand of all our employees.

7. 某一時間至另一時間或某一日至另一日之間, 用破折號分隔以示其起迄。

例: Our personnel offices are open from 8:00 a.m.—4:30 p.m.

8. 用於表示說話時的遲疑。

例: I hope so—but—that is—well, I will give it a trial.(我希望是那樣—但是—那就是—啊, 我要來試一下)。

十一、Parenthesis (( ))——圓括號

1. 以插入句, 說明或補充解釋時, 用圓括弧隔開。

例: This report (I think Mr. Smith submitted it last week) covers the period from March 1 to June 15.

2. 用於括住參考或指示。

例: The revised price pages have been prepared for all our accounts (see pages 8 to 16 for wholesaler discounts) effective May 1.

3. 用於括住同位格數字(figures)。

例: Our wheelbarrows list for twenty dollars ($20) in lots of one (1) dozen.

(2) We received fifty (50) orders for our TV sets this month.

4. 用於括住一系列項目之前的數字或字母(letters)。

例: After the orders have been registered, they are classified as follows:

⑴ F. I. (Fill Immediately);

⑵ HFC (Hold For Confirmation) and

⑶ CAR (Credit Approval Required).

十二、Bracket（〔　〕）——

1.用於作者或註釋者所加的說明或補充。

例：In that summer 〔1977〕 business dropped 20 per cent.

2.用於圓括弧句字中。

例：The creation of our research department (Mr. Wang was the man who started it 〔see page 8 of the Annual Report〕) was the principal accomplishment of our business last year.

3.說明字的發音。

例：Leicester square 〔ˈlestə skwɛr〕

透析商業英語的
語法與語感

長野格／著 林 山／譯

市面上講商業英語的書不少，但很少有一本書是專門針對商業英語語彙作解析。調查顯示，最令外語學習者感到挫折的往往不是有形的文法規則，而是難以捉摸的語感。這些平常英語學得再好也無法一窺堂奧的問題，本書中都有解答。書中除了針對單字的商業用法與語感作解析外，更有類義字的比較及出現頻率，並特別針對語意上有細微差異，或是使用不當、不得體的危險用字作提醒。商業英語不只是F.O.B.等基本商業知識，潛藏在你我熟悉的字彙中的微妙語感與語法才是縱橫商場的不二法門，掌握了它，你將是名副其實的洽商高手。

活用美語修辭
——老美的說話藝術

枝川公一／著 羅慧娟／譯

belly button（肚子的鈕扣）、maggot in one's head（腦袋的蛆蟲）、mosquito cough（蚊子的咳嗽）、sell ice to Eskimos（賣冰給愛斯基摩人），這些實際意指什麼呢？作者引用英文書報雜誌的巧言妙句為你解說，帶你徜徉於美國人的想像天地裡，享受自由聯想的語言趣味。

本書詞條採用英文字母排序，共有400個以上的詞條及例句，不限於美語的俚語，亦包括諸多英語慣用語的譬喻用法，是作者累積多年收集成冊的精華。透過書中的解說和例句，既能記憶修辭用法，使自己的用字遣詞更為豐富生動，並能刺激想像力，盡情享受英語文字的魅力。

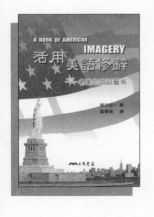

掌握英文寫作格式

笹井常三／著　林秀如／譯

英文寫作時，除了應注重內容外，是不是更應懂得標點符號的用法，以及寫作的基本要點，才能得體、不失禮呢？本書即以具體例句詳細介紹14種「標點符號」的使用法則，並介紹「大寫字」「斜體字」等的用法，以及說明「拼字的基本法則」。另外，作者更根據長年在報社擔任英文記者，以及教學累積的經驗，提供「英文寫作的小祕方」，特別介紹寫作時應注意的文法要點。相信本書不僅可以鞏固學習者寫作的基礎，更可協助寫出自然、流暢的英文，是一本適合初學寫作者最佳的參考手冊。

商用英文書信

高崎榮一郎／著　篠田義明／監修　林秀如／譯

本書收集商業人士的實際範例，內容共分15章，涉獵範圍廣泛，舉凡交涉、通知、產品簡介等基本商務書信皆包含在內；各章節結構編排嚴謹，不僅針對原文範例剖析問題點、討論例文的句型與語法，更列出改善範例以供對照，最後並點出「段落大綱」，提供讀者安排書信架構順序時的參考。例文皆附中文翻譯，以及語句註釋或「注意事項」，期望省卻讀者查閱辭典的辛苦，能完全專心針對本書中「邏輯和架構」的問題點，提高寫作能力，是一本即將出社會工作的年輕學子們最佳的商用英文書信指南。

社交英文書信

長野格、城戶保男／著　羅慧娟／譯

欣聞友人獲獎，你會用英文書寫恭賀信函嗎？商務貿易關係若僅止於商業上的書信往返，彼此將永遠不會有溫暖的互動。若想要進一步打好人際關係，除了訂單、出貨之外，噓寒問暖也是必須的。本書特別針對商業人士「社交」上的需求，編寫實用的社交英文書信範例，每一篇均附有詳細的中譯及語句註釋，讓你既學習英文也學習待人處事的技巧。內容包括了慶賀書信、邀請函、介紹信、感謝函、訪問、致歉、請求、委託、徵信查詢、遷移通知……等，包羅萬象，是你最佳的社交英文書信指南。

同步口譯教你聽英語

斎藤なが子／著　劉明綱／譯

「每次和老外說話，只要把學過的單字串在一起，再配合手勢，大多可以把自己的意思表達出來，但是對方說的話卻經常有聽沒有懂，造成對話中斷。到底有沒有什麼方法可以增強英語聽力呢？」

事實上，正確地聆聽、理解英語，是進行流利會話的第一步，而且遠比將自己的意思傳達出去還難。本書為日本名同步口譯者斎藤なが子的力作，書中提到的聽力技巧，不僅對日本讀者受用，對於國內有心增強英語聽力的讀者而言，一樣受益無窮。文中所穿插的Let's take a break!小單元，是作者多年來從事口譯的心得與見聞，有志走上口譯路途的人士務必一睹為快。

That's it!就是這句話！

宮川幸久、Diane Nagatomo／著　劉明綱／譯

英語要說得嚇嚇叫，不能靠死背文法，而是要會說、敢說。從日常中的一句話培養出興趣，再到會話，慢慢求進步。日常生活中實際的用例，像是「那不急——That can wait.」「聽起來不錯——Sounds great.」「別傻了——Don't be silly.」「各付各的——Let's go Dutch.」等等，都是課本上不教的慣用句。打了一整天的報告，電腦卻突然當機，真是令人欲哭無淚，你知道這時可以用哪句英語來表達你的心情嗎？或者是覺得對方似乎誤會了自己的意思，想趕緊澄清時，又該如何聲明呢？許許多多你絕對用得到的例子，都在這本發燒狂賣的精裝CD版中。

老外會怎麼說

各務行雅／著　鄭維欣／譯

作者以留美多年所接觸的英語為基礎，指出以文法為中心的學校英語在實際會話上有哪些不自然的表達方式，並從各個層面帶領學習者瞭解最生動自然的英語。本書內容共分成三章。第一章列出國、高中所學的英語和實際上英美所用的英語之相異處。第二章探討英語學習者在發音、文法以及字彙上常見的典型錯誤。第三章則舉例說明會話用語與書面用語兩者之間的不同處。各章下並條列介紹數十項問題的內容，指出你我常會犯的不自然語句及如何說出道地流利的英語。

英語大考驗

小倉弘／著　本局編輯部／譯

別看這本書輕薄短小，其中所收錄的114條（語意篇48條，文法篇66條）問題，都是大家最常犯錯的學校式英語。例如語意篇的「都是譯文惹的禍」所收錄的24條問題，便是受到譯文的誤導而產生的誤用。「只知其一不知其二」說的是許多常見的字其實另有含意。「超級比一比」提供了一個類義字的比較擂臺。「用英語思考」提點的是英語的語序與語感其實息息相關。

文法篇則分為「時態」「助動詞」「準動詞」「冠詞」「限詞」「介系詞」「文體」七大主題，共66條。每條一開始均先提示主題句，提供讀者找出問題點，然後再去驗證解答。你認為你的英語基礎紮實嗎？請來接受補教名師小倉弘的考驗！

打開話匣子
——Small Talk一下

L. J. Link、Nozawa Ai／著　何信彰／譯

一般人都知道可以跟外國人談一談自己的名字、工作、家庭狀況、家住哪裡等，不過接下來要聊些什麼呢？仔細學習過本書對話之後，你的交談技巧必能有所精進，也能和外國人打開話匣子！

書中的對話臨場感十足，並時而語帶詼諧。您可能會看到一些省略的措辭，甚至一些極簡的文法句構，但這些都是在現實生活中真正會被使用到的詞句與對話，而並非只能在書中被找到，讓你輕鬆掌握在地人怎麼說英語。

書末針對每一則對話，設計簡易的Comprehension Quizzes（牛刀小試），可用來評量或作為重點提示；另有Activities（建議活動）單元，極適合用在作文課或會話課，必能引起學生熱烈迴響。

英語會話從聽開始

英語聽&說

一套主要訓練英語聽和說能力的語言學習書，搭配CD做完整紮實的聽說訓練。會話的主題內容，具生活化、實用性，測驗及練習步驟明確，循序漸進、深具效果，最適合想要輕鬆應付英語大小聽力測驗的讀者。

請參考下表選擇適合自己程度的《英語聽&說》：

	多益（TOEIC）	全民英檢（GEPT）
入門篇	300～450 分	初　級
初級篇	400～600 分	初級～中級
中級篇	500～750 分	中級～中高級
高級篇	700 分以上	高　級

英語聽&說（入門篇）

白野伊津夫、Lisa A. Stefani／著
沈　薇／譯

「入門篇」是本系列的第一冊，全書著重在「提升聽力的發音技巧」，扼要說明英語的發音原則，介紹哪些音是「弱讀音」，以及「縮略字」、「同化」、「連音」等發音訣竅。建議您多聽CD並確實做好聽寫練習，必定可以提高耳朵的靈敏度。

English test

英語聽&說（初級篇）

白野伊津夫、Lisa A. Stefani／著
沈　薇／譯

「初級篇」是本系列的第二冊，全書著重在「聽力技巧」，扼要說明了掌握數字、專有名詞、片語的唸法，以及如何抓住重音位置、語調、節奏等。建議您多聽CD並確實做好聽寫練習，必定可以增強聽力，理解日常生活的必備英語。